COMPREHENSIVE ANALYTICAL CHEMISTRY

VOLUME 55

ADVISORY BOARD

Wilson & Wilson's
COMPREHENSIVE ANALYTICAL CHEMISTRY

Edited by

D. BARCELÓ
Research Professor
Department of Environmental Chemistry
IIQAB-CSIC
Jordi Girona 18-26
08034 Barcelona
Spain

Elsevier
Radarweg 29, PO Box 211, 1000 AE Amsterdam, The Netherlands
Linacre House, Jordan Hill, Oxford OX2 8DP, UK

First edition 2009

British Library Cataloguing in Publication Data
A catalog record for this book is available from the British Library

Library of Congress Cataloging-in-Publication Data
A catalogue record for this book is available from the Library of Congress

ISBN: 978-0-444-53237-4
ISSN: 0166-526X

For information on all Elsevier publications
visit our website at books.elsevier.com

Printed and bound in Hungary

09 10 11 12 13 10 9 8 7 6 5 4 3 2 1

Wilson & Wilson's

COMPREHENSIVE ANALYTICAL CHEMISTRY

COMPREHENSIVE TWO DIMENSIONAL GAS CHROMATOGRAPHY

VOLUME **55**

Edited by

LOURDES RAMOS

Department of Instrumental Analysis and Environmental Chemistry, Institute of Organic Chemistry (IQOG-CSIC), Juan de la Cierva 3, 28006 Madrid, Spain

ELSEVIER

Amsterdam • Boston • Heidelberg • London
New York • Oxford • Paris • San Diego
San Francisco • Singapore • Sydney • Tokyo

CONTENTS

PART III: ANNEX: TROUBLESHOOTING

CONTRIBUTORS TO VOLUME 55

Mohamed Adahchour
Omegam Laboratories, P.O. Box 94685, 1090 GR Amsterdam, The Netherlands

Jan Beens
Department of Analytical Chemistry and Applied Spectroscopy, de Boelelaan 1083, 1081 HV Amsterdam, The Netherlands

Jan Blomberg
Shell Global Solutions International B.V. Shell Research & Technology Centre, Amsterdam, P.O. Box 38000, 1030 BN Amsterdam, The Netherlands

Udo A.Th. Brinkman
Department of Analytical Chemistry and Applied Spectroscopy, de Boelelaan 1083, 1081 HV Amsterdam, The Netherlands

Paola Dugo
Dipartimento Farmaco-chimico, Facoltà di Farmacia, Università degli Studi di Messina, Viale Annunziata, 98168 Messina, Italia

Giovanni Dugo
Dipartimento Farmaco-chimico, Facoltà di Farmacia, Università degli Studi di Messina, Viale Annunziata, 98168 Messina, Italia

Jean-François Focant
University of Liège, Mass Spectrometry Laboratory, Biological & Organic Analytical Chemistry, Allée du 6 août, B6c, B-4000 Liège (Sart-Tilman), Belgium

Jamin C. Hoggard
Department of Chemistry, University of Washington, Box 351700, Seattle, WA, 98195, USA

Tuulia Hyötyläinen
VTT Technical Research Centre of Finland P.O. Box 1000, FI-02044 VTT, Finland

Hans-Gerd Janssen
Unilever Research and Development, PO Box 114, 3130 AC Vlaardingen, The Netherlands; University of Amsterdam, Nieuwe Achtergracht 166, 1018 WV Amsterdam, The Netherlands

Erwin Kaal
University of Amsterdam, Nieuwe Achtergracht 166, 1018 WV Amsterdam, The Netherlands; ATAS GL International, PO Box 17, 5500 AA Veldhoven, The Netherlands

Minna Kallio
Laboratory of Analytical Chemistry, Department of Chemistry, University of Helsinki, P.O. Box 55, FIN-00014, Helsinki, Finland

Sjaak de Koning
LECO Instruments GmbH, Marie-Bernays-ring 31, 41199 Mönchengladbach, Germany

Luigi Mondello
Dipartimento Farmaco-chimico, Facoltà di Farmacia, Università degli Studi di Messina, Viale Annunziata, 98168 Messina, Italia

Miren Pena-Abaurrea
Department of Instrumental Analysis and Environmental Chemistry, Institute of Organic Chemistry (IQOG-CSIC), Juan de la Cierva 3, 28006 Madrid, Spain

Lourdes Ramos
Department of Instrumental Analysis and Environmental Chemistry, Institute of Organic Chemistry (IQOG-CSIC), Juan de la Cierva 3, 28006 Madrid, Spain

Juan Jose Ramos
Department of Instrumental Analysis and Environmental Chemistry, Institute of Organic Chemistry (IQOG-CSIC), Juan de la Cierva 3, 28006 Madrid, Spain

Stephen E. Reichenbach
Computer Science and Engineering Department, University of Nebraska–Lincoln, Lincoln NE 68588-0115, USA

Jesus Sanz
Department of Instrumental Analysis and Environmental Chemistry, Institute of Organic Chemistry (IQOG-CSIC), Juan de la Cierva 3, 28006 Madrid, Spain

Gaëlle Semard
CIRSEE (Centre International de Recherche Sur l'Eau et l'Environnement) – 38 rue du président Wilson – 78230 Le Pecq, France; Laboratoire des Sciences et Méthodes Séparatives (EA 3233), Université de Rouen, IROCF, F-76821 Mont Saint Aignan cedex, France

Robert A. Shellie
Australian Centre for Research on Separation Science (ACROSS), School of Chemistry, University of Tasmania, Private Bag 75, Hobart TAS 7001, Australia

Robert E. Synovec
Department of Chemistry, University of Washington, Box 351700, Seattle, WA, 98195, USA

Peter Quinto Tranchida
Dipartimento Farmaco-chimico, Facoltà di Farmacia, Università degli Studi di Messina, Viale Annunziata, 98168 Messina, Italia

VOLUMES IN THE SERIES

Vol. 1A	Analytical Processes
	Gas Analysis
	Inorganic Qualitative Analysis
	Organic Qualitative Analysis
	Inorganic Gravimetric Analysis
Vol. 1B	Inorganic Titrimetric Analysis
	Organic Quantitative Analysis
Vol. 1C	Analytical Chemistry of the Elements
Vol. 2A	Electrochemical Analysis
	Electrodeposition
	Potentiometric Titrations
	Conductometric Titrations
	High-Frequency Titrations
Vol. 2B	Liquid Chromatography in Columns
	Gas Chromatography
	Ion Exchangers
	Distillation
Vol. 2C	Paper and Thin Layer Chromatography
	Radiochemical Methods
	Nuclear Magnetic Resonance and Electron Spin Resonance Methods
	X-ray Spectrometry
Vol. 2D	Couiometric Analysis
Vol. 3	Elemental Analysis with Minute Sample
	Standards and Standardization
	Separation by Liquid Amalgams
	Vacuum Fusion Analysis of Gases in Metals
	Electroanalysis in Molten Salts
Vol. 4	Instrumentation for Spectroscopy
	Atomic Absorption and Fluorescence Spectroscopy
	Diffuse Reflectane Spectroscopy
Vol. 5	Emission Spectroscopy
	Analytical Microwave Spectroscopy
	Analytical Applications of Electron Microscopy
Vol. 6	Analytical Infrared Spectroscopy
Vol. 7	Thermal Methods in Analytical Chemistry
	Substoichiometric Analytical Methods
Vol. 8	Enzyme Electrodes in Analytical Chemistry
	Molecular Fluorescence Spectroscopy
	Photometric Titrations
	Analytical Applications of Interferometry
Vol. 9	Ultraviolet Photoelectron and Photoion Spectroscopy
	Auger Electron Spectroscopy
	Plasma Excitation in Spectrochemical Analysis

Vol. 10	Organic Spot Tests Analysis The History of Analytical Chemistry
Vol. 11	The Application of Mathematical Statistics in Analytical Chemistry Mass Spectrometry Ion Selective Electrodes
Vol. 12	Thermal Analysis Part A. Simultaneous Thermoanalytical Examination by Means of the Derivatograph Part B. Biochemical and Clinical Application of Thermometric and Thermal Analysis Part C. Emanation Thermal Analysis and other Radiometric Emanation Methods Part D. Thermophysical Properties of Solids Part E. Pulse Method of Measuring Thermophysical Parameters
Vol. 13	Analysis of Complex Hydrocarbons Part A. Separation Methods Part B. Group Analysis and Detailed Analysis
Vol. 14	Ion-Exchangers in Analytical Chemistry
Vol. 15	Methods of Organic Analysis
Vol. 16	Chemical Microscopy Thermomicroscopy of Organic Compounds
Vol. 17	Gas and Liquid Analysers
Vol. 18	Kinetic Methods in Chemical Analysis Application of Computers in Analytical Chemistry
Vol. 19	Analytical Visible and Ultra-violet Spectrometry
Vol. 20	Photometric Methods in Inorganic Trace Analysis
Vol. 21	New Developments in Conductometric and Oscillometric Analysis
Vol. 22	Titrimetric Analysis in Organic Solvents
Vol. 23	Analytical and Biomedical Applications of Ion-Selective Field-Effect Transistors
Vol. 24	Energy Dispersive X-ray Fluorescence Analysis
Vol. 25	Preconcentration of Trace Elements
Vol. 26	Radionuclide X-ray Fluorecence Analysis
Vol. 27	Voltammetry
Vol. 28	Analysis of Substances in the Gaseous Phase
Vol. 29	Chemiluminescence Immunoassay
Vol. 30	Spectrochemical Trace Analysis for Metals and Metalloids
Vol. 31	Surfactants in Analytical Chemistry
Vol. 32	Environmental Analytical Chemistry
Vol. 33	Elemental Speciation – New Approaches for Trace Element Analysis
Vol. 34	Discrete Sample Introduction Techniques for Inductively Coupled Plasma Mass Spectrometry
Vol. 35	Modern Fourier Transform Infrared Spectroscopy
Vol. 36	Chemical Test Methods of Analysis
Vol. 37	Sampling and Sample Preparation for Field and Laboratory
Vol. 38	Countercurrent Chromatography: The Support-Free Liquid Stationary Phase
Vol. 39	Integrated Analytical Systems
Vol. 40	Analysis and Fate of Surfactants in the Aquatic Environment
Vol. 41	Sample Preparation for Trace Element Analysis
Vol. 42	Non-destructive Microanalysis of Cultural Heritage Materials
Vol. 43	Chromatographic-mass Spectrometric Food Analysis for Trace Determination of Pesticide Residues
Vol. 44	Biosensors and Modern Biospecific Analytical Techniques
Vol. 45	Analysis and Detection by Capillary Electrophoresis
Vol. 46	Proteomics and Peptidomics: New Technology Platforms Elucidating Biology

PART I
Principles, Basic Concepts and Instrumentation

CHAPTER 1

Multidimensionality in Gas Chromatography: General Concepts

Lourdes Ramos and **Udo A.Th. Brinkman**

Contents

1. INTRODUCTION

Gas chromatography with open-tubular capillary columns (GC) is a powerful separation technique that is particularly suitable for determining (semi-)volatile compounds. Application to less volatile analytes is also possible provided that these analytes are first transformed into volatile derivatives. Since its introduction by Golay [1], many examples have illustrated the potential of GC for accurate identification and quantification of individual analytes in many types of difficult real-life mixtures. The excellent resolution provided by present-day one-dimensional (1D) GC, combined with its accuracy and robustness, makes this technique the preferred separation approach in a variety of application areas. However, the improved detection capabilities provided by state-of-the-art detectors have also shown that, in many cases, aroma, food, petrochemical, and environmental samples are much more complex than was assumed a decade ago. Improved knowledge about the composition of such samples demands enhanced resolution — that is, adding an extra separation/identification capability over that achieved with 1D GC. For example, considering that 1D GC separation often

Comprehensive Analytical Chemistry, Volume 55
ISSN: 0166-526X, DOI 10.1016/S0166-526X(09)05501-9

relies on a single ("bulk") separation criterion (e.g., the different volatilities of the target analytes), if the vapor pressures of several analytes in a mixture do not differ sufficiently, co-elution will occur. Separation of these co-eluting analytes from each other will require the application of another separation criterion through which these analytes can be differentiated, for example, their different polarities. In principle, if co-elution remains, a third separation mechanism can subsequently be applied until complete resolution is achieved. Proper combination of the information provided by the several separation procedures will yield accurate information about the composition of the extracts.

If each separation mechanism applied to the resolution of a mixture is defined as a separation "dimension," the subsequent application of different separation mechanisms can be called a multidimensional approach. This chapter briefly introduces the concept of multidimensionality in the context of GC separations and discusses how switching from traditional multidimensional GC (MDGC or GC–GC) to comprehensive two-dimensional GC (GC×GC) dramatically enhances the potential of the technique.

2. BASIC CONCEPTS OF MULTIDIMENSIONALITY

An enormous variety of combinations of different separation mechanisms can be used to create multidimensional separation systems. A significant number of these combinations have already been implemented successfully, and the experimental results nicely illustrate the potential of high separation power that is typically associated with these techniques. For other combinations, especially those involving the cross-coupling of widely different separation techniques (e.g., liquid to gas; severe miniaturization; different flow regimes), development remains much more limited. That is, despite the promising advances made in the past two decades, in practice we are certainly far away from being able to set up the 10^4 to 10^6 binary combinations initially estimated by Giddings as possible multidimensional combinations based on known separation techniques and their variations [2]. The main stumbling block no doubt is the proper (on-line) coupling of the various dimensions or systems required for the multidimensional approach. Even for the relatively straightforward coupling of two modes of GC operation — and specifically for comprehensive GC — designing and properly using coupling devices or interfaces has been found to be a challenging task (see Chapter 2 for full details). The instrumental setup will rapidly become more complex when more than two separation systems are combined, and two-dimensional gas chromatography (2D GC) — whether of the MDGC or GC×GC type — can be considered the only combination of practical value.

The basic requirements for a multiple separation to be considered multidimensional were already discussed by Giddings in 1990 [2]. Two conditions should be fulfilled:

- The components of a mixture should be subjected to two (or more) separation steps in which their displacement is governed by different factors.

– Analytes that have been resolved in an earlier step should remain separated until the separation process is completed.

If, as the first condition requires, two (or more) independent separation mechanisms have to be used, this will result in a similar number of parameters to define the identity of an analyte [3]. That is, compared to 1D GC, each analyte in 2D GC is characterized by two independent retention times rather than by a single one, and characterization is done by indicating the proper analyte location in a 2D plane in which the two axes correspond to the retention time scales of the two separations. Obviously, preliminary identification of analytes becomes correspondingly more reliable.

The second condition requires the separate analysis of relatively small fractions of eluate from the first column on the second one, in order to maintain the resolution already achieved on the first column. Preferably, a cryogenic device should be used in between the two columns, acting as an interface to trap, focus, and release the subsequently arriving fractions onto the second column.

3. 2D GC: FROM MDGC TO GC×GC

The standard procedure in gas chromatography is to perform 1 D GC on a 30- to 60-m-long capillary column, under temperature-programming conditions, thereby achieving a peak capacity of some 100–150. If more resolution is desired, the higher potential of 2D GC is invoked (Figure 1).

The basic setup of the multidimensional system that is now required is, in principle, fairly simple. The 30- to 60-m GC column is coupled in series, via a cryotype interface (cf. above), to a second column packed with a stationary phase

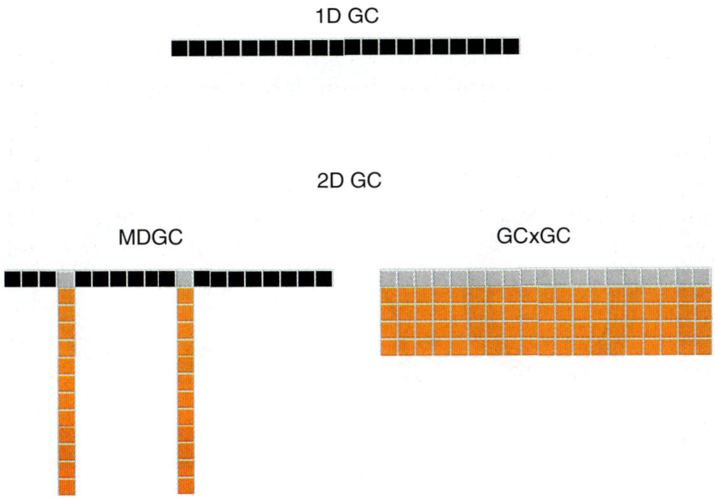

Figure 1 Comparison of the one-dimensional and the main types of two-dimensional GC.

providing another type of selectivity. The second column is often somewhat shorter than the first one (e.g., 10 to 20 m) in order to limit the overall runtime while, of course, effecting the desired resolution. The typical setup for the first class of 2D separations, introduced by Deans in 1968 [4], MDGC, is shown in Figure 2.

The typical procedure in MDGC is to select the fraction — or, the few fractions — of interest eluting from the first column — that is, those that contain the target compounds — and to subject these, one after the other, to the second, independent, separation (see Figure 1, bottom left). MDGC was used quite widely in the final decades of the twentieth century, but it never became very popular. It probably did not catch on because the instrumental setup, though considered "fairly simple" today, was thought to be too complex for routine use at the time. (For one important exception, see the discussion of the PIONA analyzer in Chapter 7.)

The main advantages of the MDGC approach are that, in principle, the most powerful second-dimension column can be selected for each individual target-analysis problem, and that there are no time constraints: as regards duration of the run, the second separation is not "coupled" to the first one. Successful applications include the separation of several groups of polychlorinated biphenyls (PCBs) co-eluting on a conventional nonpolar column and, then, adequately separated from each other, and from interfering sample constituents, on a more polar second column [6]. Admittedly, this frequently had the disadvantage that several heart-cuts had to be taken during one run. Another application of interest was, and remains, the use of enantioselective MDGC in food and fragrance analysis for determining enantiomeric ratios, for example, in order to detect

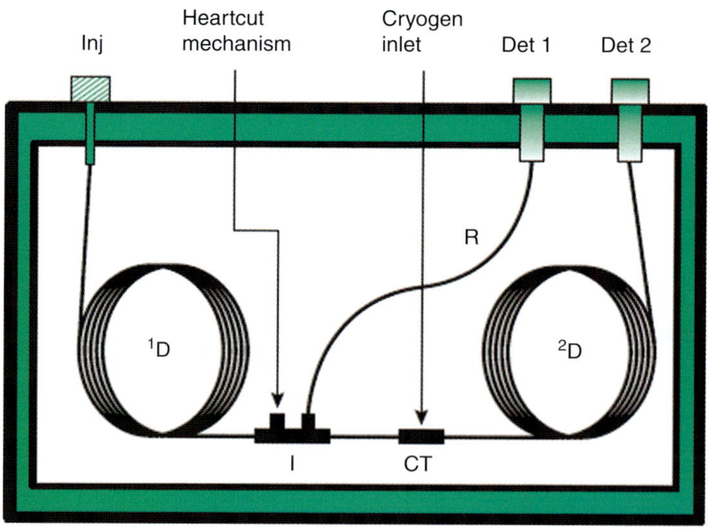

Figure 2 Schematic of the basic setup for MDGC [5]. Note that a large majority of the first-column eluent is sent to waste via capillary R (with added monitoring detector).

adulteration (see, e.g. [7]). In this case, the absence of time constraints in the second dimension is a distinct advantage: because of their rather poor selectivity, chiral separations almost always require long columns. As an illustration, Figure 3 shows the (enantioselective) analysis of CBs 153 and 105, and the two

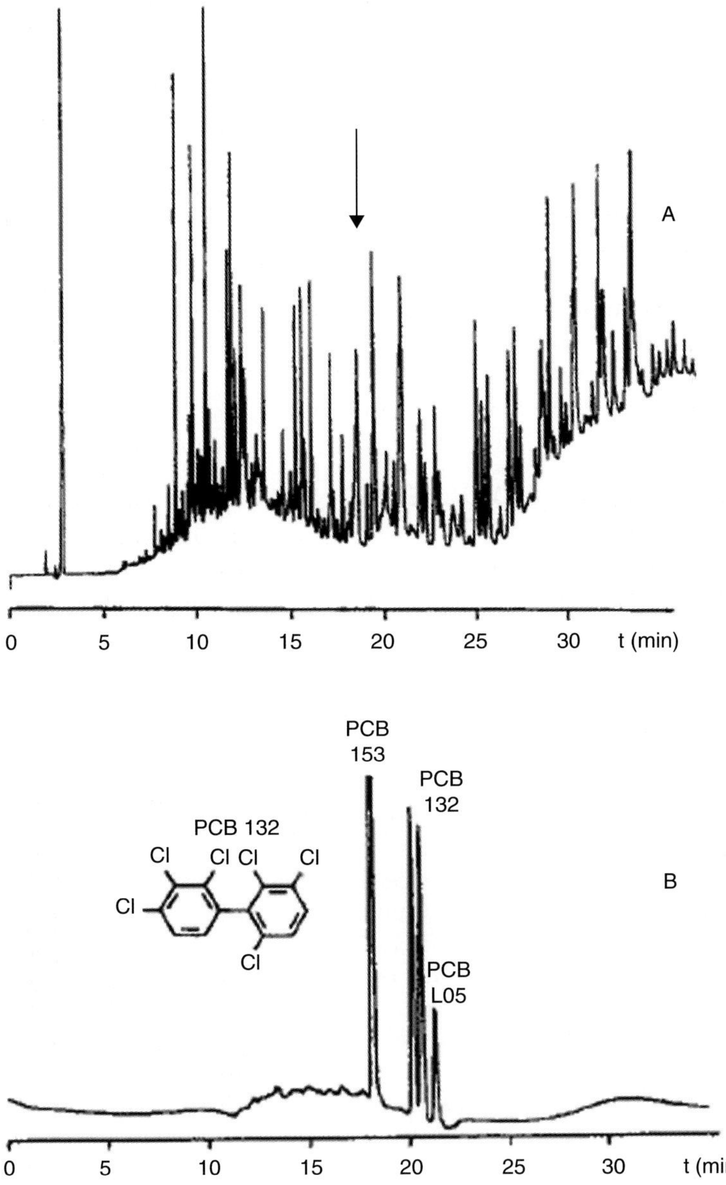

Figure 3 (A) 1D GC separation of PCBs on a DB-5 column and (B) enantioselective MDGC separation of CBs 153 and 105, and the atropisomers of CB 132 in a sediment extract [8]. The arrow in frame A indicates the position of the fraction transferred to the chiral column.

atropisomers of CB 132, in a purified sediment extract using a 30 m DB-5–10 m Chirasil-Dex column set [8].

MDGC also has serious limitations. Its main problem is that subjecting more than one or two fractions to a second analysis dramatically increases the total runtime. After all, each second-dimension analysis will easily add 30 to 45 minutes to the runtime. Even if first-dimension eluate fractions of 30- to 60-s width are used (which is, actually, a violation of the second of Giddings's conditions!), the analysis of an entire mixture will involve some 30 to 60 reanalyses. In other words, they will require something like 25 to 30 hours! Highly accurate timing during the entire process and the final precise reconstruction of the chromatogram are additional practical problems.

In summary, the general conclusion regarding the merits and demerits of MDGC is rather obvious: this technique is well suited for analysis of a limited number of target compounds in highly complex samples — what is, in practice, frequently called "heart-cutting" — but it is unsuitable for the general monitoring of entire samples and/or the search for unknowns. It is specifically for these problems that a truly comprehensive approach should be designed, which will provide second-dimension information across the board of the entire first-dimension chromatogram.

4. COMPREHENSIVE TWO-DIMENSIONAL GAS CHROMATOGRAPHY: GC×GC

In the previous section, it was concluded that MDGC cannot solve problems when total profiling of complex samples is required. One therefore has to return to the 2D GC part of Figure 1 and devise, as an alternative to MDGC, a comprehensive design in which the entire eluent from the first-dimension column is subjected to reanalysis on a second column. As indicated in the bottom right-hand side of that figure, this will have to be done in a continuous — and real-time — mode in order to (i) maintain essentially all of the first-dimension resolution, thereby meeting Giddings's second condition, and (ii) design an analytical procedure that is not excessively time consuming. In order to attain these goals, both theory and experience show that at least three to four heart-cuts should be taken across each first-dimension peak, and that reanalysis of each fraction on the second column should be completed before the next fraction arrives at the top of that column. This is the problem that, in the early 1990s, was solved by Phillips and his co-workers [9].

A typical setup for GC×GC analysis is shown in Figure 4, which at first glance seems to be little different than the MDGC schematic of Figure 2. However, this is not the case. First, no waste line is inserted in between the two columns because the *entire sample* is subjected to reanalysis and *no fractions* go unexplored. Second, because the second-dimension separation of each fraction has to be completed before the next fraction is subjected to reanalysis, the second column has to be an extremely short (and small-bore) one, with a length of, typically, 0.5 to 2 m. Such a setup makes it possible to have a 2- to 8-s analysis time for each of the three to four

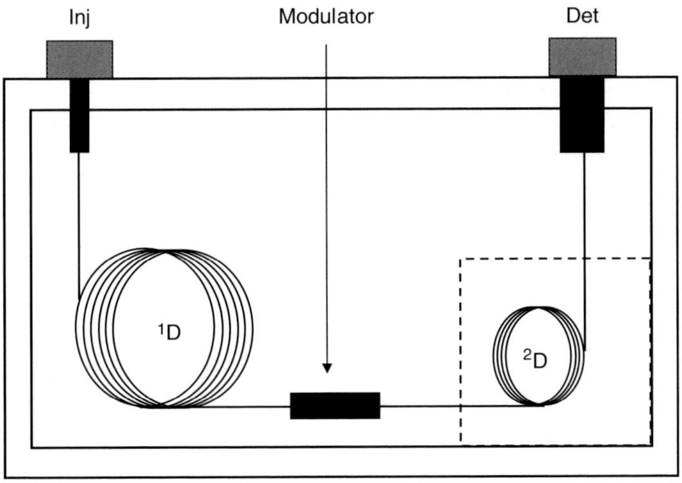

Figure 4 Schematic of the basic setup for GC×GC. Note that the whole sample elutes through both columns and that only one detector is used. Dotted line: optional secondary oven, placed inside the main oven, for independent temperature control of the second column.

heart-cuts (or *modulations*, as they are usually called in GC×GC) across a first-dimension peak with a typical baseline width of 5 to 30 s. (Experience shows that in order to really have three to four modulations per peak, the temperature-programming rate of GC×GC should be somewhat slower than that of 1D GC, viz. some 1 to 3°C/min.) The modulation process itself is of crucial importance, and, indeed, in the early years of GC×GC much effort was devoted to designing robust as well as user-friendly interfaces — invariably called *modulators* here. Today, all technical problems in this area have been solved, with cryogenic jet-based modulators with either carbon dioxide or liquid nitrogen cooling dominating the market. For a detailed discussion of the highly efficient trapping/refocusing/rapid release process (which marks a third key difference between MDGC and GC×GC) of the analytes, the reader is referred to the discussion in Chapter 2. Relevant here is to observe that the refocusing, that is, peak compression, in the modulator effects a three- to fivefold gain in analyte detectability compared to 1D GC.

With very fast separations on the second column such as mentioned above and, consequently, merely 50- to 600-ms second-dimension baseline peak widths, it is of course necessary to use detectors that have short risetimes and small internal volumes. In the early years of GC×GC, detection therefore had to be done with flame-ionization detectors (FIDs), which have negligible internal volume and data acquisition rates of 50 to 300 Hz. Today, next to this general workhorse (which is especially useful in petrochemical studies), two selective detectors are frequently used. One is the micro electron-capture detector (μECD) which is essential for all studies dealing with organochlorine and organobromine micro-contaminants (see, e.g., [10,11] and Figure 6 below). The other is the sulphur chemiluminescence detector (SCD), which is used successfully to solve a variety of problems requiring S-selective detection [12,13]. Their practicality will be amply demonstrated in the application-oriented chapters of the present book.

In the early years of GC×GC, no commercial mass-spectrometric (MS) detector was available that could cope with the required high data-acquisition rate of the comprehensive technique. Fortunately, a time-of-flight MS (ToF MS) was marketed by Leco (St. Joseph, Missouri, USA) in the early 1990s. Such an instrument — in this case the Pegasus II and, more recently, the Pegasus III and Pegasus IV — can acquire some 100- to 500-mass spectra per second. With such high rates, proper reconstruction of even very rapid second-dimension peaks does not create problems, and neither does the subsequent deconvolution of overlapping peaks. For all studies that require the provisional identification or identity confirmation of sample constituents, use of the extremely powerful GC×GC–ToF MS combination is essential. This is convincingly demonstrated by the rapidly increasing number of published papers that utilize this technique and are quoted in most of the subsequent chapters. Here, it may suffice to give a fairly simple but illustrative example that deals with the GC×GC–ToF MS determination of chlorpyrifos in tobacco [14]. In Figure 5, the vertical black line indicates the position of three peaks with the same first-dimension retention time as chlorpyrifos, but clearly separated from each other on the second column. Neither identification nor quantification of the pesticide — and, for that matter, of either of the three interfering sample constituents — will cause any problems

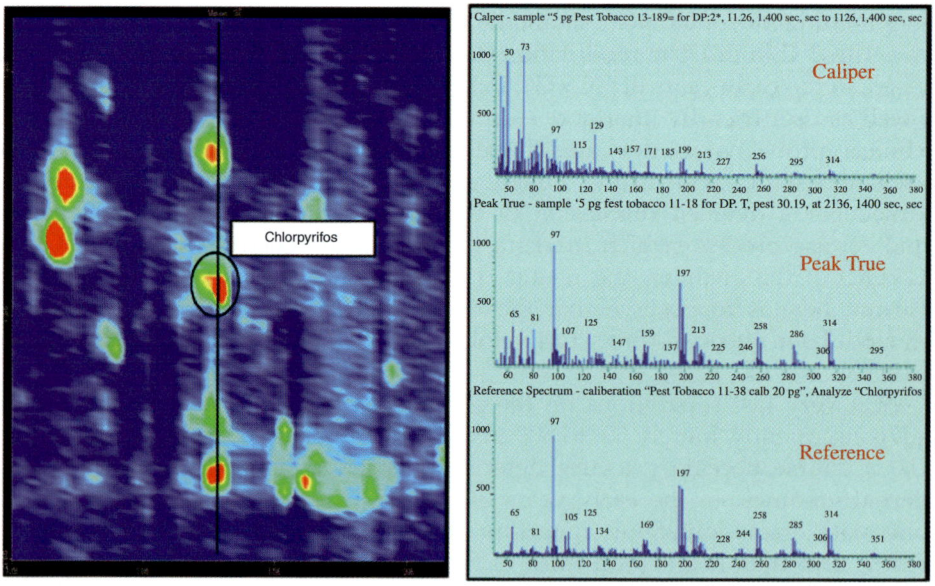

Figure 5 (Left) GC×GC–ToF MS contour plot for *m/z* 197 for a tobacco extract containing chlorpyrifos (black circle) using an Rtx-1×Rtx-200 column set. (Right) Caliper spectrum: unresolved spectrum at the peak apex where chlorpyrifos elutes. Peak true: deconvoluted spectrum. Reference: spectrum of chlorpyrifos standard [14].

now. This is in marked contrast with 1D GC where (i) the unambiguous recognition of four compounds would have been essentially impossible and (ii) the quantification of chlorpyrifos would have been adversely affected even with MS detection because all compounds have an m/z 197 fragment in their mass spectra.

In recent years, so-called rapid-scanning quadruple MS instruments have been marketed. They are much less expensive than ToF MS machines, and they also meet the data-acquisition demands of GC×GC, though over somewhat restricted mass ranges of, typically, 50 to 200 amu. As a consequence, they can be used for a variety of more or less targeted applications, and then are a valuable alternative to ToF MS. However, they cannot be recommended for wide-ranging screening studies or when sophisticated deconvolution of overlapping peaks is required [15,16]. Selected examples included in the second part of this volume will illustrate these aspects.

In GC×GC, columns have dimensions of, typically, (15–30) m×(0.25–0.32) mm ID×(0.1–1) μm d_f (film thickness) in the first, and (0.5–2) m×0.1 mm ID×0.1 mm d_f in the second, dimension. Usually, both columns are installed in the same oven. However, if there is the risk of too strong analyte retention on the second column, overlap of the GC profiles of subsequent modulations may occur (so-called *wrap-around*). A separate oven can then be installed (cf. Figure 4) to allow independent temperature control as a means of speeding up the separation.

In most applications, a nonpolar column×(medium-)polar or shape-selective column combination is the preferred setup. The main advantage is that the use of a nonpolar first column allows the virtual direct transfer of methods that have already been developed in the context of conventional 1D GC. In addition, while the analytes are separated in the first dimension on the basis of their different volatilities, the fast — and, consequently, essential isothermal — second-dimension separation neutralizes any further boiling-point contribution, and separation is only governed by specific interactions. In other words, the two chromatographic processes are independent; that is, the separation is *orthogonal* (see Chapters 2 and 3). Under these conditions, the peak capacity of the 2D separation is at its maximum, and, even more importantly, the chromatograms often display *ordered structures*, with related (classes of) compounds showing up as clusters or bands. In experimental practice, this significantly simplifies analyte or class recognition and provisional identification of unknowns. A typical example is shown in Figure 6, which features the analysis of a PCB mixture by means of GC×GC–μECD. Next to the easily recognizable separation according to the number of chlorine substituents per compound there is an additional classification based on the number of ortho chlorines present. Similar observations have been made for virtually all classes of organohalogen aromatic and aliphatic micro-contaminants (see [16,17] and Chapter 11).

More recently, it has been found that reverse-type column combinations — that is, those having a (medium-)polar first-dimension and a nonpolar second-dimension column — can also provide valuable results. This is especially true when highly polar or ionogenic analytes are among the key components of a mixture. For typical examples and a detailed discussion, the reader should consult e.g. Chapters 2 and 7 and Figures 6 and 3, respectively, in these chapters.

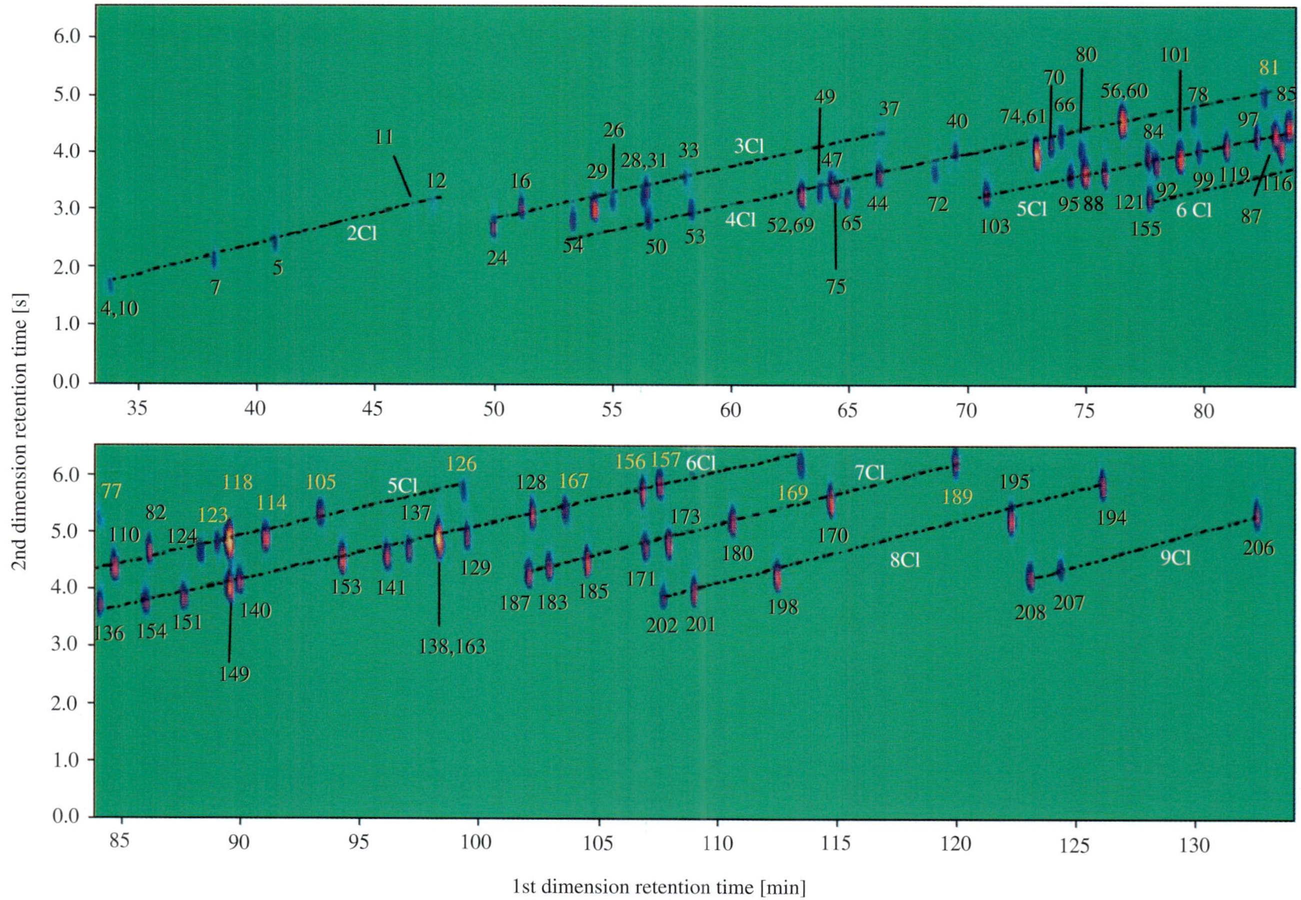

Figure 6 GC×GC–μECD chromatogram of a mixture of 90 PCBs with HP-1×HT-8 column combination. Dotted lines connect CB congeners with the same number of chlorine substituents in the molecule; the position of each CB congener within a group is determined by the substitution pattern on the biphenyl skeleton [10].

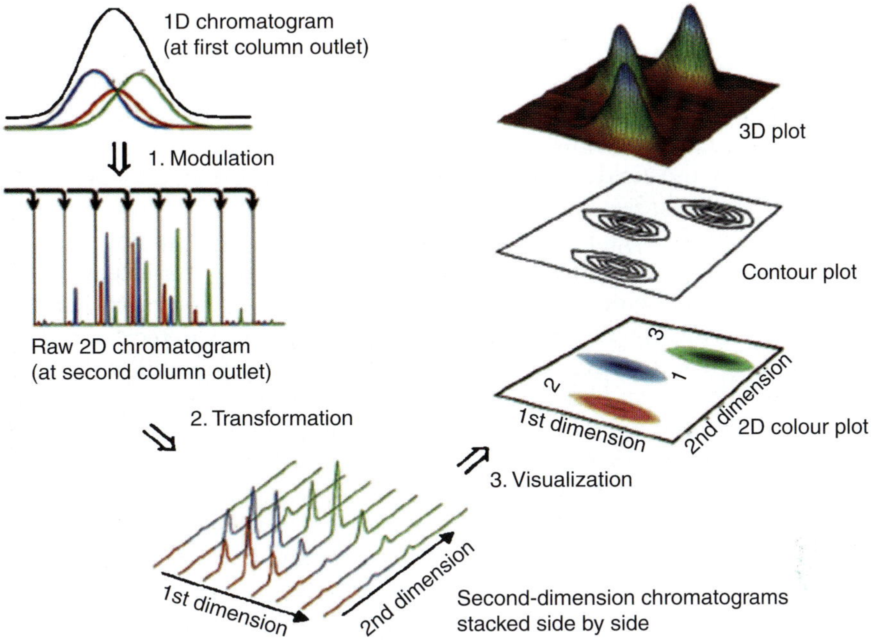

Figure 7 Data generation and visualization in GC×GC [18].

To round off the present discussion, we must finally consider the several steps of the data generation and visualization process; these are depicted in Figure 7. In brief, three analytes, which virtually co-elute after the first separation, are modulated (step 1) and transferred as narrow bands to the second column. The detector sitting at the outlet of that column registers the fractions containing the analytes, which now show distinct resolution, as a continuous sequence of high-speed second-dimension chromatograms. These raw data are then transformed (step 2) into a 2D representation in which the second-dimension chromatograms are stacked side by side, with the axes representing the retention on the two columns. After further treatment, the 2D representation is usually visualized (step 3) by means of a contour plot with, usually, a full-colour display of the signal intensities or, alternatively, as a 3D plot. Today, commercial software packages are available which include algorithms that perform the transformation of the raw data into 2D and 3D representations and incorporate, when applicable, additional relevant chemical information available in manageable data files and/or figures. Frequently, other basic data-treatment operations, such as baseline correction, peak detection, and peak integration, can also be performed in an automated way. However, despite the advances made in recent years, it should be emphasized that, specifically when complex chromatograms have to be evaluated and/or wide-range screening is a key aspect of a study, data processing and, especially, data interpretation clearly are the most time-consuming part of an analysis (see, e.g., [15,16] and Chapter 5). It is in this area that much effort is still required to arrive at robust, user-friendly, and validated solutions that analysts with limited expertise in chemometrics can routinely use.

5. CONCLUSIONS

In the past decade, GC×GC has become increasingly popular, and today, over 400 published papers are available in the literature. From an instrumental point of view, GC×GC can be called an established technique. Even so, further developments in the field of data processing and interpretation are required, specifically for all studies that involve highly rewarding, but also demanding, ToF MS detection. A large number of applications convincingly demonstrate the huge potential of the comprehensive separation technique in areas as diverse as petrochemical, organohalogen, food, flavors/fragrances, air/aerosol, and biological analysis. Typically, the number of peaks/compounds detected in complex samples is up to five- to tenfold higher with GC×GC as compared with MDGC or 1D GC. An additional benefit is that, under properly selected orthogonal separation conditions, many classes of compounds display ordered structures. This facilitates the recognition of structural relationships and the provisional identification of trace-level components and/or unknowns.

In summary, we feel that GC×GC — with GC×GC–ToF MS as its main representative — is here to stay and can be expected to have a very bright future. The contents of this book, which extensively discusses the principles, instrumentation, and many applications of the technique should convince the reader that this is indeed true.

ACKNOWLEDGMENTS

LR acknowledges MICINN for financial support via grant CTQ-2006-14993/BQU.

REFERENCES

1 M.J.E. Golay, *Gas Chromatography*, Academic Press, New York, 1958.
2 J.C. Giddings, Use of multiple dimensions in analytical separations. In: H.J. Cortés (Ed.), *Multidimensional chromatography. Techniques and applications*, Chromatographic Science Series, Vol. 50, Marcel Dekker, New York, 1990, pp. 1–27.
3 P. Schoenmakers, P. Marriott and J. Beens, *LC-GC Eur.*, 16 (2003) 335.
4 D.R. Deans, *Chromatographia*, 1 (1968) 18.
5 P.J. Marriott, P.D. Morrison, R.A. Shellie, M.S. Dunn, E. Sari and D. Ryan, *LC-GC Eur.*, 5(December) (2003) 2.
6 P. Hess, J. de Boer, W.P. Cofino, P.E.G. Leonards and D.E. Wells, *J. Chromatogr. A*, 703 (1995) 417.
7 G. Schomburg, *J. Chromatogr. A*, 703 (1995) 309.
8 A. Glausch, G.P. Blanch and V. Schurig, *J. Chromatogr. A*, 723 (1996) 399.
9 Z. Liu and J.B. Phillips, *J. Chromatogr. Sci.*, 29 (1991) 227.
10 P. Korytár, P.E.G. Leonards, J. de Boer and U.A.Th. Brinkman, *J. Chromatogr. A*, 958 (2002) 203.
11 P. Korytár, P.E.G. Leonards, J. de Boer and U.A.Th. Brinkman, *J. Chromatogr. A*, 1086 (2005) 29.
12 F.C.Y. Wang, W.K. Robbins, F.P. Di Sanzo and F.C. McElroy, *J. Chromatogr. Sci.*, 41 (2003) 519.
13 J. Blomberg, T. Riemersma, M. van Zuijlen and H. Chaabani, *J. Chromatogr. A*, 1050 (2004) 77.
14 J. Cochran, *J. Chromatogr. A*, 1186 (2008) 202.
15 M. Adahchour, J. Beens, R.J.J. Vreuls and U.A.Th. Brinkman, *Trends Anal. Chem.*, 25 (2006) 540.
16 M. Adahchour, J. Beens and U.A.Th. Brinkman, *J. Chromatogr. A*, 1186 (2008) 67.
17 P. Korytár, P. Haglund, J. de Boer and U.A.Th. Brinkman, *Trends Anal. Chem.*, 25 (2006) 373.
18 M. Adahchour, J. Beens, R.J.J. Vreuls and U.A.Th. Brinkman, *Trends Anal. Chem.*, 25 (2006) 438.

CHAPTER 2

Basic Instrumentation for GC×GC

Gaëlle Semard, Mohamed Adahchour and
Jean-François Focant

Contents

1. THE GC×GC TECHNIQUE

Comprehensive two-dimensional gas chromatography (GC×GC) has been developed to meet an increasing need for complex sample analysis and to address limitations such as peak capacity, dynamic range, and restricted specificity of one-dimensional (conventional) GC systems (1D GC), that is, to improve the global efficiency of the separation. GC×GC can be defined as a chromatographic technique in which a sample is subjected to two different separation processes coupled on-line [1]. As explained in Chapter 1, the more different the separation mechanisms applied, the more *orthogonal* is the system and the higher is its separation power. (The orthogonality principle will be discussed in Section 2.1.)

One can schematically illustrate the usefulness of such a system for the separation of a complex mixture of analytes (Figure 1). Let's consider a hypothetical

Comprehensive Analytical Chemistry, Volume 55
ISSN: 0166-526X, DOI 10.1016/S0166-526X(09)05502-0

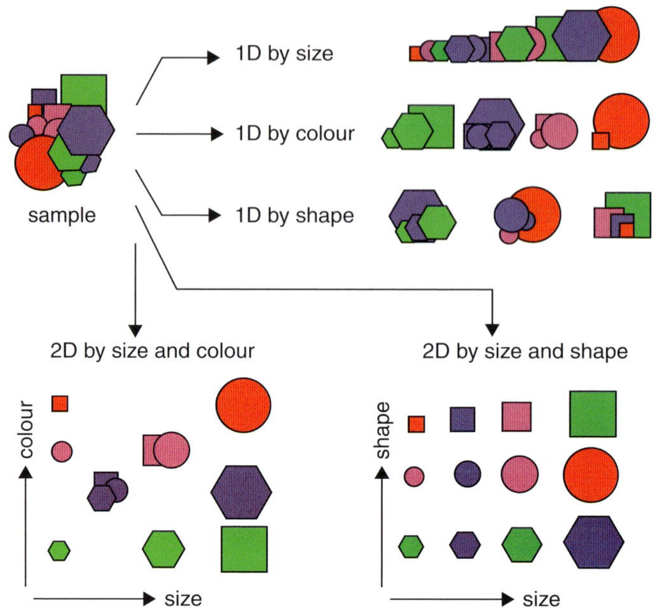

Figure 1 Match between separator and sample dimensionality in GC×GC.

sample that contains a large number of analytes that differ in shape, colour, and size. Following Giddings's guidelines [2], the sample can be characterized by a dimensionality of three. In these conditions, there is virtually no chance to separate all the analytes using a conventional single dimension system. With such a 1D system, the separation can either be performed according to size, but then colour and shape will remain unseparated; or it can be performed according to colour, but then size and shape will remain unseparated. Or, finally, the separation can be performed according to shape, and then size and colour will remain unseparated. A viable approach to achieve the separation of all the constituents of this sample is to use an orthogonal two-dimensional (2D) separation system with a dimensionality that can match the dimensionality of the sample [2]. In that case, one can use most of the available separation space very efficiently to accommodate separated analytes and create a highly structured elution pattern.

In practice, in GC×GC, everything starts by injecting and mixing the sample with the carrier gas in the injector, as in 1D GC. However, rather than entering the detector when exiting the GC column, solutes arrive in an interface named the *modulator*, placed between the two separation dimensions (columns) connected in series. The modulator ensures high sampling rates and the transfer of the sample from the first dimension [3] (^{1}D) to the second (^{2}D) while respecting Giddings's conservation rules (Chapter 1). The modulator acts as an on-line injector that produces very narrow injection pulses (down to 50 ms peak width) on the second column head, accounting for a fast sampling of compounds eluting from ^{1}D. The entire ^{1}D chromatogram is thus "sliced" following a modulation period (P_M) of a few seconds and reinjected into ^{2}D for a fast GC-type separation (Figure 2A) [4].

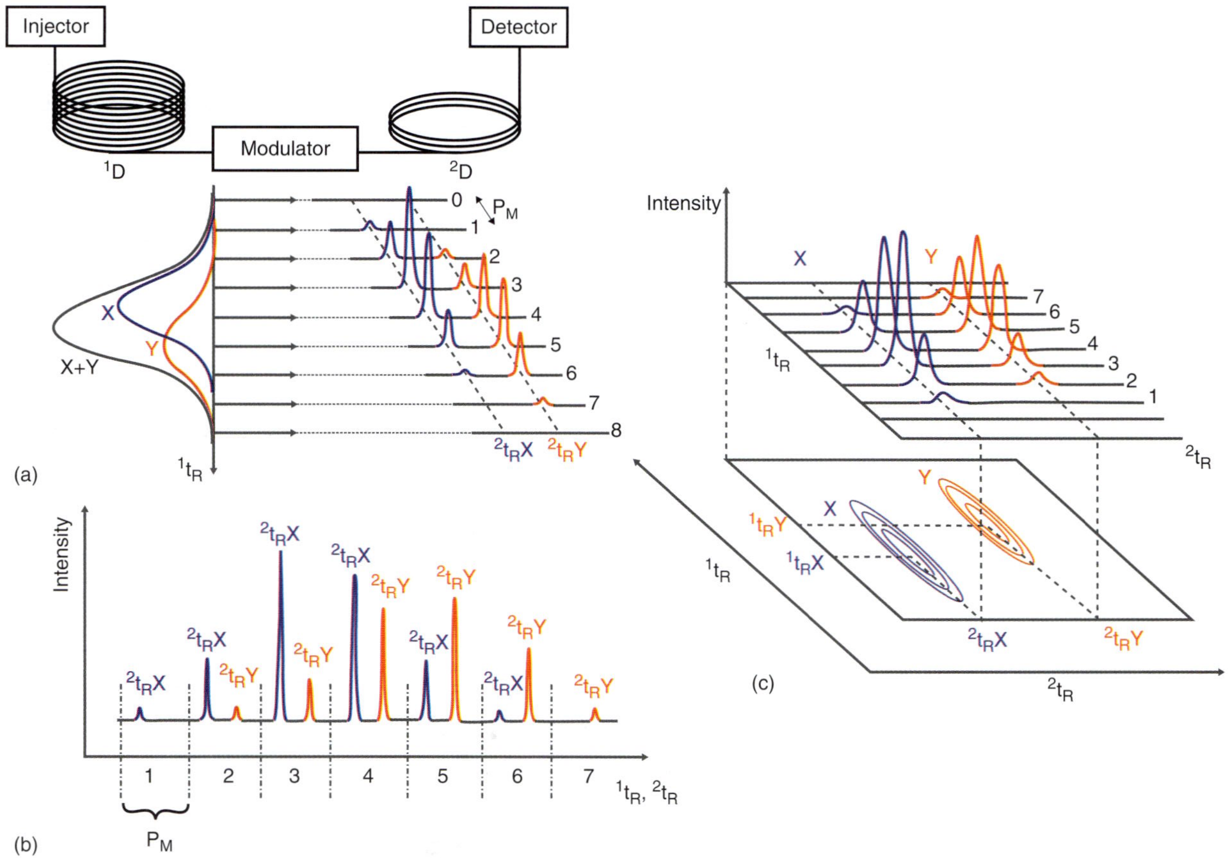

Figure 2 Scheme of the column coupling in the GC×GC setup and of how data are handled (not to scale) [4]. (A) The modulator allows rapid sampling of the analytes eluting out of ^1D and reinjection in ^2D. The modulation process is illustrated for two overlapping compounds (X and Y) coming out of ^1D at a defined first-dimension retention time (^1t_R). As the modulation process occurs during a defined P_M, narrow bands of sampled analytes are entering ^2D and appear to have different second-dimension retention times (^2t_RX and ^2t_RY). (B) Raw data signal as recorded by the detector through the entire separation process. (C) Construction of the two-dimensional contour plot from the collected high-speed secondary chromatograms of (B), in which similar signal intensities are connected by contour lines.

Ideally, the separation of analytes reinjected in ^{2}D by the modulator pulse has to be completed before another pulse is injected in the ^{2}D to avoid overlap of peaks issued from different modulation cycles (an effect called *wraparound*; see Chapter 3). That is why the second GC column is normally much shorter than the first one and why a secondary GC oven is sometimes used to heat the ^{2}D column. The use of short, narrow columns as a second-dimension results in a separation time in ^{2}D that is about 100 times more rapid than that in ^{1}D. Because sampling (modulation) occurs during the ^{1}D separation, the total GC runtime of a GC×GC separation is about the same as in conventional GC. For the detector responsible for recording the signal, everything occurs as in classical GC and a trace is monitored continuously. Actually, a series of high-speed secondary chromatograms of a length equal to P_M (3–10 s) are recorded one after another (Figure 2B). They consist of slices that can be combined to describe the elution pattern by means of 2D contour plots in the chromatographic separation plane (Figure 2C). A devoted software is responsible for processing the collected raw data and extracting the multidimensional information.

As an example, in Figure 3A, the inside of the ^{1}D oven of a GC×GC system equipped with a quad-jet dual stage cryogenic modulator is pictured (see later in this chapter for further details on the modulation process using this particular modulator). A GC split-splitless injector is located on the top left side, the ^{1}D column (30 m) is located on the right part, and the end of ^{1}D is visible on the bottom left, where one can see the press-fit connector to the ^{2}D column (2 m). The ^{2}D column enters the modulator that is located in the bottom part of the piece of hardware mounted on the left wall of the GC oven. Figure 3B shows the inside of the modulator. It is made of a bottom part and a top part that are identical. The bottom part is used for trapping the analytes, while the top part is used for refocusing. In each part, the exhaust pipe is the cold jet, and the narrow slit on the right of the column is the hot jet. In this system, a liquid nitrogen dewar is used to cool the dried nitrogen passing through the cold jets and used to cryotrap the effluent from the first-dimension column. The "modulator block" is programmed at a temperature above those of the primary and the secondary oven in order to facilitate the release of analytes in the second dimension. One can note that in this setup the modulation takes place on the ^{2}D column; that is, no modulation tube is used. The ^{2}D column then enters the ^{2}D oven where it is coiled (Figure 3C). The ^{2}D column finally exits the ^{2}D oven through the left wall of the ^{1}D oven to the detector. In this case, it exits to a time-of-flight mass spectrometer (ToF MS). If any other detector was used, the ^{2}D column will exit the secondary oven through its right wall.

Depending on the modulator used, various connecting setups can be adopted to link the ^{1}D to the ^{2}D. In most heated modulators, a piece of uncoated capillary tubing follows a piece of thick film capillary column to ensure phase termination and decrease the retention of analytes before creating the band transferred into the ^{2}D column. For cryogenic modulators, modulation can take place directly on the ^{2}D capillary column. For valve-based modulation, the valve sample loop acts as a connection between the two dimensions. In other cases, the modulation process can simply take place at the beginning of the ^{2}D column. In the early

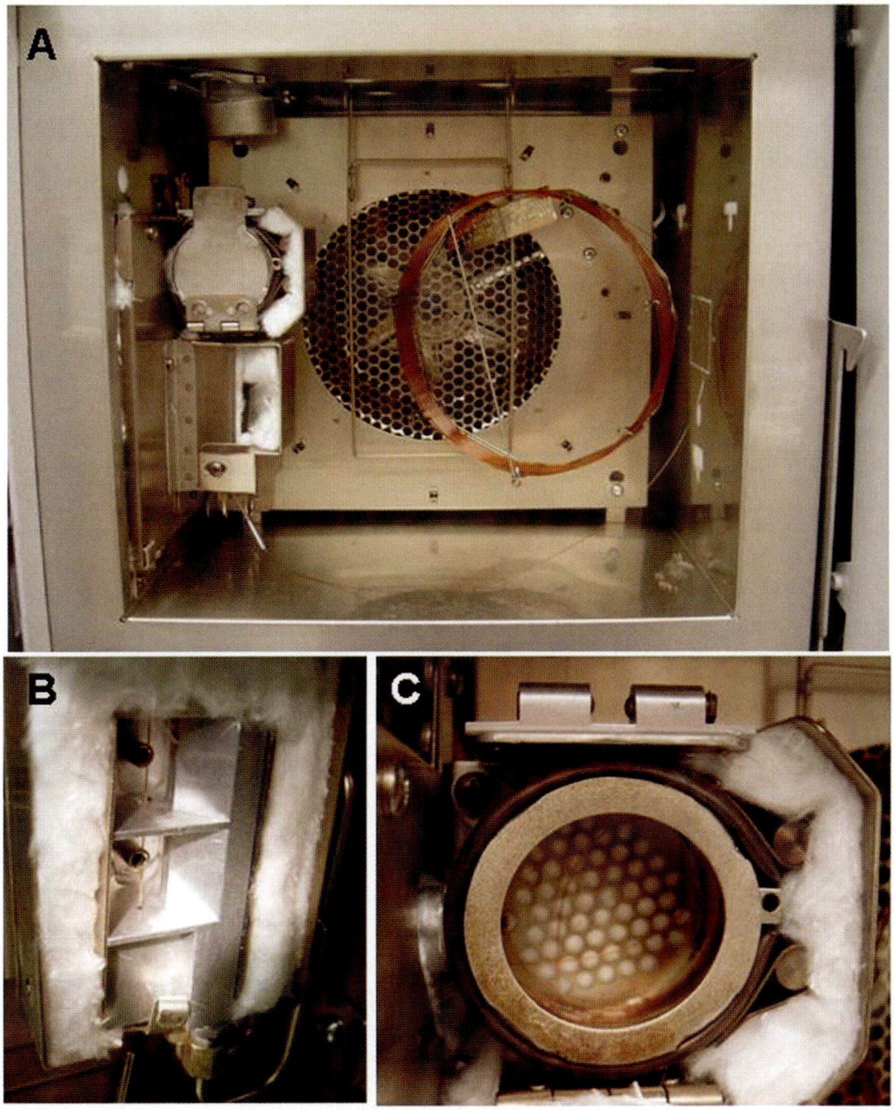

Figure 3 Picture of a typical GC×GC setup. (A) Inside of the ^{1}D oven of a GC×GC system equipped with a quad-jet dual-stage cryogenic modulator. (B) Detail of the quad-jet dual-stage cryogenic modulator. (C) Detail of the ^{2}D oven with the coiled ^{2}D column.

days, a rather complex procedure had to be used to ensure proper connection between ^{1}D column, the modulation tube, and ^{2}D column. The availability of simple glass press-fit connectors [5] greatly eased the coupling operation, which takes just a couple of minutes to be carried out either manually or with the help of an automated system.

2. COLUMN COMBINATIONS

2.1 The orthogonality principle

To satisfy the orthogonality principle, two different and independent separation mechanisms must be used in the two GC columns that are connected in series together. In most cases, GC×GC systems associate a first column having an non-polar stationary phase, for example, dimethylpolysiloxane, with a second column whose stationary phase is more polar, for example, polyethylene glycol, phenylmethylpolysiloxane, or cyclodextrine [6,7]. In this configuration, solutes are generally separated as a function of increasing boiling point in ^{1}D. Meanwhile, separation in ^{2}D is carried out under essentially isothermal conditions (a few seconds analysis for a temperature rate of 3–10°C/min) and governed by the activity coefficient of the analytes. Analytes are then separated using two independent separation mechanisms, and the process is referred as orthogonal [8]. When a first column with a non-polar stationary phase is coupled with a second more polar column, the configuration is usually defined as that corresponding to normal orthogonality (Figure 4). Conversely, a reverse orthogonality associates a first (semi-)polar column with a second non-polar column. This terminology will be adopted throughout this chapter. According to Adahchour et al. [9], in 80% of cases, normal orthogonality conditions (non-polar×medium-polar) were used in papers reported up to 2005.

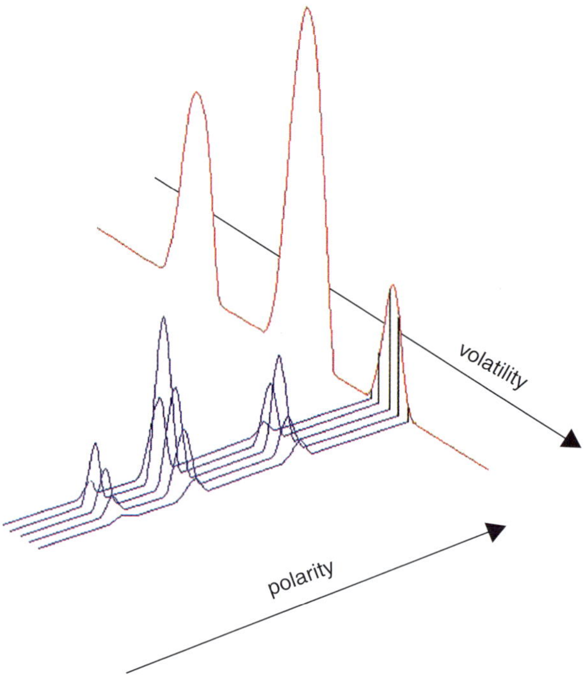

Figure 4 Orthogonality in GC×GC.

The more orthogonal the system, the more independent the retention mechanisms of the two columns will be, and thus, theoretically, the more efficient the separation of the compounds will be (Chapter 1). However, another important feature of GC×GC is the feasibility of obtaining structured chromatograms in which all compounds are gathered by chemical classes in the retention space. Structured chromatograms are indeed an important part of GC×GC, most notably in the petroleum world. However, for trace-level environmental analysis, the most important goal is to separate the analytes of interest from potential matrix interferences, and structure takes a back seat. Indeed, the peaks of compounds from a homologous class will be distributed along an elution band allowing them to be gathered together. Nevertheless, one can conclude that orthogonality and structure are thus not goals in themselves; separation is the goal. As Giddings reported [2], "structure" can be obtained under non-orthogonal conditions if the separation dimensions and sample dimensionality are properly matched.

Schoenmakers et al. [10] describe this phenomenon as a valuable tool when performing group-type identifications. Structured chromatograms can yield a fingerprint of the sample. As previously mentioned, this is particularly useful when we are considering petroleum characterisation (Chapter 7) and environmental screening for persistent organic pollutants such as polychlorinated biphenyls (PCBs), which contain a high number of congeners (Chapter 11). Focant et al. reported a highly structured separation of PCBs, using a column set far from fully orthogonal, that allows contamination pattern recognition based on the *ortho*-chlorine substitution degree of the congeners (Figure 5) [11].

Although orthogonality offers various advantages as discussed above, the reversed orthogonal approach (polar×non-polar) has also been proved to be useful in some studies. The complementarity of the "normal" and "reverse" orthogonal approaches in GC×GC is important. As an example, Figure 6 shows two GC×GC chromatograms of the same olive oil extract [12]. The normal

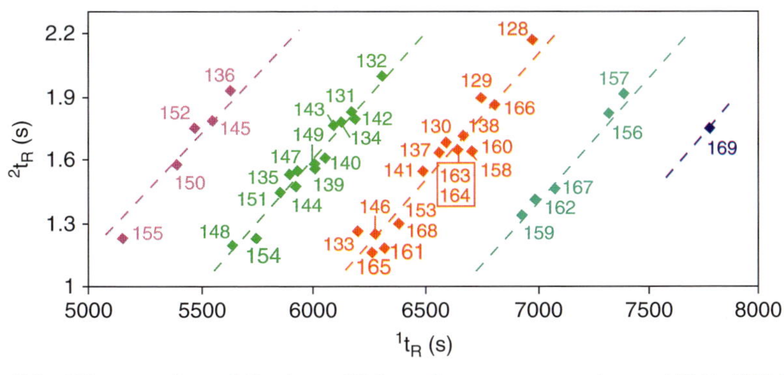

Figure 5 GC×GC separation of the hexa-CB homologue groups using an HT-8×BPX-50 column set. The *ortho*-chlorine substitution degree is described as follows: tetra-*ortho*-CBs are in pink, tri-*ortho*-CBs are in green, di-*ortho*-CBs are in red, mono-*ortho*-CBs are in light blue, and the non-*ortho*-CB is in dark blue. Boxes represent co-eluting congeners [11].

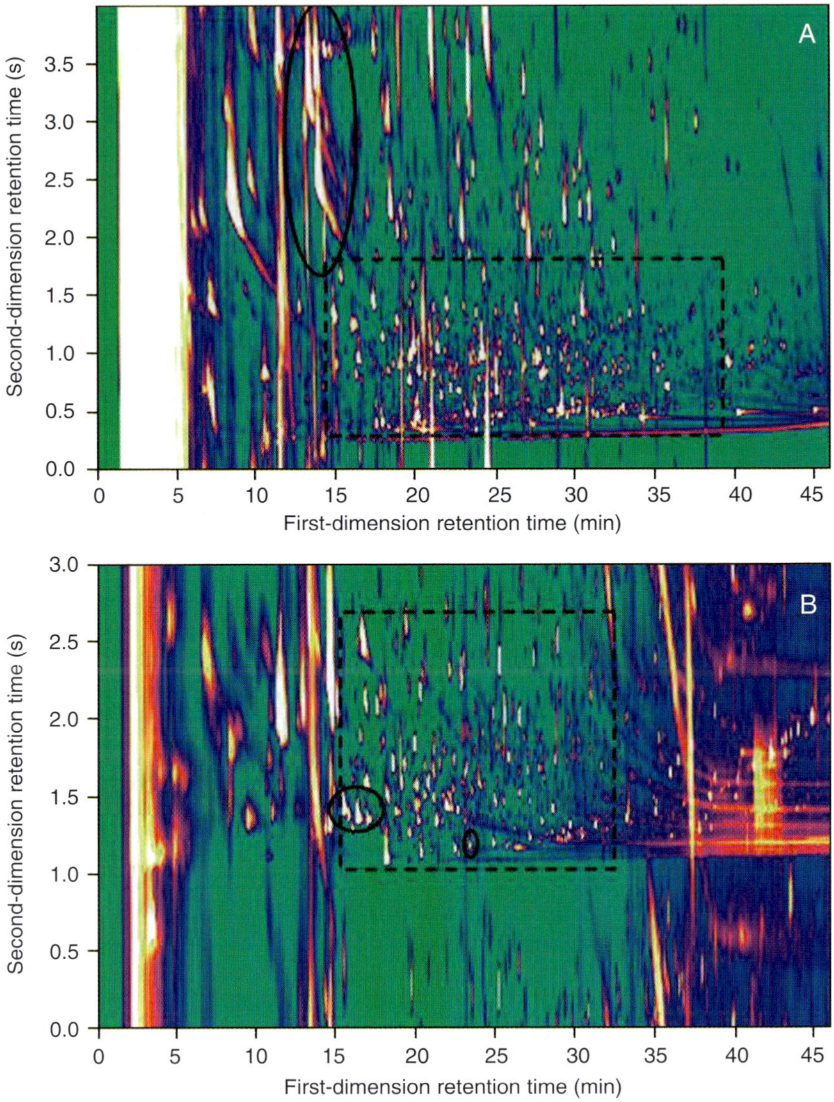

Figure 6 GC×GC–FID chromatograms of an olive oil extract obtained with (A) normal orthogonal and (B) reverse orthogonal approach. The circled spots show 3-methylbutanoic acid and the three alcohols, 1-hexanol, *cis*-3-hexenol and *trans*-2-hexenol. The zones marked by dashed lines delineate mainly non-polar analytes in (A) and polar analytes in (B) [12].

orthogonal approach yields better results for relatively non-polar analytes. These are retained less strongly than other sample constituents in the second dimension and show up as a band between 0.3 and 1.5 s in the chromatogram. In most instances, this creates an efficient separation from the polar matrix. However, as Figure 6A shows, co-elution with polar compounds showing wraparound, is a distinct drawback (see, e.g., the elongated spots eluting in the 20–25 min first-dimension time window). The reverse orthogonal approach, on the other hand, is more suitable for more polar analytes. These compounds are retained relatively strongly on the first column, which, in most instances, causes separation from the non-polar sample constituents (see Figure 6B, zone with 17–32 min first-dimension, and 1.0–2.5 s second-dimension retention times). Compared with the normal orthogonal approach, there is now no problem of wraparound (now observed for non-polar analytes), and the peak shapes of the polar analytes are fully satisfactory.

2.2 GC column dimensions

In GC×GC, the first-dimension columns are commonly 15 to 60 m long, with an inner diameter in the range of 0.25–0.53 mm and a film thickness of the order of 0.25–1 μm. These columns provide conventional peak widths of 5 to 30 s [13]. The first dimension is in fact not different from those typically used in conventional 1D GC. The main reason is that it is desirable to have rather large peak widths when entering the modulator to ensure proper sampling of potentially co-eluting analytes prior to their separation in the ^{2}D column. This point is discussed in detail in Section 3.

Ideally, the ^{2}D retention times of compounds pulsed out of the modulator must be less than or, at the most, equal to the modulation period (i.e., the duration of a complete cycle of modulation) [13,14] in order to prevent the wraparound effect. Wraparound of a compound is observed when the $^{2}t_{R}$ of the compound exceeds the P_{M} of the GC×GC system, with the result that this compound occurs in the next or subsequent sequences of modulation. Minimizing this effect is the reason the second column is usually shorter, 0.5 to 1.5 m long and has a smaller inner diameter than the first-dimension column. Film thickness is also normally reduced, in the 0.1–0.25 μm range, which allows increased separation efficiencies. In general, it may be true that narrower columns provide higher peak capacities than wider columns. In GC×GC, however, when using a narrow second column, the pressure drop across the second column is much higher than when using a wider second column. That means that the chromatographic process in the first column is much slower. The solution here is probably to use a wider second column that is somewhat longer. In addition, the second column can be placed in a separate oven in order to allow separate temperature control and help to prevent wraparound. Separations showing wraparound have long been considered as neither fully optimized nor ideal separations. It is interesting to note, however, that, as far as no new co-elutions are created due to the wrapping around of compounds, there is no major reason to spend time and effort to avoid wraparound in a separation procedure. And even if a specific chromatographic structure "wraps around," a trained eye or a dedicated software would easily reestablish a clear picture of the separation space [15]. The major disadvantage of wraparound is the slight increase of ^{2}D peak widths that could ultimately reduce the peak capacity. Another drawback of wraparound is the possibility of peak overlap (or co-elution) when single-dimensional detectors are used.

The vast majority of the applications in GC×GC were adapted from conventional GC classical methods. However, additional optimisation of column combinations, flow rate of carrier gas, and temperature program rate can bring improvement in GC×GC. Beens et al. [16] developed an Excel Calculator allowing prediction of the optimal conditions of separation for a selected set of columns. This model has highlighted that ^{2}D columns with an inner diameter of 0.1 mm were perhaps not the best choice for separation optimisation. During their study the authors obtained good results with ^{1}D columns with inner diameters of 0.25 to 0.32 mm connected with ^{2}D columns of 0.15 to 0.18 mm. When using a narrow second column, the flow rate should be far lower than the flow that is used in 1D GC. This is because the diffusion is strongly correlated with pressure in the column. So, in general, in these cases the optimal flow in the first column should be about half the flow that is used in 1D GC. Consequently, the total analysis time of GC×GC is twice the time used in 1D analysis.

Most of the studies in GC×GC are designed to obtain separation with a degree of selectivity, while conserving structure with a good sensitivity [17]. This choice causes a long runtime analysis. Harynuk and Marriott proposed another optimisation approach based on sacrificing selectivity in favour of speed using a fast GC column in first dimension. The authors used a BPX-5 (5 m×0.25 mm× 0.25 μm) coupled to BPX-50 (0.3 m×0.15 mm×0.15 μm). In this configuration, the elution temperatures of compounds were considerably reduced, thus allowing analysis of relatively high molecular compounds in short analytical time.

Zhu et al. [18] have tested several column combinations with different film thickness in first and second dimension. The authors concluded that when resolution is not necessary in the second dimension but speed is a priority, it is preferable to use a first column with a thin film thickness and a narrower second column. Conversely, when resolution is an important parameter, the use of a first column with a film thickness of about 0.25–0.5 μm and a second column with a large inner diameter (150 μm) seems to be a good compromise. An interesting possibility in GC×GC column combinations is the use of a dual parallel secondary column system by splitting the focused pulse and directing it to two parallel secondary columns instead of a single one. This was first described by Seeley et al. [19,20], who used an effluent splitter just behind the modulation valve for directing the selected fractions to two parallel and different secondary columns. The resulting technique, called dual-secondary column GC×GC (GC×2GC), produces a pair of two-dimensional contour or colour plots in a single run.

2.3 Stationary phases

Any existing stationary phases that can be used in GC can also be used in GC×GC. Table 1 lists some stationary phases that could be, and in most instances have been, used in GC×GC. A variety of stationary phases can be selected according to the intended analyte–stationary phase interaction. Although the number of commercially available columns with very narrow bore dimensions

Table 1 List of some stationary phases, with names and producers, which could be used in GC×GC

Stationary phase	Commercial code	Producers
100% dimethylpolysiloxane	HP-1, DB-1	J&W
	Rtx-1	Restek
	BP-1, BPX-1	SGE
Low polarity, proprietary	DB-XLB	J&W
	Rxi-XLB	Restek
(5% phenyl)-polycarborane-siloxane	HT-5	SGE
(8% phenyl)-polycarborane-siloxane	HT-8	SGE
5% phenyl-methylsilphenylene	BPX-5	SGE
5% diphenyl-dimethylpolysiloxane	HP-5, DB-5	J&W
	XTI-5	Restek
	SPB-5	Supelco
	Rtx-5	Restek
	BP-5	SGE
6% cyanopropylphenyl-dimethylpolysiloxane	DB-1301	J&W
	Rtx-624	Restek
35% phenyl-methylpolysiloxane	HP-35, DB-35	J&W
	Rxi-35	Restek
14% cyanopropylphenyl-dimethylpolysiloxane	BP-10	SGE
	Rtx-1701	Restek
	DB-1701	J&W
	OV 1701	Quadrex
	BGB-1701	BGB
	Mega1701	Mega
50% phenyl-methylpolysiloxane	DB-17	J&W
	Rtx-17	Restek
	Mega17	Mega
(50% phenyl)-polysilphenylene-siloxane	BPX-50	SGE
25% cyanopropyl-25%phenyl-methylpolysiloxane	Mega25	Mega
50% cyanopropylphenyl-dimethylpolysiloxane	DB-225	J&W
	Rtx-225	Restek
70% cyanopropyl polysilphenylene-siloxane	BPX-70	SGE
100% bicyanopropyl polysiloxane	SP-2340	Supelco
	Rt-2560	Restek
polyethylene glycol	BP-20	SGE
	DB-Wax	J&W
	Stabilwax	Restek
	SupelcoWax-10	Supelco
polyethylene glycol in a sol-gel matrix	SolgelWax	SGE
	MegaWaxHT	Mega
(50% liquid crystalline/50% dimethyl) siloxane	LC-50	J&K
	Rt-LC50	Restek
Permethylated β-cyclodextrin in OV 1701	Cyclodex-B	SGE
	Rt-βDEXm	Restek

has increased in past years, the offer is still quite limited regarding more polar, thermally stable phases that can be used in ^{2}D. This definitely reduces the potential of GC×GC that can be attained today. Nevertheless, several laboratories specialized in producing customized columns have used more exotic stationary phases adapted to GC×GC. Recently, Cordero et al. [21] studied the orthogonality and degree of spatial occupation for columns of mixed stationary phases. The authors tested stationary phases coated with mixture ratios of 25 to 75% of polyethylene glycol and dimethyl polysiloxane in ^{2}D. They showed that this type of column improves the resolution of natural volatile compounds. In addition, medium polarity phases in ^{2}D are very useful for analyzing polar compounds with high efficiencies (narrow peaks and less tailing) in contrast to more polar columns (polyethylene glycol) that can produce distortion in retention that affects their quantification precision. A recent report from Sidisky et al. [22] highlighted the promising thermal properties of ionic liquid stationary phases that were used at temperatures up to 240°C. Also, Seeley et al. [23] used a high-temperature phosphorium ionic liquid column in a GC×GC application proving that these phases have to be considered as good candidates for the second dimension of a GC×GC setup.

The use of conventional columns in the first dimension and a carrier gas-flow rate similar to that used in 1D GC enables all injection techniques to be used (e.g., split, splitless, large volume, PTV and SPME) [7]. That further makes any method based on a conventional GC separation amenable to a GC×GC approach without reconsideration or reoptimization of the existing sample introduction technique.

3. MODULATORS

As we saw earlier, the key point of GC×GC is the interface between the two columns, the modulator. This device ensures high sampling rates and the transfer of the sample from ^{1}D to ^{2}D while respecting Giddings's conservation rules. To guarantee conservation of the first-dimension separation achieved, the fraction eluted from the modulator should be no wider than about one-quarter of the ^{1}D peak width [13]. Recommended is the ensurance of at least three to four cuts per peak of the first dimension by the modulator. Thus, a P_M of 2–8 s is generally chosen.

Focusing effect on the modulator also produces a gain in sensitivity in GC×GC. Lee et al. [24] estimate a four- to fivefold gain in GC×GC–FID compared with GC–FID, and a two- to fivefold gain in GC×GC–ToF MS compared to GC–MS. Similar results are obtained with GC×GC–μECD (three- to fivefold) [25]. In fact, the focusing step reduces the peak widths of the analytes. As a consequence, the height of the peak is increased as shown in Figure 7.

The several different modulators that are commercially available can be classified in two main categories: thermal modulators and valve-based modulators. Thermal modulators are the most frequently used and in turn can be broken down into two categories: those whose principle involves a temperature increase and, inversely, cryogenic modulators.

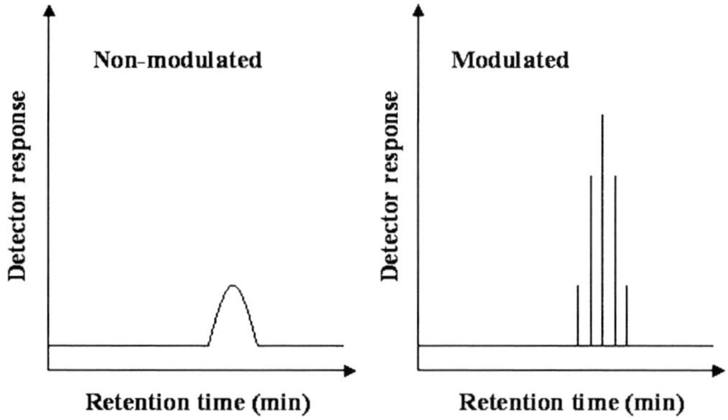

Figure 7 Gain in sensitivity by modulation in GC×GC.

3.1 Thermal modulators

3.1.1 Heating modulators

The first report on the concept of thermal modulation was published in 1985 by Phillips's group [26]. By then, the modulation principle was not yet involved in GC×GC but rather in the so-called multiplex gas chromatography concept that was used in the field of headspace sampling without preconcentration for low-level analytes. On-column thermal desorption modulation was then applied in the field of sample introduction as a device capable of narrow injection pulses for high-speed GC [27].

The concept of thermal desorption modulation (TDM) finally found its own specific area of application as the essential interface device between the two columns of a comprehensive two-dimensional separation system [28]. The TDM device consisted of a short section of a column that could be rapidly and reproducibly heated by an electrical current passing through an electrically conductive film. The authors used a segment of a capillary column with a narrow film thickness, covered with a layer of electrically conductive paint as the modulating interface between the two columns (Figure 8). Substances retained on the stationary phase of this modulator section were pulsed into the second column by resistive heating.

The low thermal mass of the conductive paint had the advantage of rapid heating and cooling of the device. A series of electrical pulses thus resulted in a series of concentration pulses in the sample stream of the second column. The main drawback was that those concentration pulses were not focused and that the analytes arriving in the modulator directly entered the second column as a relatively large injection plug.

In order to solve this refocalisation problem, a two-stage trap was later developed [29]. The two traps alternated their operation; that is, when one was hot, the other was cold, and vice versa. The reliability of this type of modulator was found to be insufficient, however, mainly because of the lack of reproducibility of the cover painting and the tediousness of proper operation.

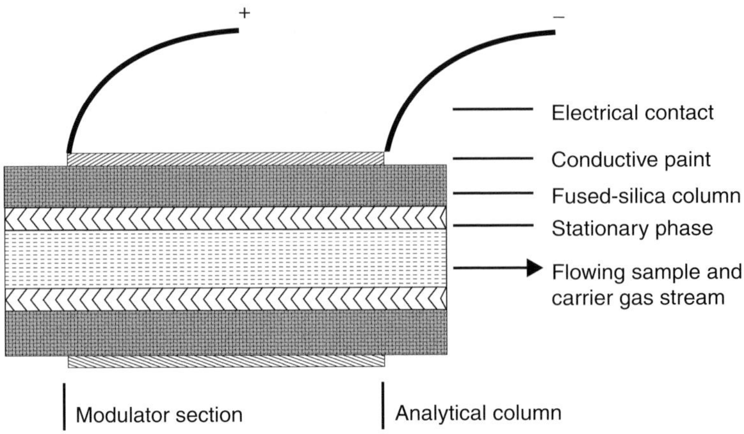

Figure 8 On-column thermal desorption modulator [27].

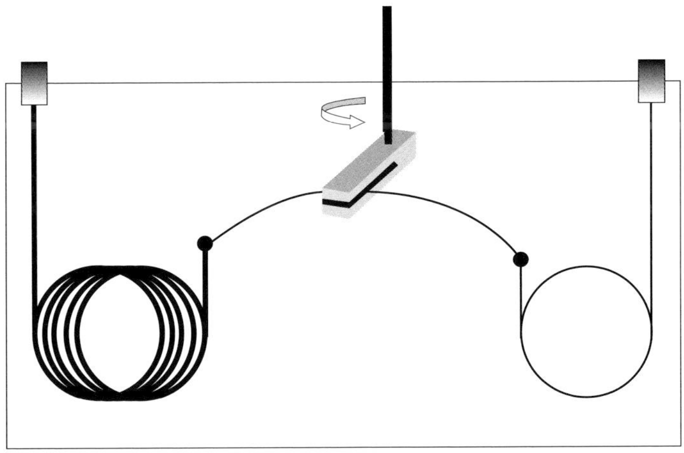

Figure 9 Diagram of the rotating sweeper modulator [31].

Further investigations by Phillips et al. eventually led to an improved thermal modulator known as the sweeper [30]. In that approach, the heating and cooling steps of the modulator tube involved applying and removing a heat source rather than directly heating the column. Sharp chemical pulses were obtained by mechanically sweeping a slotted heater (100°C higher than oven temperature) over a length of capillary column exhibiting a thick stationary phase for the trapping of analytes, the modulator tube. This multistage rotating thermal modulator allowed both the trapping and refocusing of the analytes before introduction into the second dimension. This was the first commercially available modulator (Figure 9). Independent temperature control of the modulation area and the secondary dimension was possible. It was effective for many types of analyses, but some mechanical weaknesses prevented its use as a routine tool.

Both modulators had significant disadvantages. It was practically impossible to collect volatile compounds with the heating trap. In addition, in order to prevent the thermal degradation of the stationary phase in the capillary used as a trap, the final oven temperature had to be 100°C lower than the upper working temperature of the stationary phase. Consequently, the maximum first-dimension column temperature was 230°C [32]. The range of compounds that could be analyzed was thus limited. Moreover, optimization of parameters (sweep velocity, pause time, thickness of stationary phase, and temperature difference between oven and heater) could be tedious and time consuming [22–35]. Even if the sweeper modulator was used in some 30% of all published papers [7] prior to 2003, it is no longer in use or even not commercially available today [32]. Some other heating modulators have been designed, specifically by Harynuk and Gorecki [36] and Burger et al. [37]. These modulators still had the same drawbacks and did not find large usability.

3.1.2 Cryogenic modulators

Based on the experiences and learning from the TDM investigations, a new generation of modulators appeared in the late 1990s. This new generation of modulator uses cooling rather than heating to create the required pulsing effect. In addition, efforts have been made to reduce or eliminate the need of moving parts. Cryogenic focusing was born and eventually replaced the heated trap in all fields of application [38,39].

Longitudinal Modulated Cryogenic System (LMCS) The LMCS was reported in 1999–2000 by Kinghorn and Marriot [40–44]. It is based on the use of a cryogenic trap that can be moved longitudinally along the column to focus analytes in the first centimeter of the second column. In practice, a segment of the column is cooled down with a liquid CO_2 jet, enabling analytes to be trapped in a small region at the outlet of the first column (Figure 10, position R). The analyte trap then quickly moves away from the cooled zone (Figure 10, position T). The cooled part of the column is rapidly heated by oven temperature, causing the release of trapped analytes into ^{2}D column. Once analytes are sent to the second column, the trap returns to its initial position (Figure 10, position R) and traps a second fraction.

To prevent the buildup of ice that can cause poor modulation efficiency, a stream of nitrogen is sent between the cold trap and the capillary segment. By using CO_2 as a cryogenic fluid, solutes as volatile as hexane cannot be successfully trapped and further modulated.

This modulator was the first reliable cryogenic modulator for routine use. One of the advantages of this modulator is that analyte desorption can be done at oven temperature and does not require a higher temperature. The main disadvantage of this technique is the use of liquid CO_2 as the cryogenic agent. The temperature of liquid CO_2, about −70°C, is insufficient to efficiently trap the most volatile compounds. Finally, in terms of reliability and robustness, the use of a moving shuttle at the vicinity of the fragile GC column could potentially yield to operational problems over time.

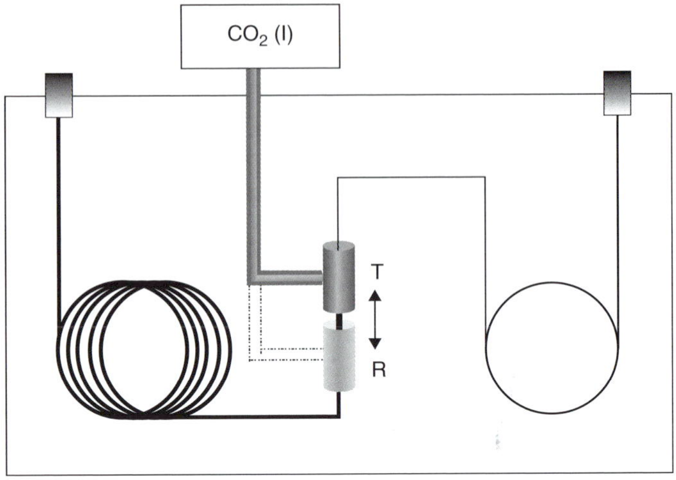

Figure 10 Schematic of the LMCS modulator. R: focusing position, T: releasing position [31].

Figure 11 Dual-stage liquid CO_2 modulator.

The dual-stage liquid CO_2 cryogenic modulator Beens et al. [45] developed another cryogenic modulator that also focuses analytes with liquid CO_2, but without undesirable moving parts. In practice, two parts of the capillary are directly and alternatively cooled in order to trap and focus each subsequent fraction, which is then remobilized by the heat of the surrounding oven air. This modulation process takes place on the first few centimeters of the ^2D. The cooling is ensured by direct spraying of expanding CO_2 onto the capillary column (Figure 11).

The principle is illustrated in Figure 12. In the first step, compounds in the effluent of ^1D are trapped and focused in the cooled section of the head of ^2D column. In step 2, by stopping the cooling, this fraction is remobilized and injected as a narrow pulse in ^2D. Meanwhile, the (continuous) eluting material

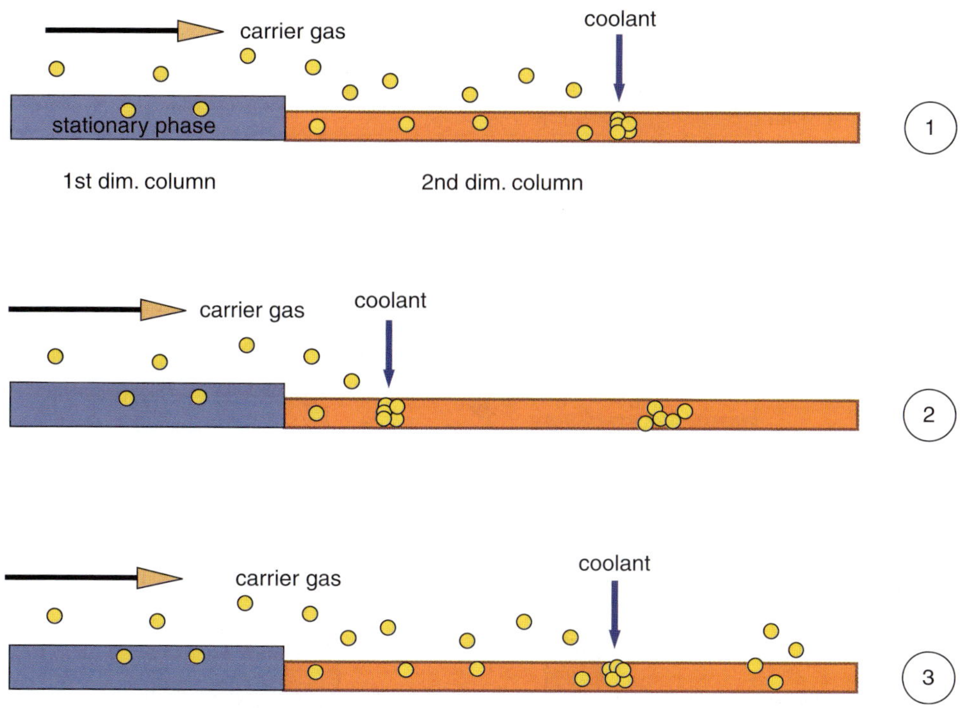

Figure 12 The modulation process in a dual-stage liquid CO_2 cryogenic modulator.

from ^1D column is temporarily stopped to avoid interference with the injected narrow band. In step 3 this cycle starts again.

This alternating cycle of two jets of CO_2 is repeated during the entire duration of the analysis, fixed by P_M that is generally between 2 and 8 s. The modulation time is the sum of the two steps shown above.

This modulator was simplified by Adahchour et al. in 2003 by introducing modulation with a single jet [46]. The main advantage of this improvement was instrumentation simplicity. The major drawback was that there was only one trapping zone, and hence proper focusing would only be attained after tedious and proper optimization of flows. The main disadvantage of cryogenic modulators using liquid CO_2 is the difficulty of focusing volatile compounds such as benzene or butadiene. The alternative for resolving this problem was modulation using liquid nitrogen, which enabled a much lower focusing temperature.

The quad-jet dual-stage modulator In 2000, Ledford and Billesbach reported on another improved CO_2 jet-cooled thermal modulator [38]. In this case, the cold jet was flowing continuously, and solenoid valves were used to pulse hot jets of gas onto a modulator tube to obtain two-stage thermal modulation without moving parts (Figure 13). The ^2D column was housed in a secondary oven allowing independent temperature control from the main oven.

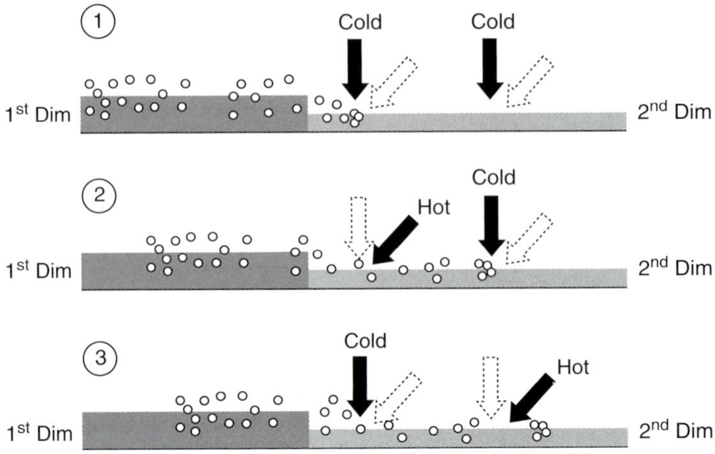

Figure 13 Sequence of events responsible for (1) trapping, (2) releasing and refocusing, and (3) reinjecting into the second column using a quad-jet dual-stage cryo-modulator [47].

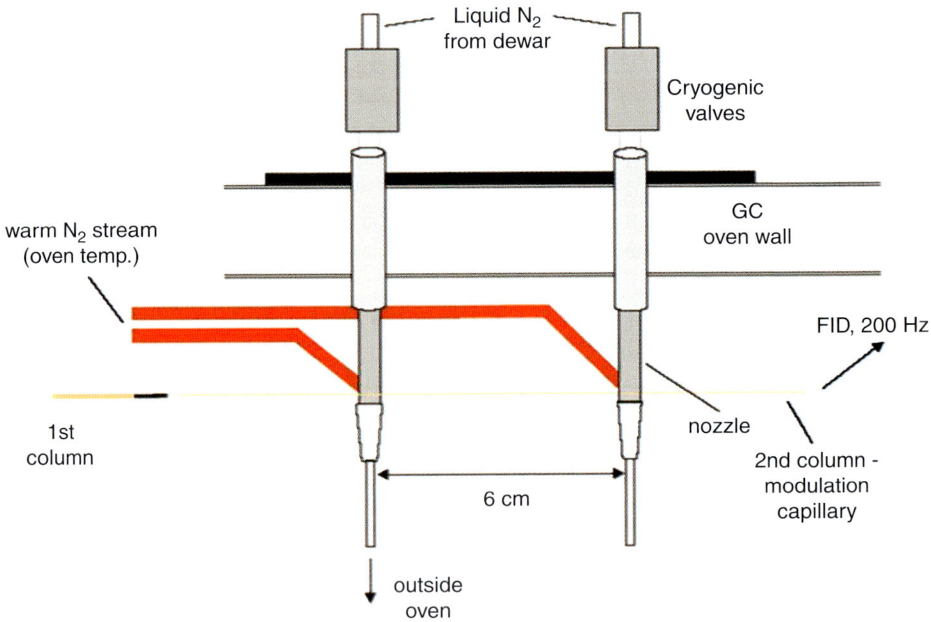

Figure 14 Design of the quad-jet N_2 modulator [48].

Three years later, Pursch et al. reported on the improvement of this modulator using liquid nitrogen instead of CO_2 for cryo-focusing [48]. They highlighted the efficiency in terms of modulation of highly volatile components such as propane and butane. The nitrogen modulator is illustrated in Figure 14.

Finally, a commercial version of the Ledford and Billesbach modulator, modified to accommodate the use of nitrogen as the cryogenic fluid, has been

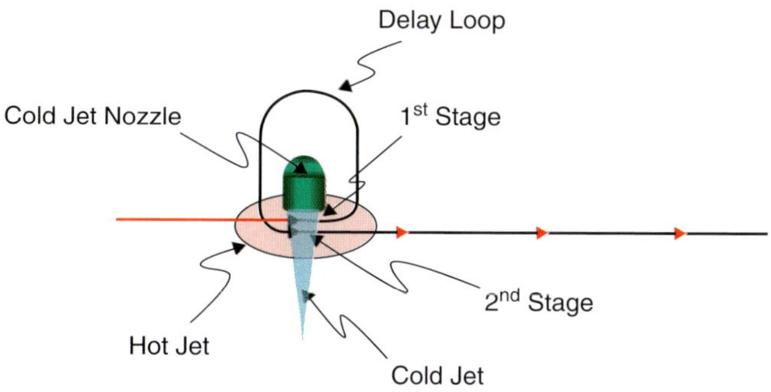

Figure 15 Design of the loop modulator [50].

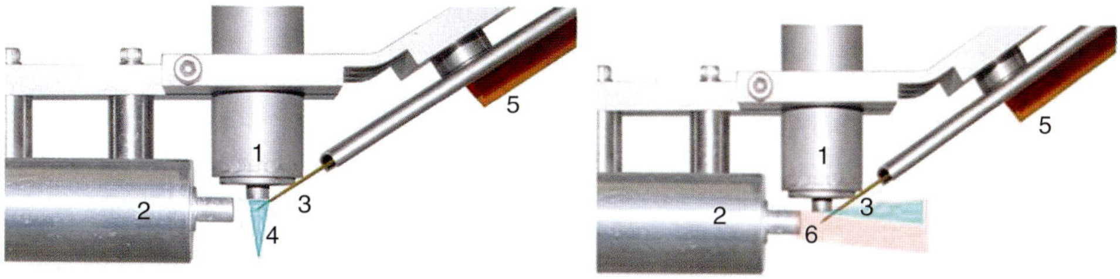

Figure 16 Jet sequence in the loop modulator [50]. 1, cold jet; 2, hot jet; 3, delay loop; 4, trap zone; 5, second column; 6, hot jet in action to release analytes.

developed and is commercialized by Zoex Corporation. LECO Corporation (GC×GC–ToF MS system) designed its own quad-jet dual-stage liquid nitrogen modulator and secondary oven under license from Zoex Corporation. This modulator is the one illustrated in Figure 3.

Ledford et al. [49] further improved Zoex's system by designing a loop modulator. The loop modulator is a dual-stage thermal modulator that employs hot and cold gas jets. The two stages are formed by looping a segment of capillary tubing through the path of a single cold jet. The piece of tubing between the two cold spots is called the delay loop [50] (Figure 15).

As illustrated in Figure 16, a hot jet is turned off when cold spots are formed, and it is turned on to inhibit the permanent cold jet for releasing the trapped analytes.

Substances eluting from the first column are trapped by the cryogenic jet and then released into the delay loop by a brief firing of the hot jet. By the time substances travel the delay loop, the cold jet is active again, and they are trapped a second time by the same cryogenic jet but at the end of the loop. The next firing of the hot jet releases a sharp chemical pulse in the second dimension while admitting the next pulse in the delay loop. The length of the loop and the velocity of the carrier gas must be chosen as a function of the modulation period used. Poor settings will lead to poor focusing. This loop modulator has the advantage of reducing the consumption of the cryogenic fluid, over dual jet systems but, as mentioned, requires fine adjustment of the delay loop to ensure optimized modulation. The system is able to modulate from C2 to C55 with a reported cold jet temperature as low as −189°C [50]. It is commercially available and compatible with most classical GC ovens. Recently, Zoex further improved the system by proposing closed-cycle cryo-refrigeration to eliminate the use of liquid nitrogen. It modulates volatile and semivolatile compounds over the C7+ range with a maximal cold jet temperature of −90°C [50].

3.2 Valve-based modulators

The first modulator operating with a valve was introduced by Bruckner et al. in 1998 [51]. The interface used four ports of a heated six-port valve. In this type of modulator, effluent from the first column is sampled periodically by the action of the valve during a very brief period before being sent to the second column. These modulators are deemed to be less sensitive than thermal modulators because only 10 to 20% of analytes eluted from the first column are really trapped and refocused before being sent to the second column.

Seeley et al. [52] proposed an alternative to this configuration by adding a valve sampling loop (Figure 17) and using its six ports. With this method about 80% of effluent from the first column is sampled. The principal practical advantage of valve-based modulators is their low construction costs. Other advantages of valve modulators is that they avoid huge temperature differences and can be used for very fast second-dimension separation of 1 s due to production of narrow injection bands in time [32]. The main disadvantage is the presence of a valve on the chromatographic way that does not withstand excessive temperature increases. Moreover, a very high flow is used to sweep the effluent out of the valve loop to the ^{2}D column to maintain a narrow second-dimension column injection pulse. One disadvantage of this is that the valve-based modulators either cannot be used, or at least have limited use, with mass spectrometers because of vacuum pumping capacity limitations.

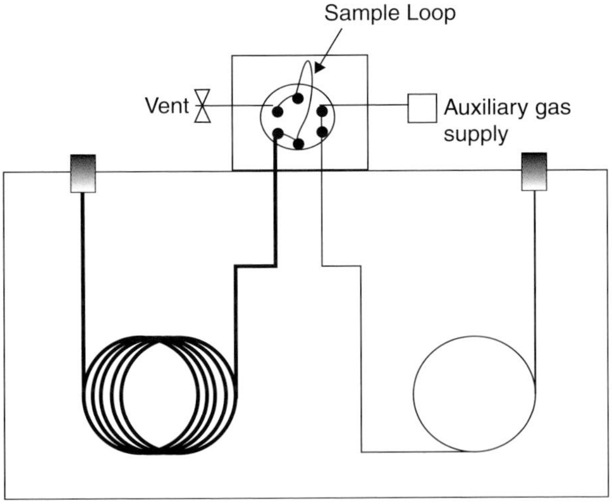

Figure 17 Example of a valve modulator: the loop modulator [31].

Gorecki et al. [53] recently improved their version of the valve modulator by proposing to periodically interrupt the flow from the first column before sampling by the modulator ("stopped flow modulation"). This approach increases analysis time in the second dimension and thus reduces wraparound phenomena without reducing first-dimension resolution. Table 2 summarizes the different modulators used in GC×GC [32].

With the number of robust modulators available today, and the different possible combinations of columns, it is difficult to find the best compromise for GC×GC separation. Jet-cooled systems have evident advantages over moving heater modulation systems. For example, the full temperature range of the GC columns can be used with no concern regarding potential overheating of the stationary phase. In addition, very long columns can be used in ^{1}D since cooling methods permit modulation at any elution temperature. When analyzing mixtures of compounds with a high boiling point, it is preferable to use a system combining a cryogenic modulator and columns whose stationary phases can resist high temperatures. For the analysis of volatile compounds, the best choice seems to be valve systems or cryogenic modulators using liquid nitrogen.

Table 2 Overview of modulators used for GC×GC

Modulator type	Focusing effect	Reference
Thermal modulator		
Dual-stage heated	phase ratio	[28,54]
Sweeper	phase ratio	[41,55]
Thermal modulator array	neg. temp. gradient	[37]
Thermal modulator	elect. heated and air cooled	[56]
Cryogenic modulator		
LMCS[a]	CO_2	[40,57–59]
Four-jet cryo[b]	N_2, liquid	[48,51]
Two-jet cryo[b]	CO_2, liquid	[45]
Three-jet cryo	N_2, liquid	[60]
Single-jet dual-stage cryo (Loop)[b]	CO_2 or N_2, gas	[50]
Microswitching and cryogenic modulation	CO_2, gas	[61]
Single-jet single-stage cryo	CO_2, liquid	[47]
Single-jet dual-stage, semirotating cryo	CO_2, liquid	[62,63]
Valve modulator		
Flow-switching modulation	valve	[52,64–67]
Diaphragm valve	valve	[68,69]
Stop-flow mode modulation	valve	[70]
Pulsed-flow mode modulation	valve	[71]

[a] Availability restricted to specific geographical regions.
[b] Commercially available.

4. DETECTORS

The detector is another important GC×GC system component. In addition to classical GC detector requirements, a GC×GC detector must offer rather high sampling rates, such as the one required for high-speed GC. The main reason is that, because of refocusing, peaks eluted from the second dimension are very narrow, typically 50 to 600 ms [7,9,40,72]. These narrow peaks require detectors with small internal volumes and fast acquisition rates in order to ensure the complete reconstruction of the second-dimension chromatograms. It is generally accepted that at least 6 to 10 acquisition points should be available to correctly define a chromatographic peak [73,74]. This is why an ideal detector should offer acquisition rates in the range of 20–100 Hz [7,9,75–77].

4.1 Flame-ionization and element selective detectors

Today, several detectors are suitable for GC×GC peak characterization. Historically, the flame-ionization detector (FID) has been the detector of choice because of its small internal volume, short risetime, and the corresponding high sampling rate that typically varies from 50 to 300 Hz [78–82]. That situation was ideal, as the major field of exploration of GC×GC in the early days was the petrochemical area, for which FID is a detector of choice owing to the high carbon content of these mixture components (Chapter 7). Another advantage of FID detectors, especially in the oil industry, is its mass response to carbon, so that no response factors of unknown compounds are needed. Since then, several other element-selective detection methods have been reported and used for GC×GC [83]. Table 3 summarizes different applications in GC×GC where FID and element selective detectors were used.

Though commercially available for quite a long time, electron-capture detectors (ECD) did not appear as a detector of choice for GC×GC before the end of the 1990s. That period actually corresponds to the market release of micro electron-capture detectors (μECDs) that offered much smaller detection zone volumes and higher scan rates (50 Hz) than classical ECDs. Coupling a μECD to GC×GC opened the field to environmental applications because of the high selectivity of this detector toward halogenated compounds such as pesticides and PCBs. In addition, μECDs are very sensitive tools (femtogram-level) well suited for ultra-trace analysis. The linear response is, however, often limited to two or three orders of magnitude. Though small, detection zone volumes of μECDs (30–150 μL) are still larger than those observed for FIDs and their use results in undesirable peak broadening potentially impacting the peak capacity of the system [25].

Other even more specific detectors can also be coupled to GC×GC. Atomic emission detectors (AEDs), and more element-selective detectors, such as sulfur compound detectors (sulfur chemiluminescence detector, or SCD), have been reported in the oil characterization area [84,85]. In these detectors, the combustion of sulfur compounds by an energetic induced plasma produces sulfur oxides that further react to and produce light at a specific wavelength that

Table 3 GC×GC applications involving the use of FID and element-selective detection [83]

Detector	Application	Detector Temperature (°C)	Acquisition Rate (Hz)	Reference
FID	Petrochemical products	NR[a]	NR	[90–94]
FID	Fat and oils	NR	NR	[95–101]
FID	Flavour and fragrances	NR	NR	[102,103]
FID	Essential oils	NR	NR	[104–109]
FID	Soil, sediment and water	NR	NR	[110–113]
FID	Air, Aerosols	NR	NR	[114,115]
μ-ECD	Toxaphene	300	50	[116,117]
μ-ECD	PCDDs, PCDFs	280–300	50	[118–120]
μ-ECD	PCBs	300	50	[121–124]
μ-ECD	Fungicides in vegetable samples	300	50	[125]
μ-ECD	Halogenated organic compounds	100–340	50–250	[126]
μ-ECD	PBDEs in dust	300	50	[127]
SCD	Sulphur-containing compounds in diesel oils	800	100	[85,128]
SCD	Sulphur-containing compounds in light catalytically cracked cycle oil–heavy gas oil mixtures	800	50	[87]
SCD	Sulphur-containing compounds in crude oils	800	50	[129]
SCD	Sulphur-containing compounds in middle distillate			[130]
NPD	Methoxypyrazines in wine	300	100	[131]
NPD	Methoxypyrazines in coffee beans	250	100	[89]
NPD	Fungicides in vegetable samples	300	50	[125]
NCD	Sulphur-containing compounds in diesel fuels		100	[88]
NCD	Nitrogen compound speciation in middle distillates	NR	100	[132]
AED	Sulphur-containing compounds in crude oils	Cavity 300	10	[84]

[a]NR = Not Reported.

is detected in a photomultiplier tube. The light emitted is directly proportional to the sulfur content of the sample. This detector offers picogram-level detection limits and rather low acquisition rates (10 Hz). AED was nevertheless reported to give satisfactory results when an extra transfer line was inserted to enhance band broadening [86]. The major drawback of the SCD is a lack of speed, which is related not to its physical dimensions, but rather to the speed of the electronics that has to be modified in order to work under suitable experimental conditions [87]. Nitrogen-specific chemiluminescence detectors (NCDs) are based on the same principle with monitoring of another specific wavelength. The use of NCD in combination with GC×GC has been reported for the analysis of neutral (indoles and carbazoles) as well as basic (pyridines and quinolines) nitrogen-containing compounds in diesel [88].

Thermionic detectors selective toward organic compounds containing phosphorus and nitrogen have also been used in GC×GC. The nitrogen-phosphorus detector (NPD) is similar in structure to FID but operates on a different principle. The NPD sensor differs from that of the FID by a rubidium or cesium chloride bead contained inside a heater coil situated close to the hydrogen jet. This detector is highly sensitive and selective to nitrogen and phosphorus. It has an acquisition rate of about 100 Hz. The analyte detectability for three methoxypyrazines present in the headspace of coffee beans was found to be 20-fold lower compared with an FID [89]. The main drawback of this detector is the high demand in terms of optimization of several gas-flow rates (hydrogen, nitrogen, air).

4.2 Mass spectrometer

One of the most powerful detectors for GC is the mass spectrometer. This is also true for GC×GC. Mass-spectrometric (MS) detectors provide structural information, which brings an additional dimension to the system. In the mid-1990 s, attempts were made to couple GC×GC to MS detectors for direct identification of separated analytes for both target compound analysis and for a broad screening of contaminants. Sector, quadrupole, and ion traps are popular MS detectors for GC that have limited use in GC×GC because of their relatively slow acquisition rates. Fast acquisition ToF MS is better suited to accommodate GC×GC ^{2}D peaks.

4.2.1 Time-of-Flight Mass Spectrometer (ToF MS)

Van Deursen et al. positively reported on the first coupling between GC×GC and ToF MS in a paper published in 2000 [133]. Other groups further investigated the coupling, and all reported on the analytical separation power of such a GC×GC–ToF MS instrument [134,135]. This interest in ToF MS was motivated by the need of high acquisition rates to characterize narrow ^{2}D peaks. Because of its principle of operation, ToF MS can attain acquisition rates up to 500 spectra per second, based on the summation of 10 transients per scan, with unit mass resolution [136]. This allows up to 50 acquisitions per 100 ms peaks, which is more than enough to describe the Gaussian shape of the peak. Contrary to sector, quadrupole, and ion traps operating in SIM mode, a full mass spectrum is collected during each ToF MS acquisition. Information on any mass included in the collected mass range is therefore available for peak identification and/or quantification. As a nonscanning MS instrument, ToF MS also has the advantage of producing nonskewed spectra because virtually all ions are collected at the same time point of the chromatogram, ensuring that ion ratios remain the same across the GC peaks. This spectral continuity allows mass spectral deconvolution of overlapping GC peaks when the fragmentation patterns of the coeluting compounds are different [137]. Deconvoluted ion currents (DICs) can then be used to solve chromatographic co-elution problems in the MS domain. Pure mass spectra can be obtained even when the purity of the compound in the chromatographic peak is poor (chromatographic co-elution). The only requirements for chromatographically co-eluting compounds to be properly deconvoluted is to have a small difference in their peak apexes' retention times and to have enough differences in their mass spectra. For example, positional isomers will not be easily deconvoluted [138].

Since then, efforts have been made to improve the coupling not only in terms of hardware, but also in terms of data-handling software. A GC×GC–ToF MS system is now commercially available (LECO Corporation) and has proven to be robust enough to be considered as a routine instrument. Up to now, GC×GC–ToF MS has been used in a wide range of applications such as analysis of petrochemical samples [139], essential oils [140], traces of pesticides in plants [141], halogenated toxicants in food [76], and cigarette smoke [142] (more examples can be found in Chapters 7–11). Improvements are still needed in terms of software to ensure user-friendly access to collected data (Chapter 4) and fast data processing in complex analytical situation like isotope dilution quantification of large groups of analytes at trace levels [143].

The hyphenation of GC×GC with ToF MS results in an analytical tool that gives an additional dimensionality to the data. It allows high separation power based on the combined use of chromatographic resolution and analytical (mass spectral) resolution. If one sees classical GC with a nonspecific detector (FID, μECD, ...) as a two-axis system (^{1}D = retention time, t_R; ^{2}D = intensity), GC×GC–ToF MS can be seen as a four-axis system allowing three separation dimensions (^{1}D = retention time in the first dimension, ^1t_R; ^{2}D = retention time in the second dimension, ^2t_R; ^{3}D = intensity; ^{4}D = mass spectral information) (Figure 18). Such an instrument is well suited to solve complex mixtures of compounds because a group of analytes will not likely have identical t_R in both dimensions (different GC phases) and identical mass spectra. In practice, compounds exit the ^{2}D column, MS acquisition takes place, and a large number of mass spectra are obtained (e.g., a 45-min GC run at an acquisition rate of 50 spectra per second produces 135,000 complete mass spectra). Recorded mass spectral data are then compared and combined following similarity criteria to identify a ^{2}D peak constituting a cluster corresponding to the same analyte. In Figure 19, peaks X3 and X4 as well as Y3 and Y4 are characterized by identical mass spectra and identical ^2t_R and are therefore recognized as part of the same peak cluster during data processing and additionally summed up for quantification purposes.

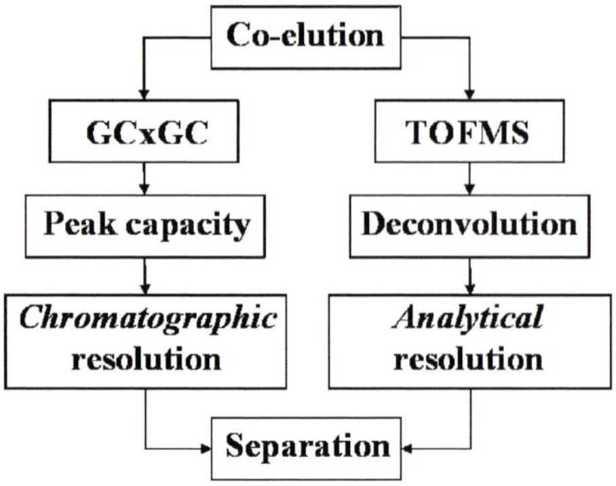

Figure 18 The two complementary options to achieve separation of co-eluting peaks using a GC×GC–ToF MS system [47].

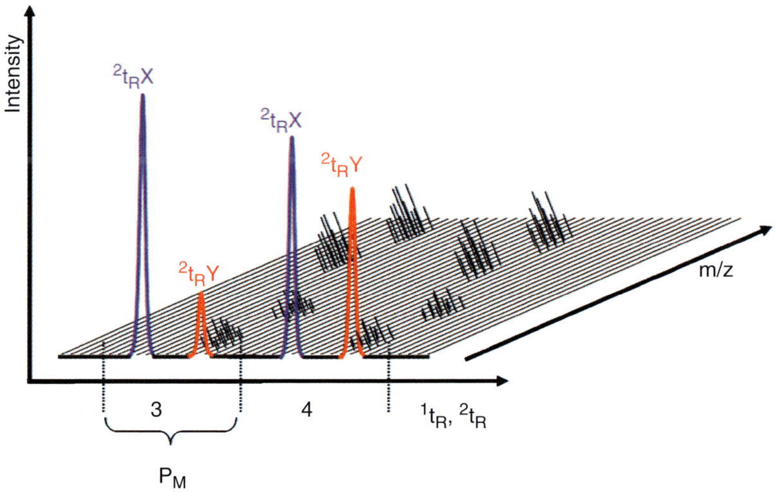

Figure 19 Expanded view of region 3 and 4 of Figure 2. The *m/z* axis lines illustrate the mass spectral acquisition process [143].

Although chromatographic co-elution problems should always be minimized, ToF MS analytical resolution power can be used to deconvolute compounds that co-elute in the GC×GC setup. Mass spectral deconvolution can be used in cases where unique ions are produced by the co-eluting species. Figure 20 presents such an example for three PCB congeners (CB-132, CB-179, CB-161) (Figure 20A), although only two were chromatographically separated in the ^{2}D. A 1D GC separation would have peaked only once (the reconstructed trace in Figure 20C), a GC×GC separation using a μECD would have peaked twice (the contour plots in Figure 20B), and the GC×GC–ToF MS permitted the identification of the three separate analytes [47].

In ToF MS technology, high mass resolution (5–10 ppm) can easily be attained but with a significant limitation on the acquisition speed. Nevertheless,

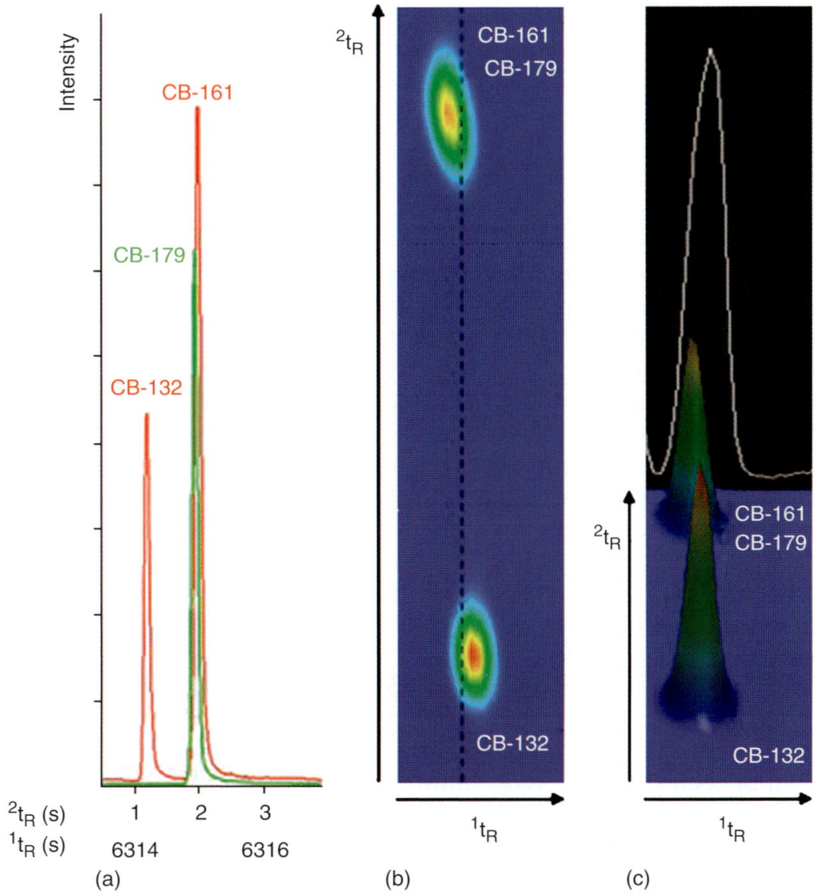

Figure 20 GC×GC–ToF MS raw data (A), contour plot (B), and three-dimensional plot (C) for the chromatographic region where CB-132, CB-179, and CB-161 were eluting using the HT-8×BPX-50 column set [11]. (A) Only one cluster of peaks, corresponding to one modulation cycle at the time represented by the dashed line on (B). The red (m/z 292×3) and the green (m/z 398) traces correspond to hexa- (CB-132, CB-161) and hepta-CBs (CB-179), respectively. (B) Contour plots corresponding to the two peak units chromatographically separated. (C) Reconstructed 1D GC–ToF MS trace (white) based on the sum of the signals recorded in ^{2}D.

attempts were carried out to evaluate the potential advantage of using a high mass resolution ToF MS coupled to GC×GC in terms of peak identification [144]. Acquisition rates of 25 spectra/s are attainable with a mass resolution of 5000. Because the acquisition speed of a ToF MS is independent of the acquired mass range, full mass spectral information is available and elemental composition estimation is possible for peak identification. No doubt, more will come regarding GC×GC–ToF MS high resolution (GC×GC–HRToF MS).

4.2.2 Quadrupole Mass Spectrometer (qMS)

The first report on the use of a qMS in GC×GC was by Frysinger and Gaines in 1999 for an application in the petrochemical field [145]. This was before the first attempt to couple GC×GC to ToF MS. The major limitation was the limited scan rate of the qMS that operated in full-scan mode (45–350 amu) for component identification. The scan rate was 2.43 scan/s, which is far too slow to characterize ^{2}D peaks with typical widths of 200 ms.

To overcome this limitation, it was necessary to slow down the GC×GC separation in order to obtain a second-dimension peak width of at least 1 s. This resulted in a situation where three data points per peak were available. That was not enough for accurate peak shape determination or for quantification. However, it allowed the use of collected mass spectra for acceptable peak identification. This was the first practical demonstration of the valuable coupling between GC×GC and MS.

New generations of rapid scanning quadrupoles offer acquisition rates of 20–35 Hz for a restricted mass range up to 200 Da [83], which permits GC×GC conditions with three to eight data points per peak [74]. However, because of the mass spectral skewing of scanning instruments, quantification and identification are seriously compromised because of variations in ion abundances at different regions of a chromatographic peak [146]. Table 4 summarizes some applications in GC×GC–ToF MS and GC×GC–qMS.

In summary, for the many applications with a limited (100–300 Da) mass range, rapid-scanning qMS is a useful alternative to ToF MS. Peak skewing issues have to be carefully monitored to ensure proper identification and may require the use of specific mass spectral libraries. Several qMS systems also offer the option of doing negative chemical ionisation (NCI) in addition to electron

Table 4 Selected applications involving GC×GC–ToF MS and GC×GC–qMS

Application	Detector	Mass range (Da)	References
Petroleum, sediments extracts	qMS	45–350	[145,147]
	ToF MS	full scan	[139]
Essential oils	qMS	41–228,5	[148]
	ToF MS	full scan	[149,150]
Allergens in fragrances	qMS	40–240	[151]
Coffee beans	qMS	40–400	[152]
VOCs in air, aerosols	ToF MS	full scan	[153–155]
Drugs Screening	qMS	42–235	[156]
	ToF MS	full scan	[157]
Pesticides	qMS	40–500	[125]
	ToF MS	full scan	[125,158]
PCAs, PBDEs, PCDD/Fs	qMS	300	[159]
	ToF MS	full scan	[15,127,160]
Cigarettes smoke	ToF MS	full scan	[161–163]

ionisation (EI) mode, the only one available with the fast acquisition ToF MS. In addition, running the qMS in selected ion monitoring mode for target analysis allows a significant increase in acquisition rate. However, whenever the target analytes cover a broad mass range or are distributed such that time scheduling offers no solution, and when searching for unknowns is a key aspect, using a ToF MS instrument is still mandatory.

4.2.3 Sector High-Resolution Mass Spectrometer (HRMS)

In the mid-1990 s, because thermal modulation coupled with a magnetic sector mass spectrometer was already expected to result in an instrument offering the ultimate attainable sensitivity for measurement of trace-level compounds, scientists attempted the coupling of GC×GC to HRMS for the measurement of polychrorinated dibenzo-*p*-dioxins and furans (PCDD/Fs) and related contaminants in human serum [163,164]. Promising data were obtained, but the use of delicate TDM and sweeper modulators limited the practical application of the method. Recently, GC×GC–HRMS using the jet-cooled loop modulator for the measurement of low levels of PCDD/Fs in human serum samples has been revisited [165]. Rather than improving chromatographic separation with GC×GC, the goal was to simply enhance the GC signal by zone compression [24,166] in order to improve limits of detection and quantification. Because of scan rate limitations of the HRMS system, a very specific tuning of the instrument was needed to ensure proper peak description and quantification. With a acquisition rate of 20 Hz (selected ion monitoring mode) and seven data points per peak, a sensitivity in the mid-attogram range was reported for tetrachlorinated dioxins. Although many challenges still remain, this clearly makes GC×GC a promising tool in the area of sensitivity enhancement. The issue is to determine whether classical ion statistics can still be applied in terms, for example, of ion ratios in such situations where the effective number of ions reaching the detectors is probably just a few thousand.

5. CONCLUSIONS

Today, GC×GC can be considered a mature technique. The instrumental setup of a GC×GC system should be in accordance with several requirements. First, all analytes undergo two independent separations. Second, compounds separated on the first dimension should remain separated in the second dimension. Finally, the elution profiles from both columns must be preserved [9].

As regards column combinations, the classical non-polar×(medium-)polar approach is the preferred way to obtain orthogonal conditions. Nevertheless, setups with a more or less polar first-dimension column attract increasing attention because of the improved peak shapes obtained for polar compounds [83]. As discussed above, one of the key GC×GC parameters is the sampling of the compounds eluted from the first column by the modulator. To succeed here, the first-dimension peaks need to be wide enough to be sampled at least three to four times by the modulator.

Concerning the modulators, both thermal or valve processes provide advantages and drawbacks. The cryogenic modulator seems to be more robust than the valve modulator, which requires further attention. The three kinds of GC×GC instruments marketed today use cryogenic modulators (dual cold jet of liquid CO_2 for TRACE GC×GC, Thermo Fisher Scientific; quad-jet dual stage of liquid N_2 for Pegasus 4D ToF MS, LECO; and one single jet of liquid N_2 with loop for QP2010MS, Shimadzu). However, Agilent has an application note with part numbers for a modulator based on the Deans Switch.

As far as detectors are concerned, an improvement has been made to give the possibility to choose the most appropriate detector according to the application, especially for the element-selective applications. Concerning mass-spectrometric detection, GC×GC–ToF MS remains the detector of choice due to its rapid acquisition rate. However, several studies showed that the qMS seems to be a good alternative, provided the mass range is reduced in order to define correctly the peaks.

In the future, to support the development of this technique, attention should be devoted to improve the data processing. Indeed, the main drawback of GC×GC and especially GC×GC–ToF MS is the time required to extract the pertinent information from a big quantity of data. Several tools such as semiautomated software need to be developed in order to make GC×GC–ToF MS a routine analytical method.

REFERENCES

1 J.C. Giddings, *Anal. Chem.*, 56 (1984) 1258A.
2 J.C. Giddings, *J. Chromatogr. A*, 703 (1995) 3.
3 P. Schoenmakers, P. Marriott and J. Beens, *LC-GC Eur.*, June, (2003) 1.
4 J.F. Focant, E. Reiner, K. MacPherson, T. Kolic, A. Sjödin, D.G. Patterson, Jr., S. Reese, F. Dorman and J. Cochran, *Talanta*, 63 (2004) 1231.
5 E.R. Rohwer, V. Pretorius and P.J. Apps, *J. High Resolut. Chromatogr.*, 9 (1986) 295.
6 P. Korytár, P. Haglund, J. de Boer and U.A.Th. Brinkman, *Trends Anal. Chem.*, 25 (2006) 373.
7 J. Dallüge, J. Beens and U.A.Th. Brinkman, *J. Chromatogr. A*, 1000 (2003) 69.
8 C.J. Venkatramani, J. Xu and J.B. Phillips, *Anal. Chem.*, 68 (1996) 1486.
9 M. Adahchour, J. Beens, R.J.J. Vreuls and U.A.Th. Brinkman, *Trends Anal. Chem.*, 25 (2006) 438.
10 P.J. Schoenmakers, J.L.M.M. Oomen, J. Blomberg, W. Genuit and G. van Velsen, *J. Chromatogr. A*, 892 (2000) 29.
11 J.F. Focant, A. Sjödin and D.G. Patterson, Jr., *J. Chromatogr. A*, 1040 (2004) 227.
12 M. Adahchour, J. Beens, R.J.J. Vreuls, A.M. Batenburg and U.A.Th. Brinkman, *J. Chromatogr. A*, 1054 (2004) 47.
13 J.V. Seeley, *J. Chromatogr. A*, 962 (2002) 21.
14 R.E. Murphy, M.R. Schure and J.P. Foley, *Anal. Chem.*, 70 (1998) 1585.
15 J.F. Focant, A. Sjödin, W.E. Turner and D.G. Patterson, Jr., *Anal. Chem.*, 76 (2004) 6313.
16 J. Beens, H.G. Janssen, M. Adahchour and U.A.Th. Brinkman, *J. Chromatogr. A*, 1086 (2005) 141.
17 J. Harynuk and P.J. Marriott, *Anal. Chem.*, 78 (2006) 2028.
18 Z. Zhu, J. Harynuk and T. Górecki, *J. Chromatogr. A*, 1105 (2006) 17.
19 J.V. Seeley, F.J. Kramp and K.S. Sharpe, *J. Sep. Sci.*, 24 (2001) 444.
20 J.V. Seeley, F.J. Kramp, K.S. Sharpe and S.K. Seeley, *J. Sep. Sci.*, 25 (2002) 53.
21 C. Cordero, P. Rubiolo, B. Sgorbini, M. Galli and C. Bicchi, *J. Chromatogr. A*, 1132 (2006) 268.

22 L.M. Sidisky, C.A. Baney, J.L. Desorcie and K.K. Stenerson, *Organohalog. Compd.*, 70 (2008) 2183.
23 J.V. Seeley, S.K. Seeley, E.K. Libby and Z.S. Breitbach, *Anal. Bioanal. Chem*, 390 (2008) 323.
24 A.L. Lee, K.D. Bartle and A.L. Lewis, *Anal. Chem.*, 73 (2001) 1330.
25 P. Korytár, P.E.G. Leonards, J. de Boer and U.A.Th. Brinkman, *J. Chromatogr. A*, 958 (2002) 203.
26 J.B. Phillips, D. Luu and J.B. Pawliszyn, *Anal. Chem.*, 57 (1985) 2779.
27 Z. Liu and J.B. Phillips, *J. Microcol. Sep.*, 1 (1989) 249.
28 Z. Liu and J.B. Phillips, *Chromatogr. Sci.*, 29 (1991) 227.
29 Z. Liu, Ph.D. dissertation, Southern Illinois University at Carbondale, Carbondale, IL, USA, 1990.
30 J.B. Philipps, R.B. Gaines, J. Blomberg, F.W.M. van der Wielen, J.M. Dimandja, V. Green, J. Granger, D. Patterson, L. Racovalis, H.J. de Geus, J. de Boer, P. Haglund, J. Lipsky, V. Sinha and E.B. Ledford, Jr., *J. High Resolut. Chromatogr.*, 22 (1999) 3.
31 T. Górecki, J. Harynuk and O. Panic, *J.Sep.Sci.*, 27 (2004) 359.
32 M. Adahchour, J. Beens, R.J.J. Vreuls and U.A.Th. Brinkman, *Trends Anal. Chem.*, 25 (2006) 540.
33 W. Bertsch, *J. High Resolut. Chromatogr.*, 23 (2000) 167.
34 R. Ong, P.J. Marriott, P. Morrison and P. Haglund, *J. Chromatogr. A*, 962 (2002) 135.
35 H.J. de Geus, J. de Boer and U.A.Th. Brinkman, *J. Chromatogr. A*, 767 (1997) 137.
36 J. Harynuk and T. Górecki, *J. Sep. Sci.*, 25 (2002) 304.
37 B.V. Burger, T. Snyman, W.J.G. Burger and W.F. van Rooyen, *J. Sep. Sci.*, 26 (2003) 123.
38 E.B. Ledford and C.A. Billesbach, *J. High Resolut. Chromatogr.*, 23 (2000) 202.
39 J. Harynuk and T. Górecki, *J. Sep. Sci.*, 25 (2005) 304.
40 R.M. Kinghorm and P.J. Marriott, *J. High Resolut. Chromatogr.*, 22 (1999) 235.
41 P.J. Marriott, R.M. Kinghorn, R. Ong, P. Morrison, P. Haglund and M. Harju, *J. High Resolut. Chromatogr.*, 23 (2000) 253.
42 P.J. Marriott and R.M. Kinghorn, *Trends Anal. Chem.*, 18 (1999) 114.
43 P.J. Marriott and R.M. Kinghorn, *J. Chromatogr. A*, 866 (2000) 203.
44 R.M. Kinghorn, P.J. Marriott and P.A. Dawes, *J. High Resolut. Chromatogr.*, 23 (2000) 245.
45 J. Beens, M. Adahchour, R.J.J. Vreuls, K. van Altena and U.A.Th. Brinkman, *J. Chromatrogr. A*, 919 (2001) 127.
46 M. Adahchour, J. Beens and U.A.Th. Brinkman, *Analyst*, 128 (2003) 213.
47 J.F. Focant, A. Sjödin and D.G. Patterson, Jr., *J. Chromatogr. A*, 1019 (2003) 143.
48 M. Pursch, P. Eckerle, J. Biel, R. Streck, H. Cortes, K. Sun and B. Winniford, *J. Chromatogr. A*, 1019 (2003) 43.
49 E.B. Ledford, C. Billesbach, and J. Termaat, Contribution No. 2262 P, Pittcon conference, New Orleans, USA, March 17–22, 2002.
50 Zoex, Technical Note KT030606-1 (www.zoex.com).
51 C.A. Bruckner, B.J. Prazen and R.E. Synovec, *Anal. Chem.*, 70 (1998) 2796.
52 J.V. Seeley, F. Kramp and C.J. Hicks, *Anal. Chem.*, 72 (2000) 4346.
53 T. Górecki and O. Panic, Conference K4-04, 4th GC×GC symposium, Dalian, China, June 4–7, 2007.
54 C.J. Venkatramani and J.B. Phillips, *J. Microcol. Sep.*, 5 (1993) 511.
55 H.J. de Geus, A. Schelvis, J. de Boer and U.A.Th. Brinkman, *J. High Resolut. Chromatogr.*, 23 (2000) 189.
56 M. Libardoni, J.H. Waite and R. Sacks, *Anal. Chem.*, 77 (2005) 2786.
57 R.M. Kinghorn and P.J. Marriott, *J. High Resolut. Chromatogr.*, 21 (1998) 620.
58 P.J. Marriott and R.M. Kinghorn, *Anal. Sci.*, 14 (1998) 651.
59 R.M. Kinghorn, P.J. Marriott and P.A. Dawes, *J. High Resolut. Chromatogr.*, 23 (2000) 245.
60 J. Harynuk and T. Górecki, *J. Chromatogr. A*, 1019 (2003) 53.
61 P.J. Marriott, M. Dunn, R.A. Shellie and P. Morrison, *Anal. Chem.*, 75 (2003) 5532.
62 T. Hyötyläinen, M. Kallio, K. Hartonen, M. Jussila, S. Palonen and M.L. Riekkola, *Anal. Chem.*, 74 (2002) 4441.
63 M. Kallio, T. Hyötyläinen, M. Jussila, K. Hartonen, S. Palonen, M. Shimmo and M.L. Riekkola, *Anal. Bioanal. Chem.*, 375 (2003) 725.
64 R. Mayadunne, T.T. Nguyen and P.J. Marriott, *Anal. Bioanal. Chem.*, 382 (2005) 836.
65 R.W. LaClair, P.A. Bueno and J.V. Seeley, *J. Sep. Sci.*, 27 (2004) 389.

66 J.W. Diehl and F.P. Di Sanzo, *J. Chromatogr. A*, 1080 (2005) 157.
67 J.N. Primeau, J.D. McCurry and J.V. Seeley, *J. Chromatogr. A*, 1086 (2005) 115.
68 A.E. Sinha, B.J. Prazen, C.G. Fraga and R.E. Synovec, *J. Chromatogr. A*, 1019 (2003) 79.
69 A.E. Sinha, K.J. Johnson, B.J. Prazen, S.V. Lucas, C.G. Fraga and R.E. Synovec, *J. Chromatogr. A*, 983 (2003) 195.
70 J. Harynuk and T. Górecki, *J. Sep. Sci.*, 27 (2004) 431.
71 H. Cai and S.D. Stearns, *Anal. Chem.*, 76 (2004) 6064.
72 A.L. Lee, A.C. Lewis, K.D. Bartle, J.B. McQuaid and P.J. Marriott, *J. Microcol. Sep.*, 12 (2000) 187.
73 K. Mastovska, J. Hajslova and S.J. Lehotay, *J. Chromatogr. A*, 1054 (2004) 335.
74 M. Adahchour, M. Brandt, H.U. Baier, R.J.J. Vreuls, A.M. Batenburg and U.A.Th. Brinkman, *J. Chromatogr. A*, 1067 (2005) 245.
75 J.B. Phillips and J. Beens, *J. Chromatogr. A*, 856 (1999) 331.
76 J.F. Focant, G. Eppe, M.L. Scippo, A.C. Massart, C. Pirard, G. Maghuin-Rogister and E. de Pauw, *J. Chromatogr. A*, 1086 (2005) 45.
77 J. Dallüge, R.J.J. Vreuls, D.J. van Iperen, M. van Rijn and U.A.Th. Brinkman, *J. Sep. Sci.*, 25 (2002) 608.
78 T.C. Tran and P.J. Marriott, *Atmosph. Environ.*, 41 (2007) 5756.
79 J. Beens, H. Boelens, R. Tijssen and J. Blomberg, *J. High Resolut. Chromatogr.*, 21 (1997) 47.
80 J. Beens and U.A.Th. Brinkman, *Analyst*, 130 (2005) 123.
81 D. Cavagnino, P. Magni, G. Zilioli and S. Trestianu, *J. Chromatogr. A*, 1019 (2003) 211.
82 J. Blomberg, P.J. Schoenmakers and U.A.Th. Brinkman, *Chromatographia*, 55 (2002) 137.
83 M. Adahchour, J. Beens and U.A.Th. Brinkman, *J. Chromatogr. A*, 1186 (2008) 67.
84 L.P.P. van Stee, J. Beens, R.J.J. Vreuls and U.A.Th. Brinkman, *J. Chromatogr. A*, 1019 (2003) 89.
85 F.C.Y. Wang, W.K. Robbins, F.P. DiSanzo and F.C. McElroy, *J. Chromatogr. Sci.*, 41 (2003) 519.
86 L.L.P. van Stee, J. Beens, R.J.J. Vreuls and U.A.Th. Brinkman, *J. Chromatogr. A*, 1019 (2003) 89.
87 J. Blomberg, T. Riemersma, M. van Zuijlen and H. Chaabani, *J. Chromatogr. A*, 1050 (2004) 77.
88 F.C.Y. Wang, W.K. Robbins and M.A. Greaney, *J. Sep. Sci.*, 27 (2004) 468.
89 D. Ryan and P. Marriott, *J. Sep. Sci.*, 29 (2006) 2375.
90 C. Vendeuvre, F. Bertoncini, L. Duval, J.L. Duplan, D. Thiébaut and M.C. Hennion, *J. Chromatogr. A*, 1056 (2004) 155.
91 C. Vendeuvre, R. Ruiz-Guerrero, F. Bertoncini, L. Duval, D. Thiébaut and M.C. Hennion, *J. Chromatogr. A*, 1086 (2005) 21.
92 C. Vendeuvre, F. Bertoncini, D. Espinat, D. Thiébaut and M.C. Hennion, *J. Chromatogr. A*, 1090 (2005) 116.
93 C. von Mühlen, C.A. Zini, E.B. Caramão and P.J. Marriott, *J. Chromatogr. A*, 1105 (2006) 39.
94 Ch.M. Reddy, R.K. Nelson, S.P. Sylva, L. Xu, E.E. Peacock, B. Raghuraman and O.C. Mullins, *J. Chromatogr. A*, 1148 (2007) 100.
95 M. Adahchour, J. Wiewel, R. Verdel, J. Beens, R.J.J. Vreuls, A.M. Batenburg and U.A.Th. Brinkman, *J. Chromatogr. A*, 1086 (2005) 99.
96 L. Mondello, P.Q. Tranchida, P. Dugo and G. Dugo, *J. Pharm. Biomed. Anal.*, 41 (2006) 1566.
97 J. Harynuk, P.M. Wynne and P.J. Marriott, *Chromatographia*, 63 (2006) S61.
98 J. Harynuk, B. Vlaeminck, P. Zaher and P.J. Marriott, *Anal. Bioanal. Chem.*, 386 (2006) 602.
99 S. de Koning, H.G. Janssen and U.A.Th. Brinkman, *LC–GC Eur.*, (2006) 590.
100 J.H. Marsman, J. Wildschut, F. Mahfud and H.J. Heeres, *J. Chromatogr. A*, 1150 (2007) 21.
101 P.Q. Tranchida, A. Casilli, P. Dugo, G. Dugo and L. Mondello, *Anal. Chem.*, 79 (2007) 2266.
102 C. Debonneville and A. Chaintreau, *J. Chromatogr. A*, 1027 (2004) 109.
103 M.S. Dunn, N. Vulic, R.A. Shellie, S. Whitehead, P. Morrison and P.J. Marriott, *J. Chromatogr. A*, 1130 (2006) 122.
104 R. Shellie, P. Marriott and P. Morrison, *J. Chromatogr. Sci.*, 42 (2004) 417.
105 X. Di, R.A. Shellie, P.J. Marriott and C.W. Huie, *J. Sep. Sci.*, 27 (2004) 451.
106 Sh. Zhu, X. Lu, Y. Qiu, T. Pang, H. Kong, C. Wu and G. Xu, *J. Chromatogr. A*, 1150 (2007) 28.
107 Ch. Ruan, G. Xu, X. Lu, R. Hua, H. Kong, K. Xiao and G. Yong, *Chromatogr. Suppl.*, 57 (2003) S-265.
108 C.A. Zini, T.F. De Assis, E.B. Ledford, Jr., C. Dariva, J. Fachel, E. Christensen and J. Pawliszyn, *J. Agric. Food Chem.*, 51 (2003) 7848.

109 M. Libardoni, E. Hasselbrink, J.H. Waite and R. Sacks, *J. Sep. Sci.*, 29 (2006) 1001.
110 M. Adahchour, E. Jover, J. Beens, R.J.J. Vreuls and U.A.Th. Brinkman, *J. Chromatogr. A*, 1086 (2005) 128.
111 R.K. Nelson, B.M. Kile, D.L. Plata, S.P. Sylva, L. Xu, C.M. Reddy, R.B. Gaines, G.S. Frysinger and S.E. Reichenbach, *Environ. Forens.*, 7 (2006) 33.
112 G.F. Slater, R.K. Nelson, B.M. Kile and Ch.M. Reddy, *Org. Geochem.*, 37 (2006) 981.
113 C.G. Johnson, G.S. Frysinger, R.K. Nelson, R.B. Gaines, N. Ohkouchi, C.M. Reddy and T.I. Eglinton, *Mar. Chem.*, 83 (2003) 5.
114 M. Shimmo, T. Hyötyläinen, M. Kallio, P. Antifa and M.L. Riekkola, *LC–GC Eur*, 17 (2004) 640.
115 X. Xu, J. Williams, C.P. Dülmer, H. Berresheim, G. Salisbury, L. Lange and J. Lelieveld, *Atmos. Chem. Phys.*, 3 (2003) 1461.
116 P. Korytár, L.L.P. van Stee, P.E.G. Leonards, J. de Boer and U.A.Th. Brinkman, *J. Chromatogr. A*, 994 (2003) 179.
117 L.R. Bordajandi, L. Ramos and M.J. Gonzalez, *J. Chromatogr. A*, 1125 (2006) 220.
118 P. Korytár, C. Danielsson, P.E.G. Leonards, P. Haglund, J. de Boer and U.A.Th. Brinkman, *J. Chromatogr. A*, 1038 (2004) 189.
119 C. Danielsson, K. Wiberg, P. Korytár, S. Bergek, U.A.Th. Brinkman and P. Haglund, *J. Chromatogr. A*, 1086 (2005) 61.
120 P. Korytar, P.E.G. Leonards, J. de Boer and U.A.Th. Brinkman, *J. Chromatogr. A*, 1086 (2005) 29.
121 L.R. Bordajandi, L. Ramos and M.J. Gonzalez, *J. Chromatogr.A*, 1078 (2005) 128.
122 L.R. Bordajandi, P. Korytar, J. de Boer and M.J. Gonzalez, *J. Sep. Sci.*, 28 (2005) 163.
123 M. Harju, A. Bergman, M. Olsson, A. Roos and P. Haglund, *J. Chromatogr. A*, 1019 (2003) 127.
124 E.M. Kristenson, H.C. Neidig, R.J.J. Vreuls and U.A.Th. Brinkman, *J. Sep. Sci.*, 28 (2005) 1121.
125 W. Khummueng, C. Trenerry, G. Rose and P.J. Marriott, *J. Chromatogr. A*, 1131 (2006) 203.
126 E.M. Kristenson, P. Korytár, C. Danielsson, M. Kallio, M. Brandt, J. Mäkelä, R.J.J. Vreuls, J. Beens and U.A.Th. Brinkman, *J. Chromatogr. A*, 1019 (2003) 65.
127 P. Korytár, A. Covaci, P.E.G. Leonards, J. de Boer and U.A.Th. Brinkman, *J. Chromatogr. A*, 1100 (2005) 200.
128 R. Hua, Y. Li, W. Liu, J. Zheng, H. Wei, J. Wang, X. Lu, H. Kong and G. Xu, *J. Chromatogr. A*, 1019 (2003) 101.
129 R. Hua, J. Wang, H. Kong, J. Liu, X. Lu and G. Xu, *J. Sep. Sci.*, 27 (2004) 691.
130 R.G. Rosario, V. Colombe, Th. Didier, B. Fabrice and E. Didier, *J. Chromatogr. Sci.*, 44 (2006) 566.
131 D. Ryan, P. Watkins, J. Smith, M. Allen and P. Marriott, *J. Sep. Sci.*, 28 (2005) 1075.
132 F. Adam, F. Bertoncini, N. Brodusch, E. Durand, D. Thiébaut, D. Espinat and M.C. Hennion, *J. Chromatogr. A*, 1148 (2007) 55.
133 M. van Deursen, J. Beens, J. Reijenga, P. Lipman, C. Cramers and J. Blomberg, *J. High Resolut. Chromatogr.*, 23 (2000) 507.
134 R. Shellie, P. Marriott and P. Morrison, *Anal. Chem.*, 73 (2001) 1336.
135 J. Dallüge, R.J.J. Vreuls, J. Beens and U.A.Th. Brinkman, *J. Sep. Sci.*, 25 (2002) 201.
136 P.A. Leclercq and C.A. Cramers, *Mass Spectrom. Rev.*, 17 (1998) 37.
137 J.W. Cochran, *J. Chromatogr. Sci.*, 40 (2002) 254.
138 J.F. Focant, J.W. Cochran, J.M.D. Dimandja, E. De Pauw, A. Sjödin, W.E. Turner and D.G. Patterson, Jr., *Analyst*, 129 (2004) 331.
139 F. Adam, C. Vendeuvre, F. Bertoncini, D. Thiébaut, D. Espinat and M.C. Hennion, *J. Chromatogr. A*, 1178 (2008) 171.
140 R. Shellie, L. Mondello, P.J. Marriott and G. Dugo, *J. Chromatogr. A*, 970 (2002) 225.
141 J. Dallüge, M. van Rijn, J. Beens, R.J.J. Vreuls and U.A.Th. Brinkman, *J. Chromatogr. A*, 965 (2002) 207.
142 J. Dallüge, L.L.P. van Stee, X. Xu, J. Williams, J. Beens, R.J.J. Vreuls and U.A.Th. Brinkman, *J. Chromatogr. A*, 974 (2002) 169.
143 J.F. Focant, A. Sjödin and D.G. Patterson, Jr., In: W. Niessen (Ed.), *Encyclopedia of Mass Spectrometry*, Vol. 8, Elsevier, The Netherlands, 2006.
144 H. Shungi, T. Yoshikatsu, F. Akihiro, I. Hiroyasu, T. Kiyoshi, S. Yasuyuki, U. Masa-aki, K. Akihiko, T. Kazuo, O. Hideyuki and A. Katsunori, *J. Chromatogr. A*, 1178 (2008) 187.

145 G.S. Frysinger and R.B. Gaines, *J. High Resolut. Chromatogr.*, 22 (1999) 251.
146 R.A. Shellie and P.J. Marriott, *Analyst*, 128 (2003) 879.
147 G.S. Frysinger, R.B. Gaines and C.M. Reddy, *Environ. Forens.*, 3 (2002) 27.
148 M.Z. Özel, F. Göğüs and A.C. Lewis, *Anal. Chim. Acta*, 566 (2006) 172.
149 M.Z. Özel, F. Göğüs and A.C. Lewis, *Food Chem*, 82 (2003) 381.
150 C. Cordero, C. Bicchi, D. Joulain and P. Rubiolo, *J. Chromatogr. A*, 1150 (2007) 37.
151 D. Ryan, R. Shellie, P. Tranchida, A. Casilli, L. Mondello and P.J. Marriott, *J. Chromatogr. A*, 1054 (2004) 57.
152 J.F. Hamilton, P.J. Webb, A.C. Lewis, J.R. Hopkins, S. Smith and P. Davy, *Atmos. Chem. Phys.*, 4 (2004) 1279.
153 W. Welthagen, J. Schnelle-Kreis and R. Zimmermann, *J. Chromatogr. A*, 1019 (2003) 233.
154 M. McGuigan, J.H. Waite, H. Imanaka and R.D. Sacks, *J. Chromatogr. A*, 1132 (2006) 280.
155 S.M. Song, P.J. Marriott and P. Wynne, *J. Chromatogr. A*, 1058 (2004) 223.
156 G. Semard, A. Bruchet, P. Cardinaël and J.P. Bouillon, *Water Sci. Technol.*, 57.12 (2008) 1983.
157 J. Zrostlıkova, J. Hajšlová and T. Cajka, *J. Chromatogr. A*, 1019 (2003) 173.
158 P. Korytár, J. Parera, P.E.G. Leonards, J. de Boer and U.A.Th. Brinkman, *J. Chromatogr. A*, 1067 (2005) 255.
159 J.F. Focant, C. Picard, G. Eppe and E. de Pauw, *J. Chromatogr. A*, 1067 (2005) 265.
160 X. Lu, J. Cai, H. Kong, M. Wu, R. Hua, M. Zhao, J. Liu and G. Xu, *Anal. Chem.*, 75 (2003) 4441.
161 X. Lu, M. Zhao, H. Kong, J. Cai, J. Wu, M. Wu, R. Hua, J. Liu and G. Xu, *J. Sep. Sci.*, 27 (2004) 101.
162 T. Pang, S. Zhu, X. Lu and G. Xu, *J. Sep. Sci.*, 30 (2007) 868.
163 D.G. Patterson, Jr., J.B. Barr, E.S. DiPietro, J. Grainger, V.E. Green, C.R. Lapeza, V.L. Maggio, S. Sirimanne and W.E. Turner, *Organohalog. Compd.*, 27 (1996) 309.
164 J. Grainger, J.M.D. Dimandja, V. Green, Z. Liu and D.G. Patterson, Jr., *Organohalog. Compd.*, 35 (1998) 28A.
165 D.G. Patterson Jr., S.M. Welch, W.E. Turner, and J.F. Focant, Riva Conference, Riva del Garda, Italy, May 27–30, 2008.
166 H.J. de Geus, J. de Boer, J.B. Phillips, E.B. Ledford and U.A.Th. Brinkman, *J. High Resolut. Chromatogr.*, 21 (1998) 411.

CHAPTER 3

Theoretical Considerations

Jesus Sanz

Contents

1. INTRODUCTION

Fundamental advances and technological developments have grown together and in a continuous way in gas chromatography (GC). The process of band elution in open tubular columns, laid out by Golay 50 years before the edition of this book [1], was the basis of the development of capillary column technology. But in some cases instrumental advances have slowed the progress in basic aspects of chromatography. Although packed columns with a low efficiency required a careful selection of the stationary phase in order to obtain a good resolution, high-efficiency capillary columns are able to solve most separation problems using general-purpose phases: their extended use was followed by a decreased interest in the development and studies of specific stationary phases.

Comprehensive Analytical Chemistry, Volume 55
ISSN: 0166-526X, DOI 10.1016/S0166-526X(09)05503-2

Although capillary columns excelled in separation power, the use of GC data for qualitative purposes was hampered by the low-information content of GC retention. Studies on selectivity tuning and on multidimensional gas chromatography, trying to solve co-elution problems, were in some way discouraged by perhaps the most important success of gas chromatography: its ability to be coupled with mass spectrometry, a technique that provides a high amount of qualitative information unrelated to that afforded by GC. Through its coupling with mass spectrometry (GC–MS), gas chromatography became a bidimensional technique that permitted obtaining qualitative and even quantitative data for compounds co-eluting in a single GC peak (Chapter 1).

The development of comprehensive two-dimensional gas chromatography (GC×GC) is also the result of basic studies and of technical improvements, but research in both fields is necessary in order to release its full power. The increase in separation power provided by GC×GC is of a higher magnitude than that resulting from the change from packed to capillary columns. When compared with GC–MS, comprehensive two-dimensional gas chromatography allows the coupling GC×GC–MS, which is a tridimensional analytical technique.

GC is based on movement of the analyte bands through the chromatographic column, which results in a response (the chromatographic peak) produced by each band at the end of the column. The basic parameters that govern the movement and dispersion of bands in one-dimensional GC are also responsible for the chromatographic signal in GC×GC, but the study of the enhanced chromatographic information obtained from GC×GC requires additional work in the study of gas chromatographic behaviour.

GC peaks can be easily shown in two dimensions, where compound retention is plotted in one dimension and quantitative data appears in the second dimension as peak size. GC×GC peaks need a completely different visual output, since two dimensions are required in the plot in order to show the retention of the compound in the two columns, while a third dimension is necessary for the representation of the quantitative information associated to the compound, as previously explained in Chapter 2.

The most important parameters related to chromatographic response (retention time, peak width, resolution), well known by GC users, are also fundamental in the characterization of GC×GC peaks. However, some of them take in GC×GC a new meaning: for instance, resolution or peak capacity must be considered in a different way. Other characteristics, such as orthogonality or structured chromatograms, are specific of GC×GC.

In this chapter, we will review the most important studies on the fundamental chromatographic parameters of GC×GC, carried out with both theoretical and applied purposes.

2. NOMENCLATURE

Gas chromatography has been a well-established technique for many years, and its nomenclature, symbols, naming conventions, and abbreviations are generally accepted. In 1993, the IUPAC published a "Nomenclature for Chromatography" [2], which recommends chromatographic terms, symbols, and acronyms, including of course those specific of GC. GC×GC is an emerging technique that has borrowed some terms from GC and from other fields, assigning a specific meaning to several of them.

A general discussion on dimensionality in analytical chemistry appears in [3]. Physical, chemical, measuring, and statistical dimensionality are considered. Multidimensional separation techniques are defined [4] as being the result of two or more independent separation steps that are linked together.

The name "comprehensive" requires a specific definition in order to avoid confusion with other GC multidimensional techniques. The requirements to define a technique as "comprehensive," which were proposed by Giddings in 1984 [5] and have already been mentioned in this volume, are:

- All sample analytes are submitted to two different separations. The system must not discriminate against any of the analytes.
- The resolution afforded by the first separation must not be lost during the second separation.

Multidimensional techniques such as heart-cutting which, after a first separation, focus only a part of the sample components and then submit them to the second separation, fail to comply with these two conditions, since not all the analytes are separated by the two procedures and since the focusing step includes loss of the resolution obtained from the first separation.

The most characteristic symbol of comprehensive two-dimensional gas chromatography (and of any other "comprehensive" coupling) is the multiplex (×), used to abbreviate its full name to GC×GC. Most of the nomenclature and symbols used in GC×GC which are now widely accepted were included by Schoenmakers et al. in a 2003 publication [6]. Their suggestions are used throughout this book.

This publication proposes different terms and definitions for the most important devices and their operation modes (modulator, modulation types), processes (focusing effect, zone compression, sensitivity enhancement), and specific properties of the two-dimensional response (orthogonality, separation space, and chromatographic structure).

Readers are encouraged to check Schoenmakers et al.'s publication for details about the different terms used in GC×GC. As an example, we list the proposed definition of three terms characteristic of GC×GC, which are frequently used to describe its graphic output:

- *Wraparound*: The occurrence of second-dimension peaks in subsequent elution sequences, caused by second-dimension retention times that exceed the modulation time of a comprehensive two-dimensional system.
- *Chromatographic structure*: The observed ordering of chemically related compounds in the plane of a comprehensive two-dimensional separation.
- *Colour plot*: Two-dimensional plot representing a comprehensive two-dimensional separation, in which the colour represents the signal intensity of the separation system.

In GC×GC graphics, the retention time of the first dimension is typically plotted in the *x*-axis, while the second-dimension retention time is represented in the *y*-axis.

While symbols for common column parameters in GC are maintained with the same meaning in GC×GC, their relationship with the columns in the first or second dimension must be distinguished in some way. The use of a superscript prefix (1 for the first column, 2 for the second column) is recommended: for instance, ${}^1t_R$ and ${}^2t_R$ are the symbols for the retention time in the first- and the second-dimension columns (^{1}D and ^{2}D columns), respectively. Other considerations useful in GC×GC nomenclature appear in "Multidimensional Confusion," an editorial of the *Journal of Separation Science* [7]. Monodimensional and bidimensional procedures can be abbreviated as 1D and 2D, or as 1-D or 2-D. The use of a hyphen (as in two-dimensional) is recommended when "two" is used to distinguish a procedure from others having a different number of dimensions.

3. PARAMETERS RELATED TO THE CHROMATOGRAPHIC RESPONSE

The response of a gas chromatographic detector to an analyte band eluting from the end of the column is the chromatographic peak. The three main characteristics that define a peak in 1D GC are its size, width, and position in the chromatogram.

GC×GC is based in the serial connection of two columns with different stationary phases. For this reason, peak size, width, and retention time have a similar meaning than in GC. As explained in Chapter 2, the modulation process divides the output from the first column into time slices, and focuses and reinjects them in the second column. However, adequate processing of the detector response allows separation of the contribution of the first and of the second columns in terms of retention time and of peak width to the final response, and it also makes it possible to present graphically the results in a two-dimensional plot.

3.1 Retention time

The time that elapses between an analyte injection in a GC column and the elution of its peak maximum is the retention time (t_R). Although retention time is related to the geometry of the column and to its operation conditions, for a given column and conditions it depends only on the interaction between analyte and stationary phase, and hence on the analyte structure. Its value contains all the qualitative information that GC affords for a compound: the coincidence of retention times between an analyte and a standard compound (analysed in the same column and conditions) is a requirement for a positive identification in GC. Differences in retention time for two analytes, together with peak width, are responsible for the chromatographic separation (resolution) between them.

In a GC×GC system, the retention of an analyte is characterized by its retention times $^{1}t_R$ and $^{2}t_R$ in the two columns. $^{1}t_R$ is defined as the time elapsed between injection of the analyte and its arrival to the modulator, while $^{2}t_R$ is the time taken for the analyte to elute between modulator and detector. In GC×GC, a coincidence of both $^{1}t_R$ and $^{2}t_R$ is required for a positive identification.

The next sections will describe the use and experimental determination in GC×GC of some retention parameters derived from retention time, the possibility of their estimation from the structural characteristics of the analytes, and the determination of the influence on their values of experimental conditions such as temperature and flow rate.

3.1.1 The retention index concept in GC and in GC×GC

The use of GC retention times with qualitative purposes for interlaboratory comparisons is very difficult because differences in chromatographic conditions (temperature, flow rate) and in column geometry (column length, internal diameter, film thickness) markedly affect their values.

Among the parameters related to the retention time, the distribution constant K — defined for a compound as the ratio between its concentrations in the stationary phase and in the mobile phase — depends only on analyte, stationary phase, and temperature, and can be used to compare retention data. K can be calculated from chromatographic measures: from its definition,

$$K = \beta \cdot k \tag{1}$$

where β is the column phase ratio (volume of gas phase divided by volume of stationary phase) and depends on the geometry and the stationary phase amount of the column, and k, the retention factor, is the ratio between the total amounts of analyte in the stationary and mobile phases. Its value can experimentally be measured as the ratio between the values of the time spent by the analyte in the stationary phase (t'_R) and in the mobile phase (t_M)

$$k = \frac{(t_R - t_M)}{t_M} = \frac{t'_R}{t_M} \tag{2}$$

where t_R and t'_R are, respectively, the retention time and the adjusted retention time of the analyte, and t_M, the holdup time, is the retention time of a nonretained compound.

But K and k depend markedly on the operation temperature, and cannot be used to compare data obtained in programmed temperature conditions. A way to increase the reproducibility of measures is to use a reference compound. In the case of retention time, flow rate and column geometry affect all the compounds in the same way: the reproducibility is increased by using retention times relative to a reference, instead of absolute values.

Use of relative values requires of a reference compound, but in GC it is not possible to find a common reference that will elute close to all the analytes. Kovats solved this problem in 1958 [8] by introducing the retention index scale, which uses as reference, instead of a single compound, the two n-alkanes of a homologous series that elute before and after the analyte. Retention indices, or

Kovats indices (abbreviated *I* as proposed in [2] or *RI*, more frequently used), are calculated by a logarithmical interpolation:

$$RI = 100\frac{(\log t'_{Ri} - \log t'_{Rz})}{(\log t'_{R(z+1)} - \log t'_{Rz})} + 100 \cdot z \tag{3}$$

where t'_{Ri} is the adjusted retention time of the analyte, while t'_{Rz} and $t'_{R(z+1)}$ are the adjusted retention times of the n-alkanes that elute before (carbon number z) and after (carbon number $z+1$) the analyte.

RI values do not depend on column dimension or flow rate. They must be obtained in isothermal mode, and they are less dependent on temperature than other retention parameters (k, K). Because there is a linear relationship between the logarithm of the adjusted retention times, obtained under isothermal conditions, of the components of a homologous series and their number of carbon atoms, n-alkanes that differ in two carbon atoms (z and $z+2$) can be used as standards in Equation (3) with only a small loss in accuracy. Other homologous series can be used instead of n-alkanes when these compounds are not appropriate — for instance, in columns with high-polarity stationary phases.

In 1963 van den Dool and Kratz [9] proposed the use of a linear interpolation (using t_R instead of $log(t'_R)$ in Equation (3)) for linear temperature-programmed operation, since in this case there is an approximately linear relationship for homologous series between retention times and number of carbon atoms z:

$$LRI = 100\frac{(t_{Ri} - t_{Rz})}{(t_{R(z+1)} - t_{Rz})} + 100 \cdot z \tag{4}$$

*LRI*s (linear retention indices) depend on programming and flow conditions, but changes in their values are not very marked, particularly when the variation of *RI* with temperature is small [9]: in this case, *LRI* and *RI* values for a compound are similar.

For both *RI* and *LRI*, the retention index of an analyte can be defined by the number (multiplied by a factor of 100) of carbon atoms of a "hypothetical n-alkane" having the same retention time than the analyte. Equations (3) and (4) can be considered as changes of a retention time scale in a 1D map, directly (*LRI*) or after applying a logarithmic transformation (*RI*), using the n-alkanes retention times in both cases as reference points.

In GC×GC, retention indices can be used as retention parameters derived from the retention times for both first and second dimensions (^1RI and ^2RI): since two values instead of one are used for characterising the retention, their usefulness for qualitative purposes is markedly enhanced. In this case, the 2D separation space defined by ^1t_R and ^2t_R is mapped by instead using *LRI* in the first dimension and *RI* in the second dimension.

The determination of retention indices in the first column is straightforward, using a suitable n-alkane mixture as reference. Equation (4) (*LRI*) is commonly used because the first column usually operates in temperature-programmed mode. The calculation of retention indices in the second column, which operates in isothermal mode, presents several problems, however. These problems will be addressed in Sections 3.1.2 and 3.4.

3.1.2 Determination of retention indices in GC×GC

The most commonly followed approach for calculating retention indices in the ^{2}D column, in order to allow retention data comparison between laboratories (with other GC×GC results or with reference 1D GC data), is to map the retention scale of the ^{2}D column using reference compounds much as has been described in the previous paragraph for one-dimensional *RI* or *LRI*. Since elution in the second column is very fast, its operation can be considered to be isothermal, and then *RI* values are more appropriate.

Application of Equation (3) for *RI* determination requires measuring the retention times of the two n-alkanes which elute respectively before and after the analyte. However, in the usual a nonpolar-polar column configuration, polar compounds will require in the second column a pair of n-alkanes with higher carbon numbers than those of the two n-alkanes that bracket the compound when elutes in the nonpolar column.

Beens et al. [10], as part of a study on selecting column sets for a given GC×GC application, proposed a solution to this problem through the use of "isovolatility" curves. If the ^{1}D injection system is not sufficiently heated, compounds evaporate slowly, producing an intense tailing effect in ^{1}D peaks. When this tail elutes at the end of the ^{1}D column, the modulator will continuously inject the compound into the ^{2}D column. The isovolatility line includes the points in two-dimensional plots that represent ^{1}D retention (^1t_R, *x*-axis) against ^{2}D retention (^2t_R, *y*-axis) at the ^{1}D elution temperature.

The scheme of Figure 1, based in [10], simulates the results of this process for an n-alkanes series in the bidimensional space. Narrow signals for nontailing

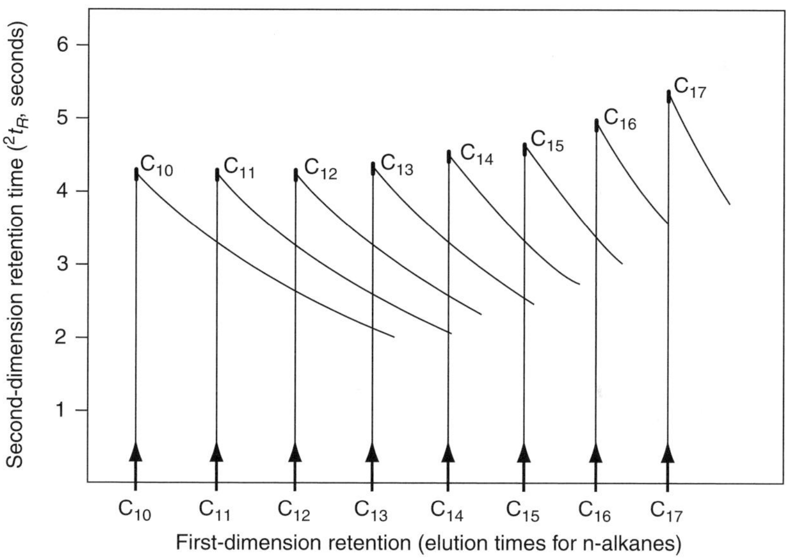

Figure 1 Simulated 2D retention behaviour for C_{10} to C_{17} n-alkanes, showing isovolatility curves resulting from a slow injection. Based on reference [10].

n-alkanes would appear at their ^{1}D retention times as marked in the x-axis. The 2D signals for n-alkane tails appear to be plotted as continuous (isovolatility) curves. The reason is that, because of temperature programming, higher retention times in ^{1}D correspond to higher elution temperatures in ^{2}D, and then to lower ^{2}D retention values: because the relationship is not linear, a decreasing curved trend appears in the plot.

The continuous introduction in ^{2}D of a controlled sample amount is difficult to attain with this procedure. Isovolatility curves for n-alkanes can also be produced using sequential injections of a n-alkanes mixture during the chromatographic run; in this case, peaks appear as discrete signals instead of as curved lines, but interpolation between peaks obtained for the same n-alkane allows to estimate for retention index calculation using Equation (3) their retention times, which would correspond to a continuous injection. Different approaches are based in this idea [11–14]. Among these approaches is a procedure proposed by Bieri and Marriott [15], which introduces the n-alkanes mixture directly into the ^{2}D column, at selected time intervals, using an SPME injector. Figure 2 presents a simulation of this procedure: in the example, injections of n-C_{10} to n-C_{17} are carried out at 10-min intervals. Circles represent the signals from n-C_{10} to n-C_{17}: gray lines (isovolatility curves) are obtained by interpolation, using, for instance, an exponential fit [15] between the different points for the same n-alkane.

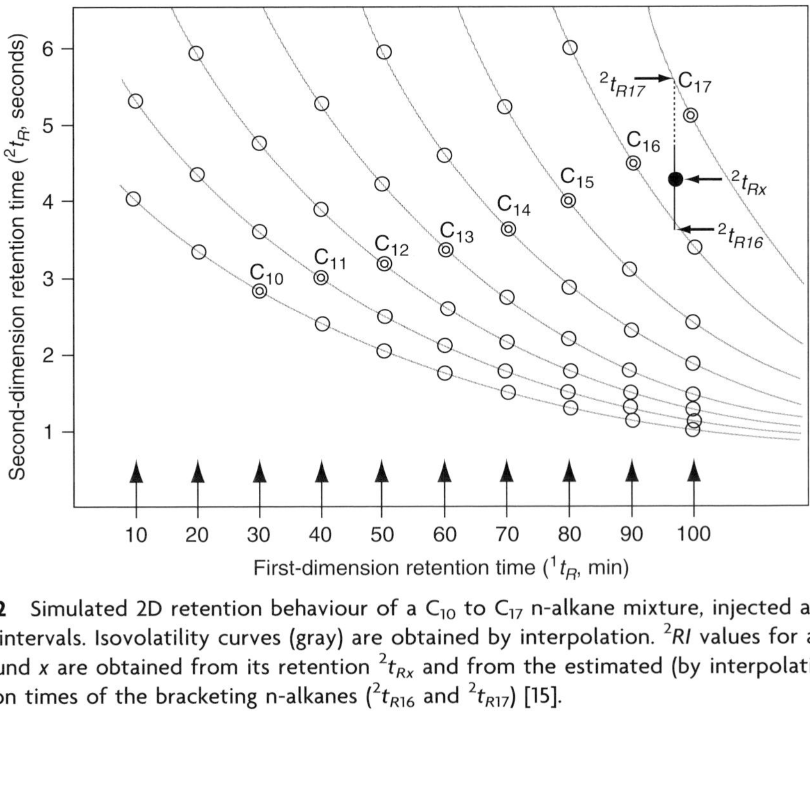

Figure 2 Simulated 2D retention behaviour of a C_{10} to C_{17} n-alkane mixture, injected at 10-min intervals. Isovolatility curves (gray) are obtained by interpolation. ^2RI values for a compound x are obtained from its retention $^2t_{Rx}$ and from the estimated (by interpolation) retention times of the bracketing n-alkanes ($^2t_{R16}$ and $^2t_{R17}$) [15].

In both cases, the mapping of the 2D space using the n-alkanes allows an easy calculation of ^{2}D RI values for a compound. Figure 2 shows the position in the 2D space of a compound x (black circle), having $^1t_{Rx}$ and $^2t_{Rx}$ as retention times, which elutes between the C_{16} and C_{17} curves. If isovolatility curves have been obtained by a mathematical fitting process, the equations involved can be used to estimate $^2t_{R16}$ and $^2t_{R17}$ at $^1t_{Rx}$. RI calculation from Equation (3) only requires, besides these values, the holdup time (see Section 3.3) in order to obtain the adjusted retention times.

The range of the isovolatility curves can be extended by extrapolation through mathematical procedures, by injections at different temperature programming rates [12] or by use of the more polar homologous series (2-methylketones and fatty acids methyl esters [11], alcohols [12]) instead of the n-alkanes.

3.2 Estimation of retention parameters

Retention time of an analyte in a given column depends on the analyte structure and on chromatographic conditions. Experienced chromatographers can estimate, for simple compounds, their comparative retention in a GC column, as well as the effect on retention of flow and temperature changes. Since the error is too high, this subjective approach cannot be used for identification purposes.

Retention behaviour in GC×GC is more complex and also far less intuitive. A very helpful paper from Ong et al. [16], on the influence of chromatographic conditions in GC×GC, describes the effects on ^{1}D and ^{2}D retention of temperature program, flow rate, stationary phase, and ^{2}D column length. In many cases, these effects are difficult to estimate even in a qualitative way. For example, an increase in flow rate will result in a lower retention time on the ^{1}D column and also in a lower ^{1}D elution temperature (T_e), which is also the run temperature for the ^{2}D column. The resulting combination in ^{2}D column of higher flow rates and lower elution temperatures could result in an increase or in a decrease of the ^{2}D retention, and it is impossible to decide intuitively between the two possibilities.

Ong et al. [16] also consider some selective effects of conditions on retention, mainly of the temperature program rate, which can even cause changes in the elution order for some compounds. The researchers also present relationships between different parameters and retention, which will be of high interest to those trying to improve their GC×GC analytical procedures. Their conclusion, "optimisation of a GC×GC analysis will be a tedious exercise," is true for present GC×GC procedures that select by trial and error column sets and operation conditions.

The solution could be a quantitative estimation of retention parameters, carried out by using objective procedures based on predictive models. The establishment in GC×GC of quantitative relationships between chromatographic retention and molecular structure, and the study of how changes in chromatographic conditions affect the retention parameters can be useful for several purposes:

- For a better knowledge of the retention process and of the physicochemical parameters involved.

- For qualitative purposes. The retention times ^1t_R and ^2t_R or the *RI* values of a compound could be checked against their calculated values, although a positive identification would always require additional information.
- As a help on the development and optimisation of a GC×GC chromatographic method. In a first step, a rough approximation could be useful to select between 2D column sets, for instance, by estimating if a low temperature limit of a given phase will result in very long runtimes or in wraparound. If values of peak widths and retention times (and then of resolution) could be accurately calculated for different conditions of flow, temperature and column geometry, the conditions for an optimum GC×GC chromatographic separation could be selected, without resorting to trial and error.

3.2.1 Estimation of the retention of a compound from its structural properties

When a compound is submitted to spectroscopic techniques, the response usually affords some information about its structure. If the retention time in GC depends, as previously mentioned, on the interactions between stationary phase and analyte, then it should be related to the chemical structure of the analyte. The presence of retention patterns correlated with similar structures in simple cases is well known for chromatographers: the regularity of the retention times for n-alkanes (and other homologous series), both in isothermal and temperature-programmed modes, is a basis of the retention index concept as already explained in Section 3.1.1).

These retention patterns have encouraged different research lines trying to find relationships between retention values and compound characteristics. Structural properties can be expressed numerically by using "descriptors" (parameters that try to describe as accurately as possible a molecular structure). Physical properties of the analyte (boiling point, vapour pressure) can also be considered as experimental descriptors.

A review [17] on quantitative structure–chromatographic retention relationships, which summarizes the work carried out in the period 1996–2006, includes a section on GC. Indications on validation (a necessary step in this type of calculations), on common mistakes, and on future trends in this field are presented.

Relatively good results have been found in the calculation of 1D *RI* values using as descriptors physical parameters of the compound, but the utility of this approach is low since these parameters are usually difficult to obtain experimentally. In many cases, it is easier to estimate values of physicochemical parameters from experimental *RI* values than the inverse process: Marriott et al. [18] present examples of how GC×GC can help in the study of chemical decompositions or transformations and other molecular processes.

Descriptors directly derived from the compound molecular structure are the easiest to obtain (molecular weight, number of H atoms). Topological parameters describing molecular size and shape are also frequently used. Also, a high number (>300) of descriptor values containing information about the atoms, bonds, size, shape, and charge distribution of an analyte can be automatically generated from its molecular structure using a suitable software such as CODESSA [19,20]. Many references related to this topic can be found in [17].

The next step consists in establishing a mathematical model (usually linear) that relates a given set of experimental GC retention times or retention indices with descriptor values. Calculation of the model parameters usually involves multiple linear regression or artificial neural networks. When descriptors are automatically generated, a previous selection in order to use only those relevant with retention properties is required [21]. After validation, retention values for other compounds can be estimated using the model. But this estimation is restricted to the original chromatographic conditions.

More interesting for practical applications is the approach of Pompe et al. [22] where two GC thermodynamic parameters (standard-state changes of enthalpy, ΔH^0, and of entropy, ΔS^0) are estimated. Since these parameters can be used to describe the variation of retention with temperature (see Section 3.2.2), estimations of retention can be extended to other temperature conditions, including programmed temperature.

The model of [22] or others described in [17] could be used in the calculation of GC×GC retention times or other parameters (e.g., *RI* values) for both ^{1}D and ^{2}D columns. However, at the time of writing this chapter, these procedures have not yet been applied in GC×GC. A publication on GC×GC retention time prediction [23] uses the previously mentioned CODESSA program package and experimental retention times obtained from the ^{1}D column in GC×GC, but the ^{2}D data are not included in the calculation.

Models that describe the relationship between GC×GC retention and structural properties should not be restricted to a fixed set of columns and chromatographic conditions. In GC×GC, any alteration in the operation parameters of the first column will cause a change in the elution conditions (flow, temperature) in the second affecting the retention in both ^{1}D and ^{2}D columns. For these reasons, practical objectives require that retention GC×GC behaviour can be predicted for different chromatographic conditions, as described in the next section.

3.2.2 Dependence of the retention with chromatographic conditions

Another approach for the mathematical modeling of the retention behaviour tries to estimate the retention times of a compound from experimental values obtained in the same system but under other chromatographic conditions. Estimation of the retention for a given compound should be easier and more accurate if we use its chromatographic data as a starting point, instead of structural descriptors or physicochemical parameters.

Equation (2) can be rewritten as:

$$t_R = (k + 1) \cdot t_M \tag{5}$$

where retention time t_R for a compound on a given column depends on flow rate through the holdup time, t_M, and on temperature through the retention factor k, although the variation of carrier gas viscosity with temperature also affects the flow conditions.

The equations that describe flow behaviour in GC are well known, but their application to GC×GC is not straightforward, mainly because of the change between the geometry of the two columns. In [24] Beens et al. list the equations necessary to describe the GC×GC flow behaviour, and present a computer programme which carries out the required calculations for different conditions and column geometries and includes the estimation of the efficiency of the columns. Harynuk and Górecki [25] developed a similar model in order to select the most appropriate conditions when using loop-type modulators.

The relationship between retention and temperature in 1D is more complex, especially when operation in temperature-programmed mode is required. The most common basis for the estimation from retention uses the thermodynamic based equation:

$$\ln k = -\left(\frac{\Delta H^0}{RT}\right) + \left(\frac{\Delta S^0}{R}\right) + \ln \beta \tag{6}$$

where ΔH^0 and ΔS^0 are, respectively, the standard enthalpy and the standard entropy of transfer of the solute from the mobile phase to the stationary phase, R is the gas constant, T (°K) is the temperature, and β the phase ratio, which is characteristic of the column. Most estimations of the retention at different temperatures use retention data at several temperatures to obtain the coefficients of Equation (6), which can be later used for the estimation of retention, although a more accurate estimation could require other equations based in three parameter models [26].

In GC×GC, the ^{2}D operates in isothermal mode, and estimation of the retention at different temperatures should be easy, although the requirement of an accurate value for t_M is a problem, as described in Section 3.3. But the ^{1}D column usually operates in programmed temperature mode. The continuous changes in temperature and in carrier gas viscosity present additional problems. Several approaches have been followed in order to estimate GC×GC retention.

The previously mentioned methods for determining retention indices in the second GC×GC column (see Section 3.1.2) are based in the assignation of *RI* values to points in the two-dimensional retention map, and can be used in an inverse design to predict the peak coordinates of compounds of known retention indices in the 2D separation space, using experimental retention data for n-alkanes as references for the mapping process, or estimating these values from 1D GC data.

A first prediction attempt of GC×GC retention using this strategy was carried out by Beens et al. [10]. Retention times in the ^{1}D column were calculated from those of n-alkanes and the analyte retention index. For the analyte retention times in ^{2}D, k was first obtained by interpolation for the n-alkane series at the elution temperatures, the retention times t'_{Rz} of these compounds were calculated using the corresponding holdup times, and then t'_{Ri} for the analyte was calculated from its *RI* at the elution temperature using Equation (3). Differences between calculated and experimental values were observed for ^{2}D retention times, although predicted elution profiles were similar to the experimental patterns. A similar approach [27] used k values measured at several temperatures to obtain a better interpolation. The accuracy of the prediction of retention times was improved: reported relative errors were lower than 20%. Zhu et al. [14] used isovolatilily curves from a mixture of n-methyl esters to calculate *RI* values and an expression derived from Equation (6) for their estimation with a high accuracy, although programming rate was the only variable used in the prediction.

Lu et al. [28] used a model based in Equation (6) to predict the retention times of a mixture of only 13 compounds which however included both nonpolar n-alkanes and polar pyridine derivatives. Agreement between calculated and experimental retention time values was good for different programming rates.

The strategy of Seeley and Seeley [29] only requires as starting data 1D GC retention times from temperature-programmed runs. After calculation of retention indices in both ^{1}D and ^{2}D columns, a transformation of these values was used to construct a plot that tries to reproduce the 2D experimental retention. The method was applied to 139 volatile organic compounds with good results in the reproduction of general patterns, although prediction of ^{2}D retention shows some errors, especially for compounds whose *RI* values depend markedly on temperature.

A computer program for the optimisation of 1D GC separations [30] has been modified to include the additional variables required for GC×GC [31]. The estimation is also based on Equation (6), but the use of the distribution constant *K* instead of the retention factor *k* allows the extension of its application to different column geometries and to mixed stationary phases. The compounds in the Grob test mixture were used in the validation. The retention times of these compounds, obtained using two different temperature programs were the only experimental data required. Results were good except for the retention of a few compounds in the ^{2}D column.

All the strategies described above result in useful estimations of retention. The reported accuracies are different, but they cannot be used for comparison among them because the objectives of the prediction, the type of available starting data, and the test compounds used are different. Future approaches should take into account for their validation the following characteristics:

- *Type of estimated retention data*. Retention times are required when the objective is to help in developing a GC×GC method, and resolution or retention patterns must be estimated for different conditions. A promising possibility consists in predicting, instead of specific retention data, parameters that allow their calculation for different columns and conditions. At the time of writing, only a simplified approach based on a solvation parameter model [32] has been presented [33].
- *Polarity and volatility of the studied compounds*. Nonpolar compounds present fewer problems, as mentioned in [34], but polar compounds cannot use the same reference compounds in two columns of different polarity. A broad volatility or polarity range will require a more careful validation in order to avoid errors for extreme values.
- *Operation modes*. Calculations must be different depending on the use of controlled flow or controlled pressure modes [24,25]. Pressure at column outlet,

which can operate at room pressure (FID) or under vacuum (MS detection), must also be taken into account.

- *Effect of the modulation process*. Although modulation is commonly not considered in the estimation of retention data, in [28,34] the effect of type and characteristics of the modulator on ^{2}D retention is studied.
- *Type of data used in the estimation*. The minimum requirement is the availability (from experimental measures or from published data) of the retention time of the analytes in GC columns having the same stationary phase used in the 2D GC×GC set object of the estimation, assuming a variation of retention with temperature similar for all the compounds. In order to obtain higher prediction accuracy, retention times should be measured at least at three different temperatures in the isothermal mode. If *RI* values are used, reference n-alkanes should be run under the same conditions. If the objective is optimisation, several runs are also required in the specific GC×GC set to be optimised, in order to correct errors related to column geometry or to the use of stationary phases from different batches.
- *Required accuracy for the estimated data*. The use of predicted retention data with qualitative purposes is still far in the future. If the objective is only a rapid screening of several two-column sets in order to check their orthogonality, to detect a possible wraparound, or to estimate analysis times, a rough approximation of retention times will be useful in most cases. For optimisation of the resolution of a complex mixture, retention time must be predicted with enough accuracy for all the components, but also peak width must be estimated (see Section 4.3).

3.3 Holdup time

Holdup time, t_M, defined as the time required to elute a nonretained analyte [2], is commonly measured in 1D GC by the injection of a compound (usually methane) having a very low retention, although different mathematically based methods have also been suggested [35]. This method can be used in GC×GC for estimating t_M in the ^{1}D column. In order to calculate ^2t_M values for the ^{2}D column, the compound should be continuously injected in the system. It should also be focused on the modulator but not retained in the ^{2}D column. In addition, programmed temperature causes viscosity, flow rate, and holdup time to change continuously during the chromatographic run in pressure-controlled mode. In order to overcome these difficulties, several procedures have been proposed to estimate holdup time in the ^{2}D column.

Beens et al. [10] describe three possible experimental approaches: (1) using a plot of retention factors k of n-alkanes against temperature and extrapolating to $k = 0$; (2) using extrapolation of the "isovolatility" curves (see Section 3.1.2); and (3) using baseline alterations caused by the modulator operation when carrier gas is continuously doped with methane. The three methods produce similar results.

In [11], a theoretical value of ^2t_M is estimated from column dimensions, gas viscosity, and pressure drop. Calculation is easy, and possible errors in variables related to column geometry (such as column length and diameter) can be corrected for a given ^{2}D column by comparing experimental 1D k values of standard compounds with the retention times measured in ^{2}D [36].

Other procedures based on approaches similar to those mentioned have been used for estimation of holdup time values when they are required for calculating retention indices *LRI* in Equation (4) (see Section 3.1.2). In [13], $^{2}t_M$ was obtained by extrapolation from the experimental retention of an n-alkanes mixture at two temperatures, which was fitted to the number of carbon atoms using its linear relationship with log k. A linear relationship with temperature, T, was assumed in order to estimate $^{2}t_M$ at other temperatures. Arey et al. [34] found that ethane and propane eluted in the ^{2}D column with the same retention as bleed from ^{1}D column, using the last as a marker through the GC×GC elution process for $^{2}t_M$ calculation.

3.4 Peak width

Band dispersion at the elution in GC is shown in the chromatographic peak as its standard deviation σ. Supposing a gaussian distribution for peak shape, we can experimentally estimate σ from peak width measures, as peak width at the baseline (w_b) or at half peak height (w_h), the latter being the parameter most commonly used in order to decrease the influence of baseline noise. In GC×GC, peak shapes are approximately elliptical, and the values of their ^{1}D and ^{2}D axis correspond to peak width in the ^{1}D and ^{2}D columns ($^{1}w_h$ and $^{2}w_h$), which can be measured from recorded data.

Peak width depends on the band dispersion at the end of the column and on the speed of the band at this point. In open tubular columns, where most of band broadening occurs in the mobile phase, band dispersion at the column end is roughly the same for all compounds [37] and peak width depends only on the band speed at the end of the column. In isothermal operation, this value is inversely related to the retention time, and peak width increases along the chromatogram. But in temperature-programmed mode operation, peak width is approximately constant when all the movement of the analyte occurs while the column is linearly programmed. Peak width for a compound that elutes in the programmed mode at a column temperature T_e can be considered to be similar to that obtained for the same compound and column in isothermal conditions at a temperature T_e [37].

For a 1D GC given isothermal run, peak width depends approximately on the retention time through a linear relationship. Flow rate has a marked effect on peak width, described for a column by the van Deemter equation [38]. The Golay-Giddings equations for band broadening, which include parameters describing column geometry, can be used for more general estimations [39]. Also, more complex models using a thermodynamic basis [40] have been proposed.

Quality of column (efficiency) and extracolumn effects such as dead volumes also contribute to peak broadening. For all these reasons, it is very difficult to make an accurate prediction of w_h unless data from the same column and compounds are used as a starting point.

But even a rough estimation of its value for both first and second columns would be of great help in GC×GC operation. In ^{1}D, peak width through the chromatogram is an important parameter in the selection of an appropriate modulation time, since at least three to four modulations are recommended to sample each ^{1}D peak (see Chapter 2). Peak capacity (see Section 4.2) in the ^{2}D columns depends on ^2w_h.

Prediction of resolution in GC×GC requires, besides calculation of retention times in both columns (which has been described in Section 3.2), estimation of peak widths (Equation (7)). Beens et al. [10] used input band width and column efficiency for ^{2}D column to estimate ^2w_h with good results. Lu et al. [28] used, as a basis for the estimation of peak width, experimental ^1w_h and ^2w_h values measured for the same columns. The van Deemter equation was used in [31].

An additional problem in estimating peak widths in GC×GC is the influence of the modulation process. It has been stated that the theoretical efficiency of the ^{2}D column cannot be attained with the broad injection pulses produced by present modulators, and that longer than optimal modulation periods also markedly reduce the efficiency of the ^{1}D column [41,42]. A few researchers [43,44] have also addressed the estimation of the effect of the modulation process in the values of ^1w_h and ^2w_h.

4. PARAMETERS RELATED TO THE SEPARATION EFFICIENCY

4.1 Resolution

In 1D GC, chromatographers frequently use some concepts because of their practical value, although they are difficult to define accurately. The common concept of "separation" between two compounds that appear as peaks in the GC profile can be quantitatively measured by using their resolution, R_s

$$R_s = \frac{t_{R2} - t_{R1}}{(w_{b1} + w_{b2})/2} \tag{7}$$

or, supposing that peaks from the two compounds have the same width, by the equivalent expression:

$$R_s = \frac{\alpha - 1}{\alpha} \frac{k}{k+1} \frac{\sqrt{N}}{4} \tag{8}$$

where it can be seen that resolution, the most important parameter in chromatography, depends not only on the efficiency of the system being used (number of theoretical plates, N), but also on the difference between the interactions of each compound with the stationary phase (separation factor, α), and on the chromatographic elution conditions (retention factor, k).

In GC×GC, definition of resolution must include the fact that separation is measured for peaks that appear in a bidimensional plot. Supposing elliptical shapes for GC×GC peaks, 2D resolution is defined as the square root of the sum of the squares of the resolution of the two (A and B) columns [45,46]:

$$2DR_s = \sqrt{R_{sA}^2 + R_{sB}^2} \tag{9}$$

For instance, resolutions of 0.8 and 0.6 in columns A and B, respectively, will result in a GC×GC resolution of 1.0. But the main advantage of GC×GC shown by Equation (9) is that the $2DR_s$ value will always be equal to or higher than the highest value of R_{sA} and R_{sB}. Since in GC×GC different stationary phases are selected for both first- and second-dimension columns, we should choose for ^{1}D or for ^{2}D a column having the adequate values for α in Equation (8).

Resolution is usually the most important criterion to be considered in the optimisation of a separation method. Its prediction from operation conditions, reducing the need for slow experimental trials, would be very useful in method development. In GC×GC, prediction of resolution requires (Equations (7) and (9)) the estimation of ^1t_R, ^2t_R, ^1w_b, and ^2w_b for the compounds involved. Research work with this objective has been previously mentioned [10,28,33] in Sections 3.2.2 (estimation of retention time) and 3.3.2 (estimation of peak width).

The number of theoretical plates N, included in Equation (8), is a measure of column efficiency, which can be individually applied to the two columns of a GC×GC set. But in comprehensive GC×GC, the use of two columns having phases with different characteristics results in the redefinition of peak capacity, a 1 D GC efficiency concept. It also results in the introduction of two new concepts related to the separation behaviour, orthogonality and chromatographic structure, which are specific to GC×GC. These three concepts are discussed in Sections 4.2, 4.3, and 4.4, respectively.

4.2 Peak capacity

Peak capacity is defined as the maximum number of components theoretically separable in a chromatographic system. In 1D GC, its value n_C can be approximated [39] from

$$n_C = \left(\sqrt{N}/4\right) \ln\left(\frac{t_R}{t_M}\right) \tag{10}$$

where N is the number of theoretical plates and t_M is the holdup time. Since n_C depends in Equation (10) on the retention time t_R, a reasonable limit for this last value must be used.

In 1D GC, values of peak capacity nearby 1000 are very difficult to attain. A 450-m column produced 5,000,000 theoretical plates, requiring 640 min for the analysis of a very complex mixture [47], which corresponds roughly to $n_C = 1000$. Peak capacities for routine 1D GC are always well below this value.

Figure 3, based in reference [48], shows how the concept of peak capacity in 1D GC (gaussian peaks at the top row of the figure, where each cell of the row holds a resolved peak) can be extended to GC×GC. In the scheme, a cell in a row (^{1}D) produces in ^{2}D a column of cells that can also be occupied by peaks.

The total number of cells is the theoretical peak capacity, ^tn. For a comprehensive two-dimensional technique such as GC×GC, ^tn is the product of the peak capacities of the ^{1}D and ^{2}D columns (^1n and ^2n, respectively):

$$^{t}n = {}^{1}n \cdot {}^{2}n \tag{11}$$

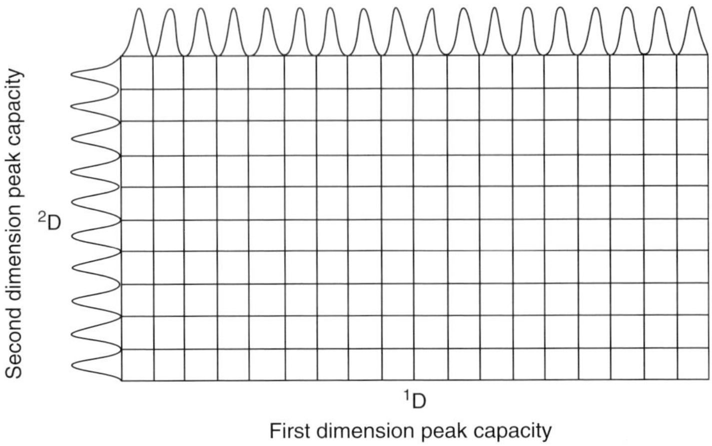

Figure 3 Theoretical peak capacity for one-dimensional GC and for GC×GC. Based on reference [48].

In GC×GC, peak capacities of 100 and 12 can be considered as average for the columns used, respectively, in the first and second dimensions. The result is a total peak capacity of 1200, which is impossible to obtain by 1D GC. Even taking into account a lower peak capacity for the ^{2}D column, the peak capacity increase from 1D GC to GC×GC is overwhelming.

But Equation (7) defines a maximum theoretical peak capacity. "Practical" peak capacity values, related to the experimental separation of components in a mixture, are lower for several reasons.

First, not all the space shown in Figure 3 is available for separation. Depending on the characteristics of the sample and of the columns, "useful" space and practical peak capacity are reduced, as shown in Figures 4 and 5, in Section 4.3.

Another reason for a reduced practical peak capacity is the loss in resolution resulting from the modulation process. As mentioned in the previous section, broad injection pulses and long modulation periods cause a loss of the optimal efficiency that can be expected from 1D GC [41,42].

Moreover, the actual number of components of a real sample that can be resolved does not depend solely on peak capacity. An important factor to take into account is peak overlap or "crowding": for compounds having a randomly distributed retention, the probability of several of them appearing at the same retention time can be statistically calculated [5]. With a random peak distribution, the number of single-component peaks is never higher than 18% of the peak capacity [49]: for any system with a peak capacity of 1000, only 180 components can be separated as single peaks. For example, the 450-m column previously mentioned [47] could not separate all the compounds in a gasoline sample. For this reason, high values of peak capacity are necessary not only for the separation of components of very complex samples, but also for simpler mixtures if we want most of their compounds to appear as single peaks. If the distribution is not random, high densities of peaks in certain areas can increase the system requirements for separation.

4.3 Orthogonality

In 1D GC, some concepts (e.g., phase polarity and selectivity) are difficult to define correctly and to measure quantitatively. Orthogonality in GC×GC presents a similar problem: it has been stated that "the definition of orthogonality, and quantification of it, is a matter of much discussion," and orthogonality has been qualified as "a slippery subject" [50]. Schoenmakers et al. [6] define orthogonality in four different fields. For analytical chemistry, orthogonality implies in a two-dimensional system that responses in the two dimensions are statistically independent. In GC×GC, it is logical to suppose that the practical separation capacity of a system will be lower when the two columns interact with sample components in a similar way (high dependence, low orthogonality), instead of as a complementary system (low dependence, high orthogonality).

Statistical definition of orthogonality [6] is clear: two data vectors are orthogonal if their values are statistically independent. But while true mathematical orthogonality requires a null correlation, in experimental techniques low-correlation values can be interpreted as having their origin in uncorrelated sources.

In GC–MS, GC partition and MS ion formation are mechanisms completely different. But even in GC–MS, for some samples the results of the two techniques could be correlated. For instance, a mixture of similar isomers can produce identical results for both retention time and mass spectra. Orthogonality depends on the sample being used for its determination, but, as for other chromatographic parameters, a general definition can be used as an approximation for practical purposes.

In 1D GC, retention data for two columns with different stationary phases are not truly orthogonal in the statistical sense, since component volatility will be in both columns a common contribution to the retention. In GC×GC operating in isothermal mode, most sample components will be distributed in a diagonal band of variable width, meaning that an important part of the two-dimensional separation space will not be used. But when a common temperature program is followed by the two columns (i.e., they are in the same oven), retention times in the second column can appear to be distributed independently of those in the first column, the correlation between them being low. The system is then almost orthogonal, according to its analytical definition. When the first column stationary phase is nonpolar, vapor pressure (volatility) being the main factor in retention, and the ^{2}D column phase presents polar interactions (the most common GC×GC set), the nonpolar factor is said to be canceled in the second column [51,52].

An explanation of this behaviour can be related to a fact described by Harris and Habgood [37]. For the GC programmed-temperature mode, each solute is largely in the gas phase at the moment of the elution from the column. Consequently, at the elution temperature, interactions responsible for the retention in the ^{1}D column will be of very low intensity. If the second-column phase presents mixed interactions, those common with the first-column phase will also be of small importance at this temperature. As part of their work in the prediction of retention times in GC×GC, Seeley and Seeley [29] present a theoretically based relationship between retention in the second column and the difference between phase-compound interactions for ^{2}D and ^{1}D columns. Experimentally, while first-dimension retention times are of course related to linear retention indices ^1LRI in the first-column phase, second-dimension retention times can be approximated as a function of the difference ($^2RI - {}^1RI$) between retention indices in both phases.

Using orthogonal column sets, most of the separation area in the 2D retention plane (the separation space) can be available for peak distribution. But while orthogonality promises the best general use of the retention plane, it cannot guarantee a correct resolution for a specific sample. As previously mentioned, orthogonality in GC×GC, as in other analytical techniques, depends on sample components and on separation conditions [48,53].

The concept of orthogonality can be used as a first step in selecting the optimum column set and the best chromatographic conditions in the development of an analytical method. This selection should take into account the "overall" properties of the stationary phases in the available columns, looking for independent phase-compound interactions. A simple approach for their estimation is to suppose that they can be explained by using two factors, one specific of the compound and the other of both stationary phase and conditions. Polarity is an intuitive concept that could be associated to these factors. But although different polarity scales have been used with quantitative purposes [54], their use presents problems, perhaps the most important being that stationary phases can interact specifically with different analytes. This selective behaviour was supposed to be described by using the retention of a small (5 to 10) number of compounds, but selection of compounds showing only "pure" interactions was difficult. For this reason, the solvation parameter model should be preferred. Poole and Poole [55] used this model to characterize capillary columns prepared with 50 stationary phases. Their database contains column parameter values at five temperatures and can be used for classification of stationary phases and as a guide for their selection in method development in GC×GC.

But in order to compare different GC×GC sets and conditions, for both method development and optimisation, a quantitative measure of orthogonality is necessary. Several methods have been proposed to quantitatively determine orthogonality and are discussed in the next section.

4.3.1 Measuring orthogonality

Slonecker et al. [56] have applied information theory in a study including different separation techniques in order to measure orthogonality. Informational similarity (IS), was used as a measure of analyte crowding in some areas of the separation space. Low IS values correspond to a low level of solute crowding: a high degree of peak overlap, inversely related to orthogonality, will be shown by IS values near to 1. The same publication describes peak scatter in the separation space using informational entropy, which is defined as the decrease in the uncertainity about the components present in the sample. The parameter used (synentropy) is the relative percent value of informational entropy contributed from each column in a two-column system. Percent synentropy (PS) values of 0% and 100% correspond, respectively, to fully orthogonal and nonorthogonal systems. IS and PS can be calculated from experimental retention times in 1D GC columns.

The method is subject to some errors if peaks appear frequently (for instance, as peak clusters) in areas far from the diagonal of the separation space, and does not take into account the operation in the programmed temperature mode.

Liu and Paterson [48] measured orthogonality by following a combined statistical and geometrical approach. When two sets of retention data for the first and second columns are considered as vectors, a mathematical orthogonality can be defined from their correlation coefficient, $C_{i,j}$. The values of this coefficient will range between 0, for a perfectly orthogonal separation, and 1, for cases of minimum orthogonality. C_{ij} is then proposed as a quantitative measure of orthogonality. From the C_{ij} value, the values of angle β in Figure 4 can be computed [48]. While in Figure 3 total peak capacity is ${}^{t}n = {}^{1}n \cdot {}^{2}n$, GC×GC peaks cannot appear in the areas marked "N" in Figure 4 and only the area marked as "A" is available for separation. From the C_{ij} value, used as an orthogonality measure, the relative amount of useful separation space can be determined [48] as a measure of "practical" peak capacity.

Orthogonality can in turn be estimated from the experimental coverage of the separation space. A geometrical approach proposed for 2D LC [57], based in dividing the separation space in rectangular bins and determining the number of bins containing at least one peak, has been discussed in [50]. The publication points out the differences between mathematical orthogonality and relative

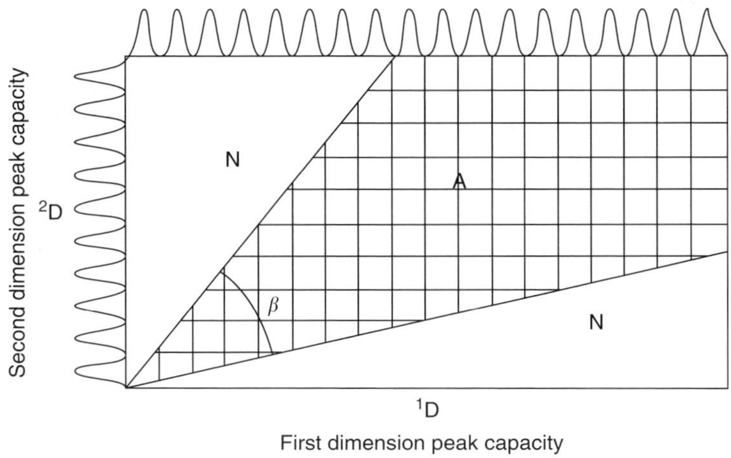

Figure 4 Orthogonality and peak capacity. β is estimated from orthogonality measures. "Practical peak capacity" available for separation is represented by the area marked "A". Based on reference [48].

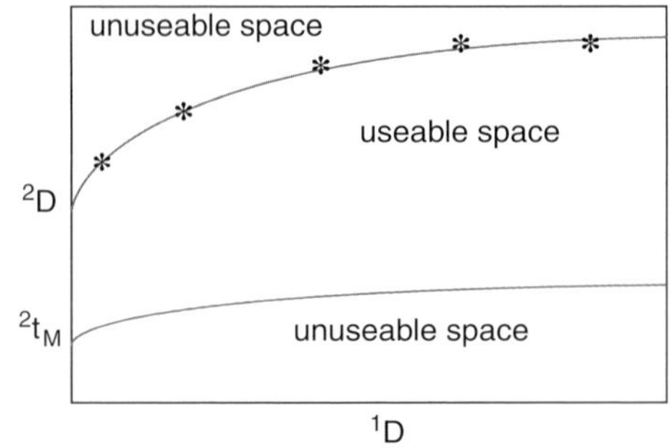

Figure 5 Diagram of the GC×GC separation space available for separation. Lower trace, holdup time ^2t_M. Symbols (*) mark the most retained compounds in ^{2}D. Adapted from [36].

coverage and recommends the use of the last parameter for a practical evaluation of the separation performance of a GC×GC separation system.

As orthogonality, coverage of separation space depends on the characteristics of sample components. Ryan et al. [36] used a test mixture with a broad range of polarities, five ^{1}D columns of increasing polarity (by combining a polar and a nonpolar columns of different lengths), and two ^{2}D columns (polar and nonpolar). The covered separation space for several column combinations was determined both experimentally and by prediction of 2D retention times (see Section 3.2.2) using retention measures in the ^{1}D columns. The approach allowed the study of specific shifts of the components caused by changes in column polarity, but it also showed the general distribution of the available separation space of the solutes (Figure 5), which lays between the holdup time (^2t_M, lower trace) and the most retained compounds (asterisks, upper trace) in the second-dimension axis.

A drawback of this approach is that, as shown in Figure 5, the calculated useful space depended on the last eluting compounds of the mixture (upper trace) more than on the overall distribution of the compounds.

Cordero et al. [53] followed an experimental approach related to that in [36]. An essential oil and a mixture of standards were run through GC×GC. Two ^{1}D columns with OV-1 and polyethyleneglycol as stationary phases were combined with ^{2}D columns coated with mixtures of these two phases and with OV-17: 10 different sets of columns were used in this way. Experimental GC×GC results should correspond to different levels of orthogonality, which Cordero et al. evaluated using the parameters previously described, based on correlation analysis [48], spreading angle, β, theory of information [56] (informational similarity, IS, and percent synentropy, PS), and percent usage of separation space [36,50,57]. Most parameters confirm that the best GC×GC separation is obtained for the ^{1}D nonpolar×^{2}D polar set, but it is worth noting the good practical orthogonality obtained with mixed phases (probably because the temperature program reduced the nonpolar interaction in the ^{2}D mixed phase column). Different parameter values were obtained from the two test mixtures. This showed how orthogonality depended on the analyzed compounds, confirming as mentioned for other procedures that, for method development purposes, the test mixture used to assess the value of orthogonality and/or of available separation space must contain all components spanning the n-dimensional range of properties that can be expected for possible analytes [51].

An alternative for a more general measure of orthogonality should use specific parameters of the columns being considered, instead of analyte retention data. The solvation parameter model [55] characterizes column retention by using five parameters. These parameters define for each column a vector in a five-dimensional space. If θ is the angle between two column vectors, $\cos \theta$ will be nearly 1 for very similar pair of stationary phases, while values of $\cos \theta$ close to 0 will correspond to column sets of high orthogonality [58].

4.4 Chromatographic structure

The 1D GC separation of a mixture of n-alkanes using a linear temperature program is an example of how common structural features can appear represented in a chromatographic profile. The 1D GC retention time depends only on vapour pressure for compounds, like n-alkanes, which have the same polarity. And since vapour pressure for n-alkanes is related to their number of carbon atoms, the GC profile shows peaks regularly spaced along the retention time scale. Figure 6 simulates (lower trace, black circle marks, corresponding to a 1D profile) the regular elution of an n-alkanes mixture. The presence of other sample components (for instance, n-alcohols) can, however, confuse the regular elution

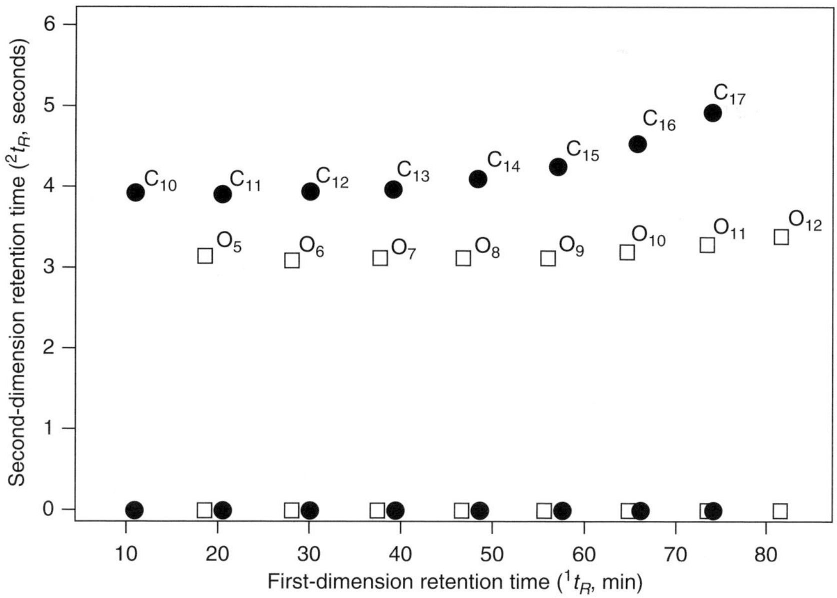

Figure 6 Simulated 1D (lower marks) and 2D (upper marks) retention behaviour of a sample containing n-alkanes (•) and n-alcohols (□).

patterns in 1D GC of n-alkanes. Later eluting peaks from n-alcohols (Figure 6, lower trace, square marks) will co-elute with those n-alkanes.

Although the n-alkanes retention can be described by assigning to each compound a single variable (e.g., number of carbon atoms), more complex mixtures will require for their description additional variables. In these cases, samples are said to have a higher dimensionality [4]. For instance, a dimensionality of 2 can be assigned to the mixture of n-alkanes and n-alcohols plotted in Figure 6, which can be described using the number of carbon atoms and the presence of a hydroxyl group.

When an orthogonal 2D set of columns is used in GC×GC, two-dimensional mixtures can be resolved showing elution patterns in the separation plane which reflect the properties of sample components. For instance, for the n-alkanes: n-alcohols mixture, if the retention of the first column is mainly related to compound vapour pressure, and the retention of the second column to compound polarity, when sample components elute their position in the 2D separation plane will be related to those two properties (see the upper marks in Figure 6). The resulting chromatograms are said to be "structured" because compounds presenting similar values for one of these properties will appear grouped or related in the bidimensional contour plot. Even in mixtures containing very different compounds, that is, having a high dimensionality, analytes with common properties can appear plotted in structured chromatograms.

Two main pattern types are produced in GC×GC plots. Globular clusters correspond to groups of isomers with very similar properties, and then with similar retention times in both ^{1}D and ^{2}D columns. In these cases, separation will require the use in 1D or 2D of a stationary phase that is more selective toward the small structural differences among isomers, or use of a separation technique of higher dimensionality (see Chapter 6).

More common, and also more interesting, are elongated clusters or linear trends, which will appear when compounds have a common property (e.g., polarity) but differ in other (e.g., volatility). These properties are related to the separation mechanism of the ^{1}D and ^{2}D columns.

In Figure 7 from [59], which shows the separation of the major components present in the GC×GC chromatogram of a lavender essential oil analysed in an nonpolar–polar two-column set, the two pattern types are represented. Several isomeric monoterpene hydrocarbons having the same retention in the two columns (peaks 16 to 20) appear clustered in a tight group, while the oxygenated terpenes appear distributed according their polarity in the second dimension above the hydrocarbons (region A, alcohols; region B, acetates).

Other examples of the last type of structures appear in the chapters of this book that detail GC×GC analyses of complex samples of different types. For instance, Figure 8 of Chapter 7 shows the structured results of the analysis of sulphur compounds in gasolines.

Structured chromatograms also appear when using reverse GC×GC sets, that is, a polar column in the first dimension and an nonpolar column in the second dimension. In Figure 2 of Chapter 9, the GC×GC results for a citrus essential oil analysed using a polyethyleneglycol×SPB-5 set are shown. Chromatographic

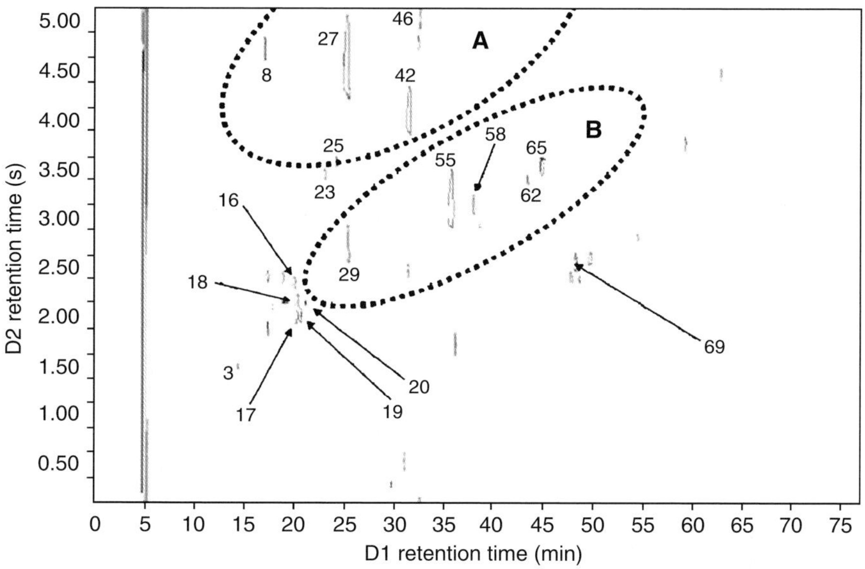

Figure 7 GC×GC plot of major components of a lavender oil. A, alcohols; B, acetates. Adapted from [59].

structures appear for alcohols (lower relative retentions in the nonpolar ^{2}D column), aldehydes, esters and hydrocarbons (higher retentions in ^{2}D).

For a given GC×GC column set and conditions, the size and shape of these structures depend on the number and type of compounds sharing the common property. For instance, a homologous series with a high range of number of carbon atoms usually appear as curves stretched along the ^{1}D elution in nonpolar-polar systems. The trend is clear: the curve can even be fitted to nonlinear expressions, and the results can be used for qualitative purposes (see for instance Figure 4 of Chapter 10).

When groups of compounds in the series differ only in minor structural characteristics having a small influence in $^{1}t_R$ and $^{2}t_R$ (for instance, when the series is formed by several groups of isomers having the same number of carbon atoms but a different substitution pattern), the band is widened by the dispersion in the 2D retention of these groups. This is a common behaviour for complex families of compounds (see Chapters 10 and 11). When this secondary characteristic affects more markedly the retention in the ^{2}D column, a new set of structures can appear, stretched mainly along the second dimension. In this way, a compound can belong to two different chromatographic structures. Examples of these cases appear in the previously mentioned Figure 8 of Chapter 7 and Figure 4 of Chapter 10.

The presence of these characteristic trends in a GC×GC chromatogram is an advantage toward its interpretation, since the overall sample type or the existence of an atypical composition can visually be checked in an easy way. Chromatographic structures also help by reducing the number of peaks that need to be checked for an MS identification. In very complex samples, the presence of elongated structures allows better use of the separation space available (see Section 4.2), since components of a group are usually spread in a long band and can be separated in the ^{2}D dimension from other groups with different properties.

Section 4.3 describes the importance of the sample being analysed in the determination of available peak capacity and of orthogonality. The presence of structured chromatograms also depends on the specific sample: complex samples that include a representative number of compounds with a common characteristic will present a higher probability of showing these groups of compounds in a 2D structure. However, experimental or estimated data are always required in order to select the "best" stationary phase set for this purpose, including normal (low polarity-high polarity) or reversed (high polarity-low polarity) column arrangements.

5. FINAL REMARKS

GC×GC can be used for basic research purposes, but the theoretical considerations presented here are related mainly to its practical aspects. As for other chromatographic techniques, the GC×GC main objective is the qualitative and quantitative analysis of complex mixtures. Chapters in the second part of this volume detail its analytical applications in several fields, usually including its use as a quantitative tool.

Qualitative analysis by GC×GC must be based on its 2D retention times (^1t_R and ^2t_R) or on derived parameters. Although the measure in the ^{2}D of these parameters presents some difficulties, several procedures developed in order to obtain precise values are described in Sections 3.1 and 3.3.

The optimisation of an analytical method based in GC×GC is perhaps one of the most important problems associated with this technique. The number of operation conditions to be optimised is higher in GC×GC than in 1D GC, but the main difficulties are the effect on the ^{2}D separation of any change in the ^{1}D conditions, and the high number of possibilities offered by the different two-columns sets.

For these reasons, research work on predictive models addressed to the estimation of the parameters responsible for the resolution is expected to increase in the near future. Methods for estimating peak retention and peak width, described in Section 3.2, are promising and have already presented good results.

Orthogonality and the presence of chromatographic structures are characteristics of a GC×GC separation which describe the overall use of system efficiency. These concepts, important in the analysis of very complex samples, are difficult to define and even more difficult to express numerically or to compare on a quantitative basis. However, the predictive models mentioned could also be applied in a rough calculation of the retention of the compounds possibly present in samples for a given application. Orthogonality, use of the separation space, and presence of chromatographic structures, could be estimated for several sets of columns and/or conditions, and the general efficiency for the application could be optimised in this way.

Since both GC×GC and GC are based in the same chromatographic processes, GC×GC will benefit from previous work on basic aspects of GC, but also more GC research will be required as a basis for the optimisation of GC×GC-based methods. Comparison of different strategies for optimisation and generalization of their results will require consideration of (1) the possible effect of the modulation process in the response parameters, which is disregarded in most approaches but appears to be of the utmost importance in others, and (2) the dependence of results in the particular composition of the test sample used for optimisation.

REFERENCES

1 M.J.E. Golay, in *Gas Chromatography 1958* (Amsterdam Symposium), D.H. Desty (Ed.), London, Butterworths, 1958, p. 36.
2 L.S. Ettre, Nomenclature for Chromatography (IUPAC Recommendations 1993), *Pure Appl. Chem.*, 65 (1993) 819.
3 K. Danzer, J.F. van Staden and D.T. Burns, *Pure Appl. Chem.*, 74 (2002) 1479.
4 J.C. Giddings, *J. Chromatogr. A*, 703 (1995) 3.
5 J.C. Giddings, *Anal. Chem.*, 56 (1984) 1258.
6 P. Schoenmakers, P. Marriott and J. Beens, *LC-GC Europe*, 16 (2003) 335.
7 T. Greibrokk, *J. Sep. Sci.*, 29 (2006) 479.
8 E. Kovats, *Helvetica Chim. Acta*, 41 (1958) 1915.
9 H. van den Dool and P.D. Kratz, *J. Chromatogr.*, 11 (1963) 463.
10 J. Beens, R. Tijssen and J. Blomberg, *J. Chromatogr. A*, 822 (1998) 233.
11 R.J. Western and P.J. Marriott, *J. Sep. Sci.*, 25 (2002) 832.
12 R.J. Western and P.J. Marriott, *J. Chromatogr. A*, 1019 (2003) 3.
13 T. Pang, S. Zhu, X. Lu and G. Xu, *J. Sep. Sci.*, 30 (2007) 868.
14 S. Zhu, X. Lu, Y. Qiu, T. Pang, H. Kong, C. Wu and G. Xu, *J. Chromatogr. A*, 1150 (2007) 28.
15 S. Bieri and P.J. Marriott, *Anal. Chem.*, 78 (2006) 8089.
16 R. Ong, P. Marriott, P. Morrison and P. Haglund, *J. Chromatogr. A*, 962 (2002) 135.
17 K. Héberger, *J. Chromatogr. A*, 1158 (2007) 273.
18 P.J. Marriott, T. Massil and H. Hügel, *J. Sep. Sci.*, 27 (2004) 1273.
19 A.R. Katritzky, E.S. Ignatchenko, R.A. Barcock, V.S. Lobanov and M. Karelson, *Anal. Chem.*, 66 (1994) 1799.
20 O. Ivanciuc, *J. Chem. Inf. Comput. Sci.*, 37 (1997) 405.
21 B. Lucic, N. Trinajstic, S. Sild, M. Karelson and A.R. Katritzky, *J. Chem. Inf. Comput. Sci.*, 39 (1999) 610.
22 M. Pompe, J.M. Davis and C.D. Samuel, *J. Chem. Inf. Comput. Sci.*, 44 (2004) 399.
23 Y. Ren, H. Liu, X. Yao and M. Liu, *Anal. Bioanal. Chem.*, 388 (2007) 165.
24 J. Beens, H.G. Janssen, M. Adahchour and U.A.Th. Brinkman, *J. Chromatogr. A*, 1086 (2005) 141.
25 J. Harynuk and T. Górecki, *J. Chromatogr. A*, 1086 (2005) 135.
26 M. de Frutos, J. Sanz, I. Martinez-Castro and M.I. Jiménez, *Anal. Chem.*, 65 (1999) 2643.
27 C. Vendeuvre, F. Bertoncini, D. Thiébaut, M. Martin and M.C. Hennion, *J. Sep. Sci.*, 28 (2005) 1129.
28 X. Lu, H. Kong, H. Li, C. Ma, J. Tian and G. Xu, *J. Chromatogr. A*, 1086 (2005) 175.
29 J.V. Seeley and S.K. Seeley, *J. Chromatogr. A*, 1172 (2007) 72.
30 F.L. Dorman, P.D. Schettler, C.M. English and D.V. Patwardhan, *Anal. Chem.*, 74 (2002) 2133.
31 F.L. Dorman, P.D. Schettler, L.A. Vogt and J.W. Cochran, *J. Chromatogr. A*, 1186 (2008) 196.
32 M.H. Abraham, A. Ibrahim and A.M. Zissimos, *J. Chromatogr. A*, 1037 (2004) 29.
33 J.V. Seeley, K. Hill-Edwards, E.K. Libby, and S.K. Seeley, Symposia HTC-10, Bruges, 2008.

34 J.S. Arey, R.K. Nelson, L. Xu and C.M. Reddy, *Anal. Chem.*, 77 (2005) 7172.
35 R.C. Castells, *J. Chromatogr. A*, 1037 (2004) 223.
36 D. Ryan, P. Morrison and P. Marriott, *J. Chromatogr. A*, 1071 (2005) 47.
37 W.E. Harris and H.W. Habgood, In: *Programmed Temperature Gas Chromatography*, John Wiley & Sons, New York, 1966.
38 J.J. van Deemter, F.J. Zuiderweg and A. Klinkenberg, *Chem. Eng. Sci.*, 5 (1956) 271.
39 C.F. Poole, *The Essence of Chromatography*, Elsevier, Amsterdam, 2003.
40 K. Krisnangkura and V. Pongtonkulpanich, *J. Sep. Sci.*, 29 (2006) 81.
41 L.M. Blumberg, *J. Chromatogr. A*, 985 (2003) 29.
42 L.M. Blumberg, F. David, M.S. Klee and P. Sandra, *J. Chromatogr. A*, 1188 (2008) 2.
43 J.V. Seeley, *J. Chromatogr. A*, 962 (2002) 21.
44 T. Gorecki, J. Harynuk and O. Panic, *J. Sep. Sci.*, 27 (2004) 359.
45 R.E. Murphy, M.R. Schure and J.P. Foley, *Anal. Chem.*, 70 (1998) 1585.
46 A.E. Sinha, B.J. Prazen and R.E. Synovec, *Anal. Bioanal. Chem.*, 378 (2004) 1948.
47 T.A. Berger, *Chromatographia*, 42 (1996) 63.
48 Z. Liu and D.O. Patterson, *Anal. Chem.*, 67 (1995) 3840.
49 J.M. Davis and J.C. Giddings, *Anal. Chem.*, 55 (1983) 418.
50 N.E. Watson, J.M. Davis and R.E. Synovec, *Anal. Chem.*, 79 (2007) 7924.
51 C.J. Venkatramani, J.Z. Xu and J.B. Phillips, *Anal. Chem.*, 68 (1996) 1486.
52 J.B. Phillips and J. Beens, *J. Chromatogr. A*, 856 (1999) 331.
53 C. Cordero, P. Rubiolo, B. Sgorbini, M. Galli and C. Bicchi, *J. Chromatogr. A*, 1132 (2006) 268.
54 W.O. McReynolds, *J. Chromatogr. Sci.*, 8 (1970) 685.
55 C.F. Poole and S.K. Poole, *J. Chromatogr. A*, 1184 (2008) 254.
56 P.J. Slonecker, X.D. Li, T.H. Ridgway and J.G. Dorsey, *Anal. Chem.*, 68 (1996) 682.
57 M. Gilar, P. Olivova, A.E. Daly and J.C. Gebler, *Anal. Chem.*, 77 (2005) 6426.
58 S.K. Poole and C.F. Poole, *J. Sep. Sci.*, 31 (2008) 1118.
59 R.A. Shellie, L.L. Xie and P.J. Marriott, *J. Chromatogr. A*, 968 (2002) 161.

CHAPTER 4

Data Acquisition, Visualization, and Analysis

Stephen E. Reichenbach

Contents

1. INTRODUCTION

An introduction to informatics for comprehensive two-dimensional gas chromatography (GC×GC) should begin with the strikingly beautiful and

Comprehensive Analytical Chemistry, Volume 55
ISSN: 0166-526X, DOI 10.1016/S0166-526X(09)05504-4

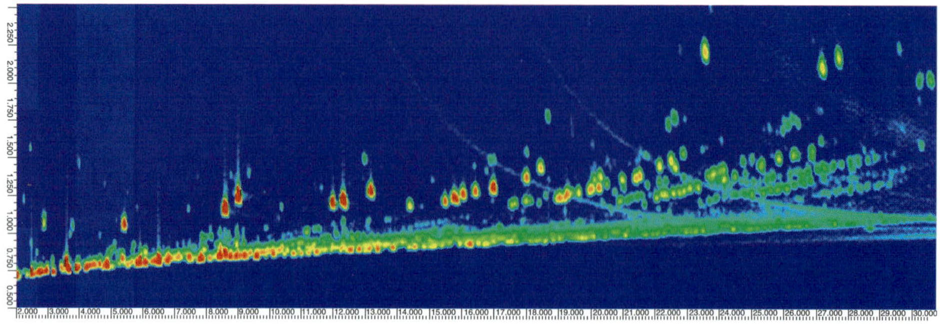

Figure 1 GC×GC data from a gasoline analysis visualized as a digital image. Only a portion of the data is shown. (This and other figures were generated with GC Image® software [1]. Data supplied by Zoex Corporation.)

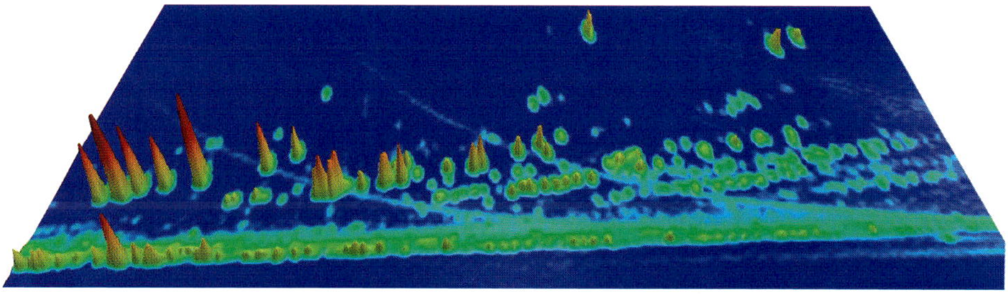

Figure 2 GC×GC data visualized as a three-dimensional surface. A subregion of the data from Figure 1 is shown.

complex pictures of data visualization. Whether viewed as a pseudocolorized two-dimensional image, as in Figure 1, or as a projection of a three-dimensional surface, as in Figure 2, GC×GC visualizations impress even observers lacking chromatographic expertise with their colorful and multitudinous features. Chromatographers recognize, within these pictures, complex patterns embedding a wealth of multidimensional chemical information. The richness of GC×GC data is immediately apparent, but the size and complexity of GC×GC data pose significant challenges for chemical analysis.

This chapter examines methods and information technologies for GC×GC data acquisition, visualization, and analysis. The quantity and complexity of GC×GC data make human analyses of GC×GC data difficult and time-consuming and motivate the need for computer-assisted and automated processing. GC×GC transforms chemical samples into raw data; information technologies are required to transform GC×GC data into chemical information.

The typical data flow is a sequence of: acquiring and storing raw data, processing data to correct artifacts, detecting and identifying chemical peaks, and analyzing datasets to produce higher-level information (including quantification) and reports. In applications for which the analysis is fairly well understood and routine, information technologies may fully automate this process.

However, because GC×GC is so powerful, it frequently is used for analyses that are not well understood or are not routine. In such cases, information technologies must support semi-automated processing, visual interpretation, and interactive analysis.

This chapter addresses the following fundamental tasks in transforming GC×GC data into chemical information:

- Acquiring and formatting data for storage, access, and interchange.
- Visualizing multidimensional data.
- Processing data to remove acquisition artifacts and detect peaks.
- Identifying chemical constituents.
- Analyzing datasets for higher-level information and reporting.

2. DATA ACQUISITION

Although GC×GC is a true two-dimensional separation, the process serializes the data — producing data values in a sequence. In GC×GC, the first column progressively separates and presents eluates to the modulator, which iteratively collects and introduces them into the second column, which then progressively separates and presents eluates to the detector. As explained in detail in Chapter 2, in the detector, the analog-to-digital (A/D) converter samples the chromatographic signal at a specified frequency. In concept, this operation is similar to how some optical systems create an image with as few as one detector by progressively scanning the detector(s) across the two spatial dimensions, but, in GC×GC, the two dimensions are the two retention times. Then, the digitized data and relevant metadata (information about the data) are stored in a file with a defined format for subsequent access.

2.1 Modulation and sampling

The modulation frequency and the detector sampling frequency typically are under user control. Setting these frequencies (subject to the limitations of the hardware) involves trade-offs between resolution and other constraints. The desire for high resolution suggests that the modulation and sampling rates should be as rapid as possible. A Gaussian peak is not band-limited, so truly sufficient sampling is not possible. Therefore, higher modulation and sampling rates provide greater information capacity and increased resolution for detecting co-eluted peaks. However, the modulation frequency must allow adequate intervals for separations in the second column, and the sampling frequency involves a trade-off in data size (i.e., higher sampling frequencies generate more data) and diminishing returns in selectivity and precision. Full consideration of these and other issues (such as duty cycle and noise) in setting the modulation and sampling frequencies involves instrumental and application-specific concerns that are beyond the scope of this chapter, but consideration of the data suggests general guidelines.

Experimental and theoretical studies [2] suggest that the modulation rate should be at least one cycle per two times the primary peak standard deviation σ_1 (i.e., the standard deviation of the peak width from the first column separation), which translates to at least four modulation cycles over $8\sigma_1$ (the effective width of peaks from the first-column separation). The considerations for GC×GC detector frequencies are similar to those for traditional one-dimensional chromatography, for which a rate of at least one sample per peak standard deviation is recommended [3,4], that is, eight samples over $8\sigma_2$ (the effective width of peaks from the second-column separation). With these considerations, Murphy et al. [5] recommend that method development begin with determining the shortest time for adequate chromatographic separation in the second column and then a first-dimension method be used that provides peak widths of at least four times the modulation interval. With the wide variety of chemical mixtures and analytical goals for GC×GC, a broad range of modulation and sampling frequencies are used. Modulation cycles from 2 to 20 seconds (s) and sampling frequencies from 25 to 200 hertz (Hz) are not unusual. Again, however, the application should be considered; slow modulation and sampling rates relative to peak width may be sufficient for applications that require only quantification of well-separated peaks, and fast modulation and sampling rates relative to peak width may be required for applications that involve compounds that are difficult to separate.

A common problem in GC×GC data processing is inadequate sampling of the first-column output; that is, the modulation period is too long with respect to the first-column peak widths, or, put another way, the first-column chromatography produces peaks too narrow for the modulation period. Of course, if the modulation period is constrained by the time required for second-column separations, then broadening the peak widths from the first column may require longer runs (thereby increasing cost). Inadequate sampling of the second-column output is less commonly problematic because most detectors used for GC×GC are fast and most laboratories typically use detector sampling rates that exceed what is required for the analysis (and so generate more data than may be necessary). However, as explained in Chapter 2, some types of detectors — for example, quadrupole mass spectrometer (qMS), atomic emission detector (AED), and electron capture detector (ECD) — may be challenged by the acquisition speeds required for GC×GC.

2.2 Digitization and coding

GC×GC systems use an A/D converter to map the intensity of the chromatographic signal to a digital number (DN). Among the many types of detectors used with GC×GC, the major distinction is between detectors that produce a single number at each time sample of the chromatogram, such as a flame-ionization detector (FID) and a sulfur chemiluminescence detector (SCD), and multichannel detectors that produce multiple values (typically, over a spectral range) for each time sample, such as a mass spectrometer (MS). In either case, each DN is represented with a limited number of bits indicating a value in a limited range with limited precision.

Because GC×GC can produce large datasets, GC×GC systems often employ data compression in their file formats. Sampling at 200 Hz, a detector for single values with a 48-bit dynamic range (as supported by Agilent's IQ data file format [6]) produces data at the rate of 4.3 megabytes/hour (MB/h). Most programming languages must perform arithmetic on 48-bit values with 64-bit long integers or 64-bit double-precision floating-point numbers. Mass spectrometers can produce data at sub-1 GHz (e.g., one 8-bit spectral intensity per nanosecond), a data rate of greater than 1 gigabyte/sec (GB/s). In order to more efficiently store data, GC×GC systems may compress the data. For example, because data values are correlated with neighboring values in the sequence, Agilent's IQ data file format implements a second-order backward differential coding that compresses values from a 48-bit range to 2 bytes. Even more aggressive compression commonly is used for MS data. For example, ORTEC's FastFlight-2TM [7] can accumulate successive spectra in hardware and output only the summed spectra for a much smaller data rate. In a MS with GHz raw speed, summing 100 transient spectra in 100 K channels generates 100 spectra per second (compared to 10,000 raw spectra per second). The FastFlight2 also offers a lossless compression mode that uses fewer bytes to represent smaller values and a lossy compression mode that detects and encodes only the spectral peaks in the MS data — a process sometimes called centroiding because each spectral peak is represented by a single centroid indicating the center, intensity, and sometimes the peak width.

2.3 File formats

Most GC×GC systems use a proprietary data file format, which affords vendors a high degree of control (e.g., to implement data compression), but which poses a barrier and inconvenience for sharing or processing data across systems. Currently, there is no standard format for GC×GC data, but GC×GC data can be shared using nonstandard text files or existing standards for gas chromatography (GC) data. GC×GC data can be converted to text, for example, ASCII-format comma-separated values (CSV), but the resulting files are non-standard and are larger than binary or compressed data files. The ASTM has issued Analytical Data Interchange (ANDI) standards for chromatography [8] and MS [9]. These standards lack some requirements for GC×GC metadata (e.g., a metadata element for the modulation cycle) but can be used to communicate raw data and other chromatographic metadata. These standards were developed primarily for data interchange and lack some desirable features for more routine use. Another limitation of the ANDI standards is that the network Common Data Form (netCDF) [10], upon which the standards are built, was defined for 32-bit computing systems, limiting their usability for data larger than 2 GB. The ASTM has sanctioned an effort to develop a new format standard for analytical chemistry data, the Analytical Information Markup Language (AnIML) [11,12], utilizing the eXtensible Markup Language (XML) [13]. Standard formats for analytical chemistry data facilitate data portability and interchange, but despite such considerations proprietary GC formats have continued to dominate the market.

3. VISUALIZATION

Visualization is a powerful tool for qualitative analysis of GC×GC data (e.g., to troubleshoot the chromatography). Various types of visualizations are useful: two-dimensional images provide a comprehensive overview, three-dimensional visualizations effectively illustrate quantitative relationships over a large dynamic range, one-dimensional graphs are useful for overlaying multivariate data, tabular views reveal the numeric values in the data, and graphical and text annotations communicate additional information. This section explores some of the methods and considerations in the various types of visualizations.

3.1 Image visualizations

3.1.1 Rasterization

A fundamental visualization of GC×GC data is as a two-dimensional image. GC×GC data, which is acquired sequentially, can be reorganized as a raster — a two-dimensional array, matrix, or grid of picture elements called pixels — in which each pixel value is the intensity of the detector signal. As a two-dimensional array of intensities, GC×GC data has many similarities with other types of digital images and so many methods and techniques from the field of digital image processing can be applied or adapted for GC×GC data visualization and processing.

The standard approach for rasterization is to arrange the data values acquired during a single modulation cycle as a column of pixels, so that the ordinate (*Y*-axis, bottom-to-top) is the elapsed time for the second-column separation, and then to arrange these pixel columns so that the abscissa (*X*-axis, left-to-right) is the elapsed time for the first-column separation. This ordering presents the data in the commonly used right-handed Cartesian coordinate system, with the first-column retention time as the first index into the array. Other orderings are possible but less commonly used. The problems of correctly synchronizing the columns of data with the modulation cycle and of modulation cycles that are not evenly divisible by the detector sampling-interval are examined in Section 4.1.

3.1.2 Colorization

For presentation as an image, the pixels are colorized; that is, the GC×GC values are mapped to colors of the display device. Scalar values, such as single-valued GC×GC data, can be colorized simply on an achromatic grayscale, familiar from so-called black-and-white images. Scalar values can be extracted from multi-spectral data in various ways, for example, by adding all intensities in each spectrum to compute the total intensity count (TIC) of the data point or by taking the value in a selected "channel" of the spectrum. A grayscale mapping typically is defined by setting a lower bound, below which values are mapped to black; an upper bound, above which values are mapped to white; and a function to map values between the bounds to shades of gray, with brightness increasing with value. Linear, logarithmic, and exponential mapping functions are useful for different effects: linear mapping treats gradations at all intensity levels similarly;

logarithmic mapping emphasizes gradations nearer the lower bound; and exponential mapping emphasizes gradations nearer the upper bound. Although grayscale colorization provides a straightforward ordering of values from small to large that is intuitively meaningful, humans may be able to distinguish fewer than 100 distinct grayscale gradations [14]. Therefore, grayscale images cannot effectively communicate many differences among values over a large dynamic range such as is common for GC×GC data.

Pseudocolorization takes advantage of the differing sensitivities in human vision for different frequencies of light [14]. These differing sensitivities enable "color" perception, with greater selectivity than for grayscale. Because humans have trichromatic vision based on three types of color receptors (cones), a trichromatic color model is sufficient for image colorization. Various trichromatic color models have been developed. RGB (with values for red, green and blue) and HSV (with values for hue, saturation, and brightness value) are widely used color models for digital imaging.

Pseudocolorization maps data values with three independent functions for the three color components. The mapping functions for the color components typically are not monotonically nondecreasing (as grayscale mapping functions typically are), so discerning relative values in a pseudocolor image is not as straightforward as with grayscale (for which brighter means larger). However, a good pseudocolor scale can communicate a clear ordering of values. For example, topographic and temperature images commonly use a pseudocolor scale sometimes called cold-to-hot, which has a mapping from small to large that progresses through blue, cyan, green, yellow, and red, with intermediate colors. In Figure 1, the color scale has the smaller values of the background colorized dark blue and the larger values of the peaks colorized with the cold-to-hot scale to show increasing values. This mapping is easily interpreted because it is familiar. Pseudocolor images can present many distinguishable colors, but there is a trade-off between having a pseudocolor scale with an ordinal progression that is simple to understand and the number of gradations that can be discerned: an easily understood scale visually differentiates a smaller number of gradations, and a scale that visually differentiates a larger number of gradations makes the value ordering more difficult to understand.

Pseudocolorization offers better visualization than grayscale for gradations across a wide dynamic range of values, but to be effective the mapping still must allocate color variations to the value range according to the presence of gradations. Specifying pseudocolorization interactively can be tedious and difficult, so automated determination of pseudocolor mapping is useful. Gradient-Based Value Mapping (GBVM) [15] is an automated method for mapping GC×GC data values onto a color scale, for example, the cold-to-hot scale. For a given dataset, GBVM builds a value-mapping function that emphasizes gradations in the data while maintaining ordinal relationships of the values. The first step computes the gradient (local difference) at each pixel. Then, the pixels (with computed gradients) are sorted by value, and the relative cumulative gradient magnitude is computed for the sorted array. The GBVM function is the mapping from pixel value to the relative cumulative gradient magnitude of the sorted array. GBVM is effective at showing local differences across a large dynamic range.

Each resolved chemical compound in a sample increases the value in a small cluster of pixels, which, if the colorization effectively shows local differences, are seen as a localized spot with different colors than the surrounding background. If the colorization is not effective over the full dynamic range, spots with small values may not be visible or spots with large values may not show significant relative differences.

3.1.3 Navigation

Standard operations for navigating digital images include panning, scrolling, and rescaling. Rescaling requires resampling the data — creating a displayed image with more pixels to zoom in or a displayed image with fewer pixels to zoom out. (Visualization does not change the underlying data used for later processing.) Enlarging an image by rescaling entails reconstruction, which is the task of rebuilding the signal at resampling points between the data values. Popular methods for digital image reconstruction include nearest-neighbor interpolation, bilinear interpolation, and various methods using cubic polynomial functions for interpolation or approximation [14]. Bilinear interpolation provides a good compromise between quality and computational overhead. It is important to remember that reconstruction estimates signal values and that large zoom factors entail numerous estimates. Therefore, although nearest-neighbor interpolation creates blocky images with less accurate reconstruction, the result makes clear the modulation and sampling rates of the data. Similarly, nearest-neighbor interpolation will show changes in the aspect ratio imposed during rescaling (e.g., to compensate for different sampling rates in the two dimensions, such as undersampling the first-column separation and oversampling the second-column separation). Figure 3 compares bilinear and nearest-neighbor interpolation. Bilinear interpolation shows a spot that more closely represents the continuous peak produced by chromatography. Nearest-neighbor interpolation shows rectangular pixels that make clear the discrete nature of the digitized signal.

3.1.4 Qualitative analysis

Visualization can quickly and clearly show important characteristics of GC×GC data, including problems related to the chromatography. Three such examples are considered briefly here. First, if the retention time of a compound in any second-column separation exceeds the length of the modulation cycle, the associated compound will elute during a subsequent modulation cycle and the peak will appear as a spot that is wrapped around into a subsequent column of pixels in the image. If the retention time is only slightly too long, the spot will appear in the otherwise blank region at the bottom of the image corresponding to the void time of the next second-column separation. This problem can be recognized upon visual inspection, and the chromatographer can change the acquisition settings, for example, lengthening the modulation cycle time or accelerating the second-column separations with a temperature program or shorter column. A second problem sometimes is seen in crescent-shaped trails

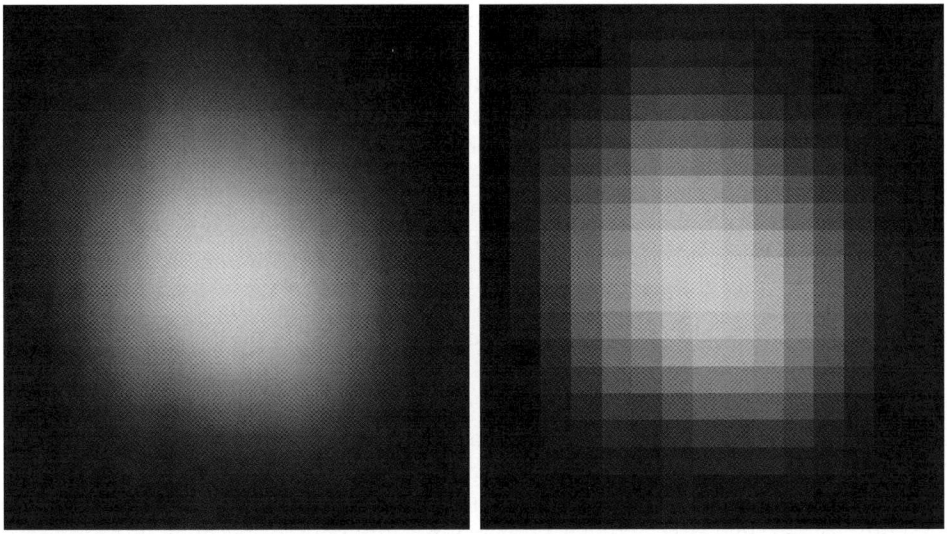

Figure 3 A single GC×GC peak enlarged by bilinear interpolation (left) and nearest-neighbor interpolation (right). Bilinear interpolation yields a truer (i.e., higher fidelity), more pleasing spot; but nearest-neighbor interpolation more clearly shows the individual data points.

that, from left-to-right, slope downward quickly at first and then level out. These artifacts indicate a continuous presentation of eluates from the first column into the second column, perhaps caused by incomplete bake-out (an unclean first column) or by incomplete modulation (i.e., a thermal modulator that is not heated sufficiently to fully release). A third problem seen in visualizations is peak tailing in the second-column separations, which can be caused by various chromatographic issues. Figure 1 illustrates small artifacts of crescent-shaped "bleed" and peak tailing. Data visualization enables quick inspection of the data for these and other qualitative issues.

3.2 Other visualizations

3.2.1 Three-dimensional visualizations

Three-dimensional visualizations use many of the same techniques as two-dimensional image visualizations, including rasterization, colorization, navigation, and reconstruction. A three-dimensional visualization is based on a surface, with the surface elevation relative to the base plane given by each pixel's value. The elevation scale can utilize a mapping function (e.g., linear, logarithmic, or exponential functions). Constructing and viewing an artificial surface utilizes many of the techniques of computer graphics. The surface can be rendered in various ways, for example, pseudocolorized at each pixel, colorized with a solid color and illuminated to provide shading, or built as a wire frame. Then, the surface is projected onto a two-dimensional viewing plane for display. A common projection is the perspective view from a single viewpoint. Additional navigation operations enable the user to rotate the surface in space, in order to view the surface from different perspectives. Figure 2 illustrates a three-dimensional perspective view of a portion of the GC×GC data shown in Figure 1 with values shown as the third dimension (i.e., elevation), with log scaling.

With the added dimension of height, three-dimensional visualizations are better able to show quantitative relationships over a large dynamic range. However, in three-dimensional visualizations, points on the surface can be obscured, and there is no correspondence between the dimensions of the data and the axes of the display, so interactive operations such as point-and-click indexing are more difficult and problematic than with a two-dimensional image. In that sense, different visualizations are complementary, each with its own utilities.

3.2.2 One-dimensional visualizations

One-dimensional graphs are useful for various purposes, including showing slices or integrations of GC×GC data in a graphical format that is familiar to traditional chromatographers. For example, the values in different secondary chromatograms (or rows along the first-column separation) can be rendered as a graph and overlaid to show whether the profiles change over time and/or the results of peak detection in one dimension. Similarly, values in different spectral "channels" of a pixel column (or row) can be graphed and overlaid to show if the multispectral profiles reveal the presence of co-eluted peaks, as illustrated in Figure 4.

3.2.3 Text and tabular visualizations

Some information is best communicated in a text format. For example, the values of the two-dimensional data array can be shown directly as a table, in which each cell displays a numeric pixel value. Visualization features available in spreadsheets are useful for tabular text visualizations. For example, colorization of the text or textboxes can be useful for highlighting different features of the data, such as peak

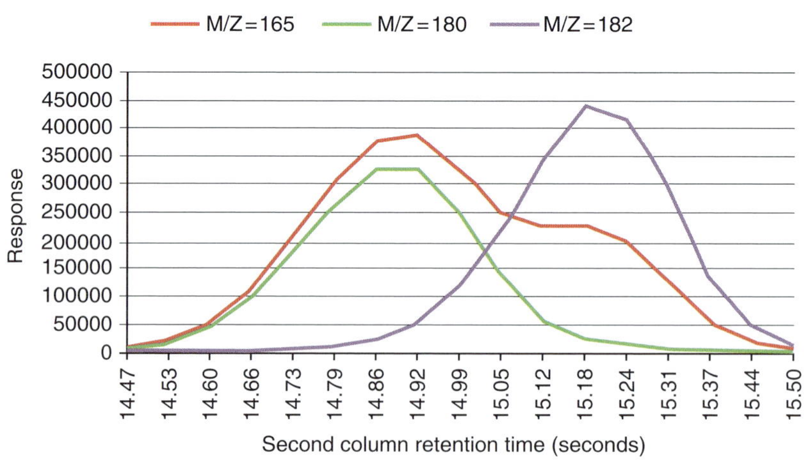

Figure 4 A one-dimensional visualization graphing values in selected-ion channels along a slice through co-eluted peaks.

	21.15	21.20	21.25	21.30	21.35	21.40	21.45
1.430	0.0571	0.1008	0.1730	0.5366	1.5977	0.5145	0.0056
1.425	0.0608	0.1814	0.2747	1.0831	3.1361	1.0161	0.0743
1.420	0.0428	0.3823	0.4438	2.1658	6.0174	1.9412	0.0978
1.415	0.0831	0.7240	0.9287	3.8826	11.2317	3.6697	0.2208
1.410	0.0841	1.3725	1.7997	6.8363	20.1688	6.6537	0.3567
1.405	0.1725	2.5766	3.4523	11.1887	34.9439	11.6780	0.6015
1.400	0.3449	4.7235	6.2445	17.6071	56.5288	19.6141	1.0032
1.395	0.7235	8.4680	11.2150	25.7758	85.6142	31.5504	1.6243
1.390	1.2746	14.2762	19.3033	35.3864	117.9551	47.0344	2.5674
1.385	2.0418	23.0246	31.9252	44.7295	147.8911	64.6653	3.7363
1.380	2.9983	34.3495	49.7217	52.4607	166.2080	80.5460	5.0810
1.375	4.0430	47.7378	71.9066	56.9784	167.6671	90.6433	6.2284
1.370	4.9709	60.2989	95.1407	57.8070	150.7534	91.0964	6.8776
1.365	5.4613	69.1711	113.9597	55.5665	120.4287	81.0968	6.7406
1.360	5.3638	70.7810	122.5308	50.6774	84.3015	63.1797	5.7978
1.355	4.6100	64.3888	117.6683	44.3232	50.4783	41.9836	4.3647
1.350	3.4788	50.9848	100.1199	35.8963	25.2809	23.4741	2.8337
1.345	2.3031	34.7295	74.5050	26.7257	10.3831	10.7412	1.6212
1.340	1.3587	19.7956	47.2256	17.3471	3.7661	4.2601	0.7790
1.335	0.7333	9.4821	24.8462	9.9335	1.2948	1.5094	0.3424
1.330	0.3598	3.8899	10.6588	4.8219	0.4114	0.5122	0.1174
1.325	0.1834	1.4643	3.8885	2.1327	0.0980	0.1950	0.0448
1.320	0.1014	0.5461	1.3180	0.8218	-0.0022	0.0817	0.0131

Figure 5 A tabular visualization of data values in two adjacent peaks with colorization to show primary peak membership.

membership, as shown in Figure 5. Statistical views of the data can be presented simply in a table, and other spreadsheet functions, such as sorting and averaging, are useful for quantitative analysis, which is the subject of the next section.

3.2.4 Graphical overlays and annotations

Graphical overlays are useful for communicating metadata — additional information about the data. For example, in Figure 6, semitransparent bubbles are used to indicate detected peaks. This analysis is for ASTM D5580 Standard Test Method for Determination of Benzene, Toluene, Ethylbenzene, p/m-Xylene, o-Xylene, C9 and Heavier Aromatics, and Total Aromatics in Finished Gasoline by Gas Chromatography [16], so bubbles are activated only for the peaks of interest. The areas of the bubbles are proportional to the peaks' total response, and the colors indicate the chemical group membership of the peak. (Peak detection and identification are described later.) Lines connecting peaks show associations with internal standards for quantitative calibration. Graphical shapes, such as polygons and polylines, are used to indicate chemical groups — in this

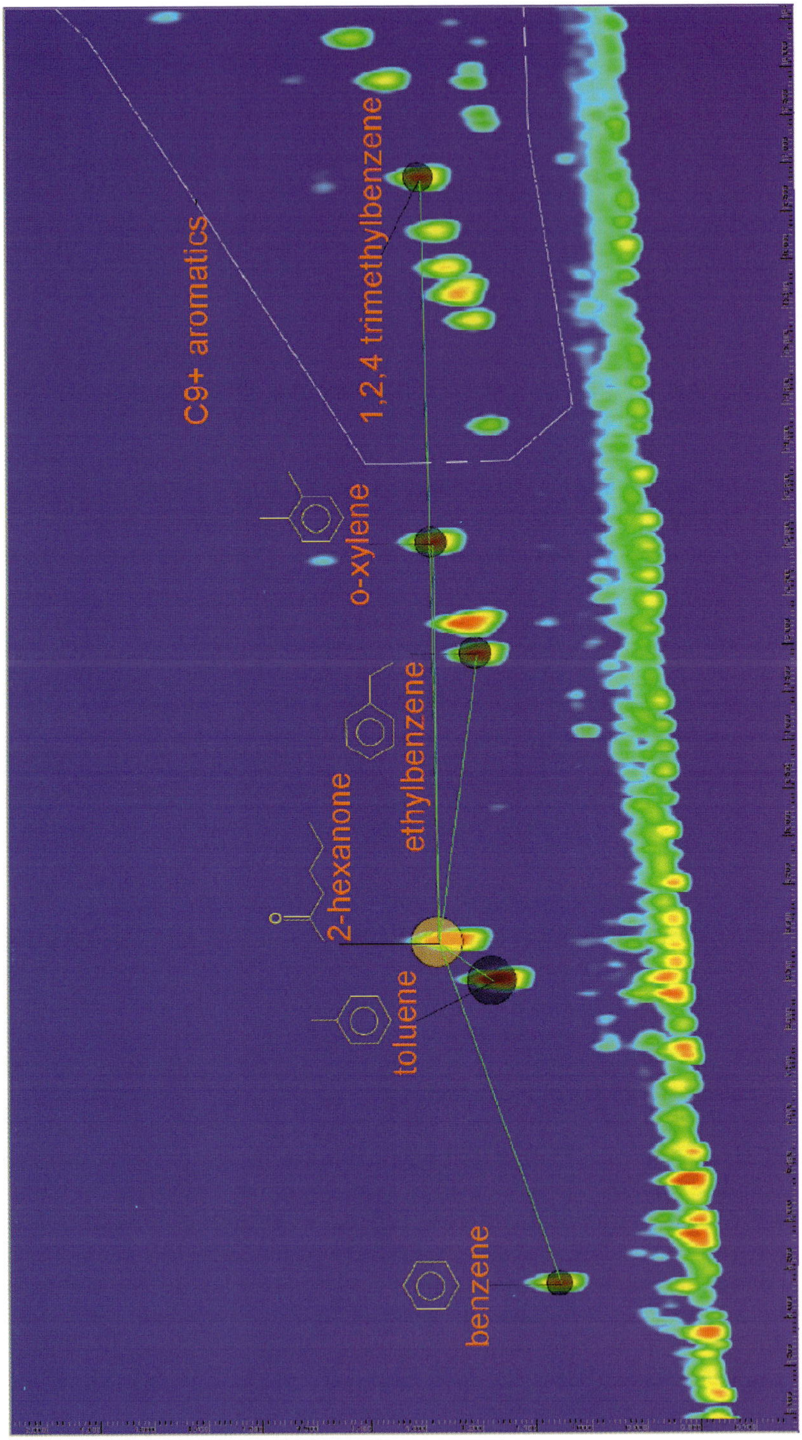

Figure 6 A graphical overlay with semitransparent bubbles for detected peaks of interest, a polygon to indicate the C9+ aromatics, text labels, and graphical chemical structures. A subregion of the data from Figure 1 is shown.

example, the C9+ aromatics. Text labels and chemical structure graphics communicate additional information.

4. DATA PROCESSING

Data processing extracts higher-level information from the raw data for further analysis. This section presents an overview of basic processing operations for GC×GC data:

1. Phase correction — synchronize the columns of data points with the modulation cycles.
2. Baseline correction — remove nonsignal baseline offsets in the data values.
3. Peak detection — detect signal peaks induced by separated compounds.

4.1 Phase correction

In rasterizing GC×GC data, it is typical that the starting data point of each secondary chromatogram in the image corresponds to the time that the modulator released its sample into the second chromatographic column. Then, the vertical axis of the image properly reflects the retention time in the second column. Typically, this is performed by the chromatographic system, but if the data acquisition is out-of-phase with the start of the modulation time, phase correction may be required.

Phase correction is the operation of shifting data in the image so that the data point acquired at the start of each modulation cycle (i.e., the start of each second-column separation) is the first pixel in each image column. (Other synchronizations, e.g., starting each column at the holdup time, are possible but less commonly used.) In the data itself, there may be no markers for the start of the modulation cycles, in which case corrective processing requires inference. (If there are such markers, phase correction is simple.) If the modulation and sampling frequencies are known accurately, then it is possible to accurately infer the first data point corresponding to the modulator release in every modulation cycle from the data point corresponding to the modulator release in just one modulation cycle by iteratively adding (or subtracting) the product of the modulation interval and the sampling rate.

For example, a modulation interval of 4 s and a sampling rate of 200 Hz mean the data point for the start of each modulation cycle follows 800 data points after the data point at the start of the previous modulation cycle. Suppose, in this example, the first data point of the first full modulation cycle is not the first pixel in the first image column but is instead the 400th pixel (i.e., in the middle of the first image column). Then, phase correction could be performed by dropping the first 400 pixels of the first image column, corresponding to the data points before the start of the first full modulation and shifting the data. So, given the modulation and sampling frequencies, it is sufficient to know the second-column retention time of any constituent compound and then to identify the peak pixel for that compound, in order to establish a known mapping between data points and the modulation cycle. From that known point in the modulation cycle, the starting data points for every modulation cycle can be inferred and shifted accordingly.

If the required phase correction is not an integer, two options are possible: (1) round the phase correction to the nearest integer pixel index and accept a timing error of not more than one-half of the sample interval or (2) resample the data so that the resample point is precisely at the start of the modulation cycle. The first option typically is preferred because it maintains the original data, without introducing resampling errors, and is computationally simpler.

A similar issue exists if the product of the modulation interval and the sampling frequency is not an integer. In this situation, each pixel column may have a different fractional offset relative to the modulation start time. Then, the fractional phase correction varies among image columns, and so rounding may result in image columns with heights that differ by one pixel. For visualization, but not for subsequent analysis, this requires that a pixel be added to shorter rows (or that a pixel be excised from longer rows), for example, in data for the void time at the start of the separation.

4.2 Baseline correction

In gas chromatography, the signal peaks, induced by constituent compounds in the sample, rise above a baseline level in the output. Under controlled conditions, the baseline level consists primarily of the steady-state standing-current baseline of the detector and column-bleed (which may cause a progressive rise in temperature-programmed runs). Figure 7 illustrates a three-dimensional

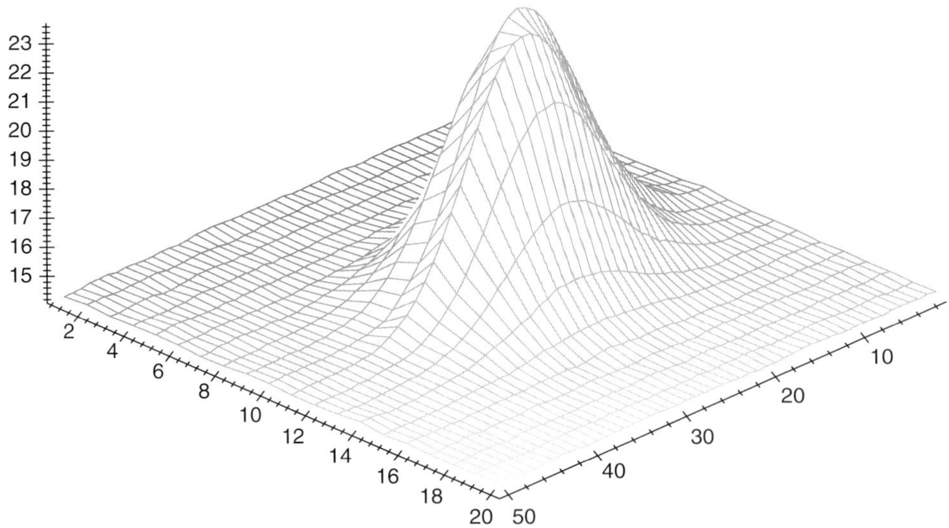

Figure 7 A GC×GC peak on a non-zero baseline.

perspective plot of an isolated peak rising to a maximum value of over 23 picoamps. However, the baseline in the region of the peak is more than 14 picoamps, so the actual maximum peak height induced by the sample compound is less than 10 picoamps. As this example makes clear, accurate quantification of the analyte peaks requires subtraction of the baseline level from the signal.

There are two general approaches for estimating the baseline for correction: (1) estimate the baseline around each peak separately and (2) estimate the baseline across the data comprehensively. The first approach requires that a data point value just outside a peak indicate the baseline level, but this is problematic in regions of the data that are crowded with peaks because the values just outside a peak may be acted upon by neighboring peaks. The second approach requires multiple data point values indicating the baseline level with sufficient frequency that the baseline can be reconstructed.

In GC×GC data, the baseline usually can be observed at many points, for example, during the void time of each second-column separation, even if other regions of the data are crowded with peaks. This is an important attribute of GC×GC for accurate quantification because if the baseline cannot be estimated, then peak integration is less accurate. Typically, the baseline does not change significantly over the brief time of a few modulation cycles, so these observations are sufficient to reconstruct the baseline in a comprehensive fashion.

In a simple model of the GC×GC process, each data point value produced by the system is the sum of:

- A nonnegative baseline offset value that is present even when there is no sample compound detected.
- The signal due to the presence of the detected sample compound(s).
- Random noise fluctuations (including digitization round-off).

Under typical controlled conditions, the baseline offset values change relatively slowly over time, and the signal and noise fluctuate more rapidly over time.

Reichenbach et al. [17] described a method for extracting the GC×GC baseline comprehensively. The first step identifies background regions (i.e., regions without analyte peaks) by locating data points with the smallest values in each second-column chromatogram (or other interval). Then, the local means of the values from data points in the background regions are taken as first estimates of the baseline, and the variances of the values are taken as first estimates of the variance of the noise distribution (which also is present in the background). Then, signal processing filters are used to reconstruct the baseline as a function of the local estimates. Finally, the baseline estimate is subtracted from the signal.

Figure 8 shows two examples of baseline correction: with a blank sample (top) and a diesel sample (bottom). On the left, images of the data before baseline correction are shown with a narrow grayscale range of 1.0 picoamp from black to white. As can be seen in both images, but especially the blank data, there is a temperature-induced increase in the baseline from left to right such that the baseline at the right is nearly 1.0 picoamp greater than the baseline at the left. On the right, images of the data after baseline correction are shown with an even narrower grayscale range of 0.1 picoamp from black to white centered about

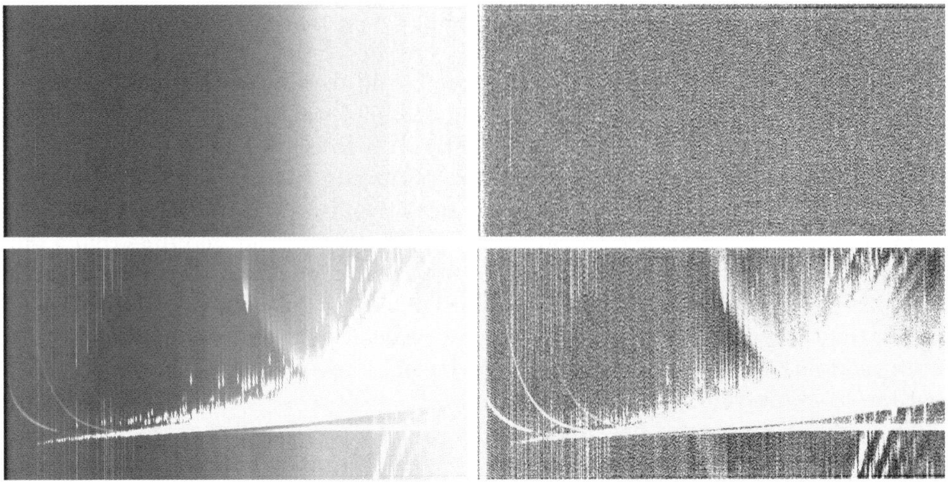

Figure 8 Baseline correction for a blank sample (top) and diesel sample (bottom), before baseline correction (left, with a grayscale range of 1.0 picoamp, 14.5 to 15.5) and after baseline correction (right, with a grayscale range of 0.1 picoamp, from −0.5 to 0.5).

0.0 picoamp. As can be seen in the images after baseline correction, the baseline is removed, and the remaining background values consist of near zero-mean noise with variance less than 0.1 picoamp. The baseline correction is successful not only for the blank run, but also for the diesel sample in which signal obscures much of the baseline.

For systems producing multichannel data, such as GC×GC–MS, the baseline can be estimated in each channel using the same method. Baseline correction for centroided multispectral data is difficult because the centroiding process removes many (or all) of the background values. Therefore, baseline correction should be performed before or at the same time as spectral centroiding (but, unfortunately, that is not always done).

4.3 Peak detection

Blob detection is the process of aggregating peaked clusters of pixels. The term *blob*, from the digital image processing literature, means a cluster of pixels that are brighter (or darker) than their surround. For GC×GC data, it is useful to distinguish blobs from analyte peaks, because a detected blob might be formed from several co-eluted analyte peaks, or a single analyte peak might be detected incorrectly as several blobs (e.g., due to false minima introduced by noise). After blob detection, peak detection may require unmixing blobs resulting from co-elution and merging blobs resulting from incorrectly split peaks.

Two alternative approaches for GC×GC blob detection are: (1) use traditional one-dimensional chromatographic peak detection along each second-column chromatogram and then form two-dimensional blobs from the unions of adjacent one-dimensional peaks [18,19] or (2) perform detection in both dimensions simultaneously. The first approach, of relying on one-dimensional chromatographic peak detection, builds on an accepted methodology but does not fully utilize all available relevant information as it detects peaks in one dimension without reference to the other dimension. The second approach requires a two-dimensional algorithm but can use all available relevant information in each step of the detection.

The drain algorithm for two-dimensional blob detection in GC×GC data [20] is an inversion of the watershed algorithm [21]. The approach is a "greedy" dilation algorithm that proceeds by starting blobs at peak tops and iteratively adding smaller pixels bordering the blobs until there are no more smaller, positive-valued pixels in the surrounds. This process can be understood conceptually by picturing the image as a relief map with larger values having higher elevation (i.e., as a three-dimensional surface as in Figure 2). The surface is placed under enough "water" to submerge the highest elevation; then, the water is progressively "drained." As the draining proceeds, peaks appear as "islands" and are distinguished with unique blob identification numbers. As more water drains, islands (blobs) expand as lower-lying pixels around the "shore" are exposed. When the water between two islands disappears, then a border between blobs is set. When the water level reaches zero, the process is stopped (as negative values are due to noise fluctuations below the baseline). In order to prevent noise from being detected as spurious peaks, blobs that are too small — either in number of data points, apex value, total blob intensity, and/or other criteria — can be ignored.

The example in Figure 9 illustrates the drain algorithm. The intensity of the data point is the base number (values up to 99), and the subscript indicates the order (1–12) in which the data points are added to a blob (dark gray for Blob 1 or light gray for Blob 2). In A, the data point with largest value, 99, starts Blob 1, and then the data points ordered by values 95, 88, and 80 are added to Blob 1 because they neighbor another data point previously assigned to Blob 1. In B, the data point with value 77 starts a new blob, Blob 2, because it is the next largest value and is not adjacent to a data point in any other blob. Then, the data point with value 72 is added to Blob 2. In C, the data points with values 63 and then 61 are added to Blob 2 and Blob 1, respectively, based on their adjacencies to previously assigned data points. In D, the data points with values 42, 38, and 34 are assigned, in order, to Blobs 1, 1, and 2. Where a data point is adjacent to more than one previously assigned data points, the data point is assigned to the same blob as its largest neighbor.

95_2	61	34	77
99_1	71	38	72
80_4	88_3	42	63

95_2	61	34	77_5
99_1	71	38	72_6
80_4	88_3	42	63

95_2	61_9	23	77_5
99_1	71_7	38	72_6
80_4	88_3	42	63_8

95_2	61_9	24_{12}	77_5
99_1	71_7	38_{11}	72_6
80_4	88_3	42_{10}	63_8

Figure 9 Data points, with intensity shown as the base number, are assigned in order of their intensity, with order shown as the subscript, to a blob (dark gray for Blob 1 or light gray for Blob 2). Snapshots of the assignment process are shown from left to right.

One type of error for any blob detection algorithm is oversegmentation — detection of multiple blobs that should be detected as a single peak. This problem can be caused by noise-induced false minima within a peak or other acquisition artifacts. Various approaches can be used to reduce or eliminate oversegmentation. For example, smoothing can be implemented before detection as a convolution with a two-dimensional Gaussian spot whose width is parameterized according to the variance of the noise: a wider blurring function is appropriate for more noise and a narrower blurring function is appropriate for less noise. Too little blurring does little to correct incorrectly split peaks, whereas too much blurring can cause the opposite problem of incorrectly merged peaks.

Undersegmentation, in which multiple analyte peaks are detected as one blob, occurs if analyte peaks are so close in time that there are no minima between them (or small minima are removed by smoothing). For example, a small co-eluted peak may appear as a shoulder on the larger peak. Even if there are minima between overlapping peaks, the watershed algorithm does not "unmix" the peaks; it simply delineates the minima between them. As described in Chapter 5, numerical methods may be used to unmix co-eluting peaks. For example, if each peak has a consistent shape with respect to every row and with respect to every column of the data, then unmixing can be seen as the task of inverting (or deconvolving) a separable, bilinear system for single-valued data or tri-linear model for multichannel data. However, the inversion problem is ill conditioned, and the peak shapes and data are subject to noise and other variables, so the unmixing problem is difficult. MS data can be especially useful for unmixing co-eluting peaks that have differing spectra. Even with MS data, unmixing nearly coincident peaks may require external information (e.g., the spectra of the coincident peaks).

Various chromatographic conditions can cause problems for peak detection algorithms. For example, if the temperature for the second-column separation changes rapidly relative to the modulation cycle, then the apexes of one-dimensional peaks in consecutive second-column separations of a single compound may be offset from one another. For a two-dimensional method such as the drain algorithm, the two modulations may be detected as two separate peaks if the shift is two or more samples. Similarly, a one-dimensional method may fail to join the two one-dimensional peaks. Smoothing, described above, may ameliorate this problem. Chromatographic solutions include more rapid modulations, a slower temperature program, and/or a slower sampling rate. As discussed in Section 2.1, long modulation cycles or slow sample rates relative to (respectively) the first-column and second-column peak widths yield narrower troughs between co-eluting (or nearly co-eluting) peaks, which can lead to undersegmentation as the separate peaks become more difficult to discern. In this case, chromatographic solutions include more rapid modulations, a slower temperature program, and/or a faster sampling rate.

After blobs are detected (or even as they are detected), important statistical features of the blobs can be computed. Most important for quantification, the integration or sum of all of a peak's intensity values is indicative of the relative amount of the compound inducing the peak (subject to the responsivity of the detector to the compound). Geometrically, the integration under a two-dimensional peak is a volume (with two retention-time dimensions and the response dimension), analogous to the integration under a one-dimensional chromatographic peak as an area (with one retention-time dimension and the response dimension). Quantification that accounts for the detector responsivity is described in Section 6.1.

Many other statistics can be computed. The number of data points (or pixels) in the peak is a measure of its retention-time footprint or area, with two retention-time dimensions. Symmetry can be measured as a ratio of the tailing and leading half-widths in each dimension. Various measures with weighted and unweighted moments indicate the center of the peak, center of gravity, variance in each retention-time dimension, orientation, eccentricity, and so on. The GC Image Users' Guide [22] documents more than 70 GC×GC peak features. These features are useful in identifying unusual and possibly problematic blobs, for example, blobs resulting from co-eluting peaks or blobs resulting from split peaks, which then can be subject to visual inspection and interactive correction.

As experienced chromatographers know, automated peak detection is sometimes erroneous, especially for small peaks that are barely detectable amid noise and co-eluted peaks that are nearly coincident. So, interactive tools are useful, but even human experts may not be able to solve difficult peak detection problems. As described in the next section, complex features can be computed as the combination of elementary features for chemical identification.

5. CHEMICAL IDENTIFICATION

A common analytical goal is an assay with individual compounds or group identities and quantitative concentrations of target constituents. (Compounds belonging to the same chemical group are related to one another in some chemical or physical way.) Accurate quantification involves not only the peak responses, but also the responsivity of the detector because detectors may have differing quantitative responses to the same concentrations of different compounds. Therefore, analyte identification (described in this section) typically is performed before quantification (described in the next section). With single-valued GC×GC data, analyte identification must be based primarily on retention time. With multi-channel data, such as from GC×GC–MS, multivariate methods can be used for chemical identification.

5.1 Chemical identification by retention time

A common method for chemical identification in one-dimensional chromatography is to define retention-time windows for peaks of interest. Under repeatable, reproducible, and tightly controlled chromatographic conditions, the peaks for target compounds will fall reliably within fixed retention-time windows. However, narrow windows may be required for peaks with nearby neighboring peaks (to avoid false positives), and, with narrow windows, even slightly different chromatographic conditions may cause a peak to drift outside its window. Here, "drift" is used to characterize a local variation that may be related to more complex systemic variations as might be caused by differing column conditions, temperatures, gas pressure, etc.

Some standard one-dimensional GC methods use reference peaks to help recognize drift [23]. For more widely varying chromatographic conditions, retention times for targets can be related using a linear retention index (LRI) [24], in which retention times are referenced relative to the retention times of marker compounds. A common LRI scheme uses the n-alkanes as marker points with indices equal to 100 times the carbon number (following the Kováts index [24]); then the indices for peaks between marker points are computed using piecewise linear interpolation. If retention-time windows are defined relative to marker peaks that can be located, then any linear retention-time transformation observed in the marker peaks can be applied to the windows used for chemical identification.

Retention-time windows can be used in two dimensions, but the problems of drift exist in both dimensions, with drift in the first dimension possibly inducing drift in the second (related to the temperature program). In an intralaboratory study of GC×GC retention times across separate column sets, chromatographs, and days, Shellie et al. [25] demonstrated highly reproducible peak positions, but with statistically significant drift over separate days and other chromatographic conditions. Ni et al. [26] showed that peak pattern variations over widely varying chromatographic conditions could be modeled well by affine transformations (i.e., translation, scale, and shear). As illustrated in Chapter 3, several approaches have been put forward for two-dimensional indexing [27–31], but none has yet achieved wide acceptance and research continues. A robust approach for dealing with two-dimensional retention-time transformations that can be tailored to specific applications is to locate and identify target peaks relative to the positions of many other peaks in the sample, not just a few standard markers. With this approach, the transformation observed in the pattern of many peaks can be applied to the windows for chemical identification.

Template matching is a powerful extension of the traditional approaches of reference and marker peaks to identify compounds by recognizing peaks in multidimensional separations subject to multidimensional retention-time transformations. A template records the pattern of peaks expected for an analysis, along with information for chemical identification, such as the compound name and/or chemical group for peaks of interest. A template can be built from prototypical data either automatically with all peaks meeting specified criteria (e.g., the largest peaks) or interactively with selected peaks. Templates can be constructed based on peak retention times in one chromatogram or based on averaged peak retention times in several chromatograms. Then, given a template and the set of peaks observed in a sample for analysis, peak pattern matching finds a subset of peaks in the sample data that forms the same pattern as the template. A template-matching algorithm establishes as many correspondences as possible between peaks in the template and peaks in the sample data subject to the allowed retention-time transformation (e.g., shifting or scaling the template) and the allowed retention-time window [32–37]. After peak correspondences are established, the annotated information (such as compound name or group) from peaks in the template is copied into corresponding peaks in the data. Consequently, all the matched compounds in the template are identified in the data.

Figure 10 illustrates a template constructed from the gasoline analysis in Figure 6. The template from the gasoline analysis is overlaid and matched to the chromatogram of a diesel analysis acquired four years later with a different chromatograph and different columns. This template is a multitype template that contains not only a pattern of expected peaks, shown with open circles, but also other information for annotating and reporting on the data. Polygons define regions in which peaks for chemical groups are expected. Text and chemical structure objects are included to provide annotations for visualizations. Graphical lines are used to visually highlight associations between compounds and the internal standards used for calibration. (However, the internal standard, 2-hexanone, is not present in the diesel sample.) The locations of the matched peaks in the diesel chromatogram are shown with filled circles connected by lines to the nearby, corresponding template peaks (shown with open circles). As can be seen, template matching is an effective method for quickly identifying peaks and chemical groups. Other objects in the template are geometrically transformed according to the transformation of the matched peaks, as can be seen for the shifted polygon and its label. Any errors in template matching can be corrected interactively.

5.2 Multivariate methods for chemical identification

Methods for identifying chemical compounds by multichannel data signatures (such as searching a MS library for a matching multispectral signature) are essentially the same for GC×GC as for GC, but GC×GC, with its superior separation power, can significantly reduce co-elution and so improve the accuracy of chemical identification. With multichannel detectors, different compounds have different multivariate signatures (although signatures of similar compounds can be quite similar). The signatures of unidentified peaks can be compared to the known signatures of compounds of interest, with a mathematical computation of difference or similarity between signatures, to find a match that identifies the compound. The National Institute of Standards and Technology (NIST) distributes a library of MS signatures for more than 163 K compounds and a program for searching the library [38]. This approach can be highly effective for chemical identification, but there are many issues that can cause misidentifications, for example, the unknown compound may not be documented in the library, observed signatures are variable, co-elution mixes signatures. In the presence of variability, co-elution, and noise, the search program may find the wrong match. GC×GC can greatly reduce co-elutions, thereby producing purer signatures that can be better identified.

Rule-based methods follow another approach for chemical identification with multichannel data. Experienced analytical chemists often use rules to deduce chemical identity [39,40]. In a computer-based system, rules express the reasons or criteria for chemical identification. Welthagen et al. [41] used a rule-based

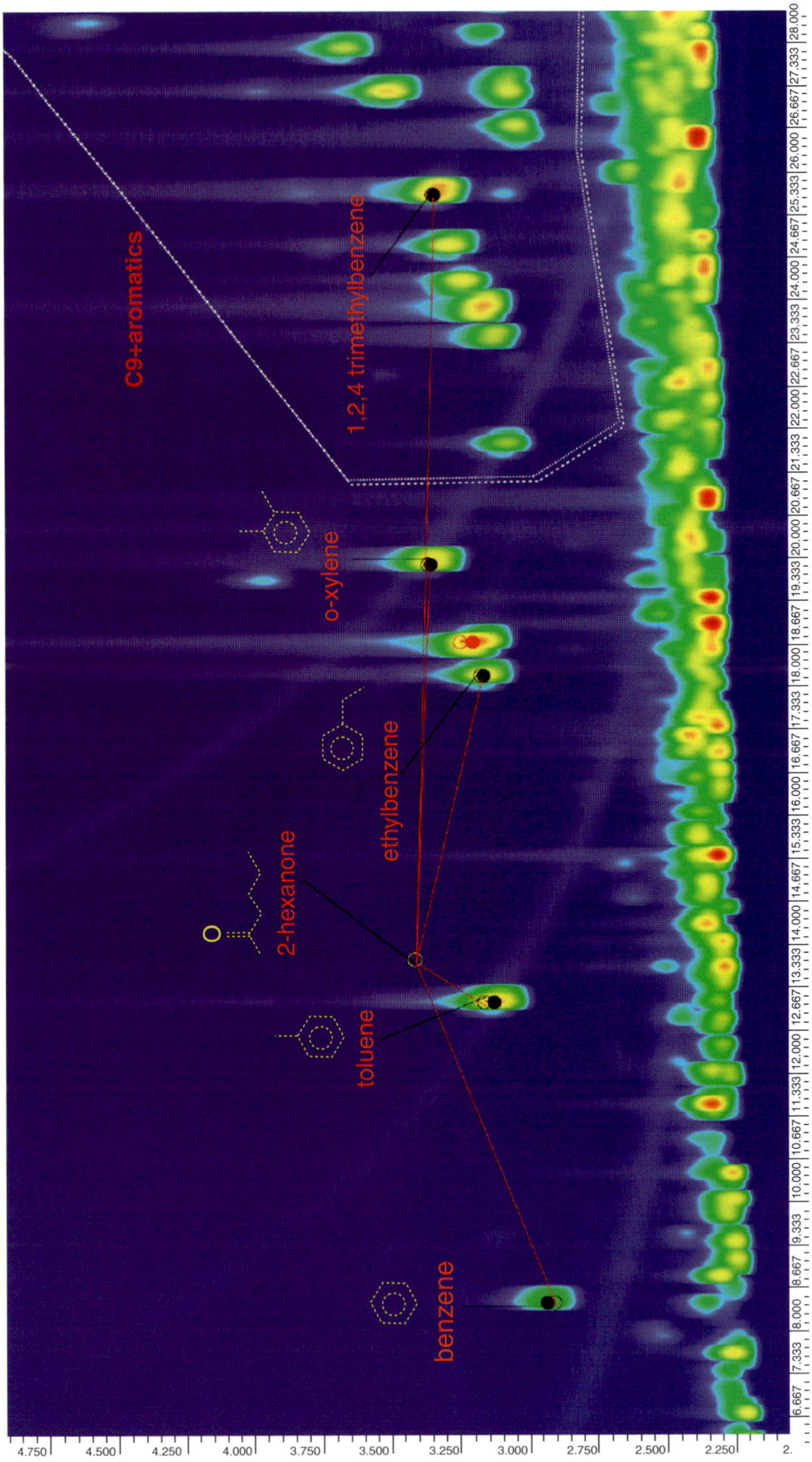

Figure 10 A template from a gasoline analysis is overlaid and matched to peaks in a diesel sample, thereby identifying the peaks and groups of interest. A subregion of the data is shown with open circles, showing the expected peak positions in the template.

approach based on GC×GC retention times and MS signatures to classify chemical groups in the analysis of airborne particulate matter. For example, the rule for polar benzenes with or without alkyl groups in the GC×GC–MS data was:

> The MS intensity at mass-to-charge ratio (m/z) 77 is greater than 25% of the intensity of the MS base peak (i.e., the largest MS intensity), and the retention time in the second column is greater than 2 s.

The Computer Language for Identifying Chemicals (CLIC) [42] defines a grammar for expressing rules for chemical identification based on multi-dimensional retention times and spectral characteristics, including library search. CLIC offers functions of multidimensional chromatographic retention times, functions of MS characteristics (such as selected-ion intensity counts), functions for MS library search, numbers for quantitative and relational evaluation, and logical and arithmetic operators. The CLIC expression for the above rule for identifying polar benzenes is:

$$(\text{Relative}(77) > 25)\ \&\ (\text{Retention}(2) > 2).$$

This rule can be applied to the spectra of all peaks to determine which are polar benzenes. Even more complicated rules involving selected-ion intensity counts can be derived using classifiers [43,44], and other features can be applied with GC×GC [45].

Rule-based identification works well for multispectral constraints but is less convenient for retention-time constraints (e.g., describing a many-sided polygon to restrict the retention times for a group of peaks in a chemical class). Complex retention-time rules can be more easily expressed graphically, for example, in templates. Rule-based constraints and templates have complementary strengths that can be combined for highly effective chemical identification (as described next).

5.3 Smart Templates

Smart Templates [46] combine retention-time templates with rule-based chemical constraints. Templates express retention-time patterns in a convenient graphical form that is highly visual; CLIC expressions efficiently define rules with an arsenal of functions, constants, and mathematical and logical operators. In complex chromatographic regions, if template matching finds several peaks in the data that are candidates to match a template peak, then a rule associated with the template peak can eliminate incorrect matches. (In this case, the CLIC expression can be applied only to the peaks that are potential matches.) Similarly, if a spectral rule to identify peaks in a chemical group identifies peaks with too widely ranging retention times, then a template polygon with the associated rule can restrict group identification using both the rule and convenient graphical retention-time constraints. The combination is a powerful methodology for chemical identification.

Ledford [47] developed a Smart Template for analysis of diesel aromatics in experimental work on a standard analytical method under development for evaluation by the ASTM. (Vogt et al. [45] developed a similar analysis.) Figure 11 shows an example group analysis with Ledford's Smart Template. The Smart Template uses a retention-time polygon with CLIC expression on the GC×GC–MS data for each chemical group, with distinctions for both chemical classes and carbon number. The result is a descriptive group analysis.

6. QUANTIFICATION AND MULTI-DATASET ANALYSES

Several important analytical problems involve multiple datasets.

- *Sample quantification*: calibrate for quantification by measuring detector responses to different levels of concentrations in multiple chromatograms.
- *Sample comparison*: characterize similarities and differences between datasets, for example, to find anomalies such as might be responsible for a desirable or undesirable trait.
- *Sample classification*: use many GC×GC datasets to characterize sample classes based on within-class commonalities and between-class differences and then classify a sample into one of the classes based on GC×GC analysis.
- *Sample recognition*: establish the identity of a sample's source by pattern recognition comparing a GC×GC dataset against many GC×GC datasets stored in a library to find the best match. This is sometimes referred to as chemical fingerprinting.
- *Sample query*: find a dataset(s) that have specified characteristics from among a repository of many datasets. Other standard database operations, such as insert and delete, are useful for maintaining and using repositories.

6.1 Quantification

After the detector responses for a peak have been integrated, accurate quantification requires consideration of the detector's responsivity to the compound inducing the peak. In this, calibration and quantification of GC×GC peak responses are performed with the same approaches as for GC (including internal calibration, external calibration, and response factors), but research surveys document that the quantitative performance of GC×GC is superior to that for one-dimensional GC [48–50]. In an early report of quantitative performance for GC×GC, Gaines et al. [51] reported two- to fourfold improvements in limits-of-detection for trace oxygenate and aromatic compounds with FID. Lee et al. [52] observed a four- to fivefold increase in sensitivity for GC×GC with FID, which was consistent with their model predicting both peak response enhancement of roughly 20-fold from peak focusing and increased noise associated with faster sampling rates. Other researchers reported detectability improvements of two- to fivefold for GC×GC–MS [53] and GC×GC–ECD [54]. Of course, the greatest benefit of GC×GC for quantification frequently is greater selectivity, which allows quantification of compounds that otherwise would be co-eluted and difficult to quantify accurately.

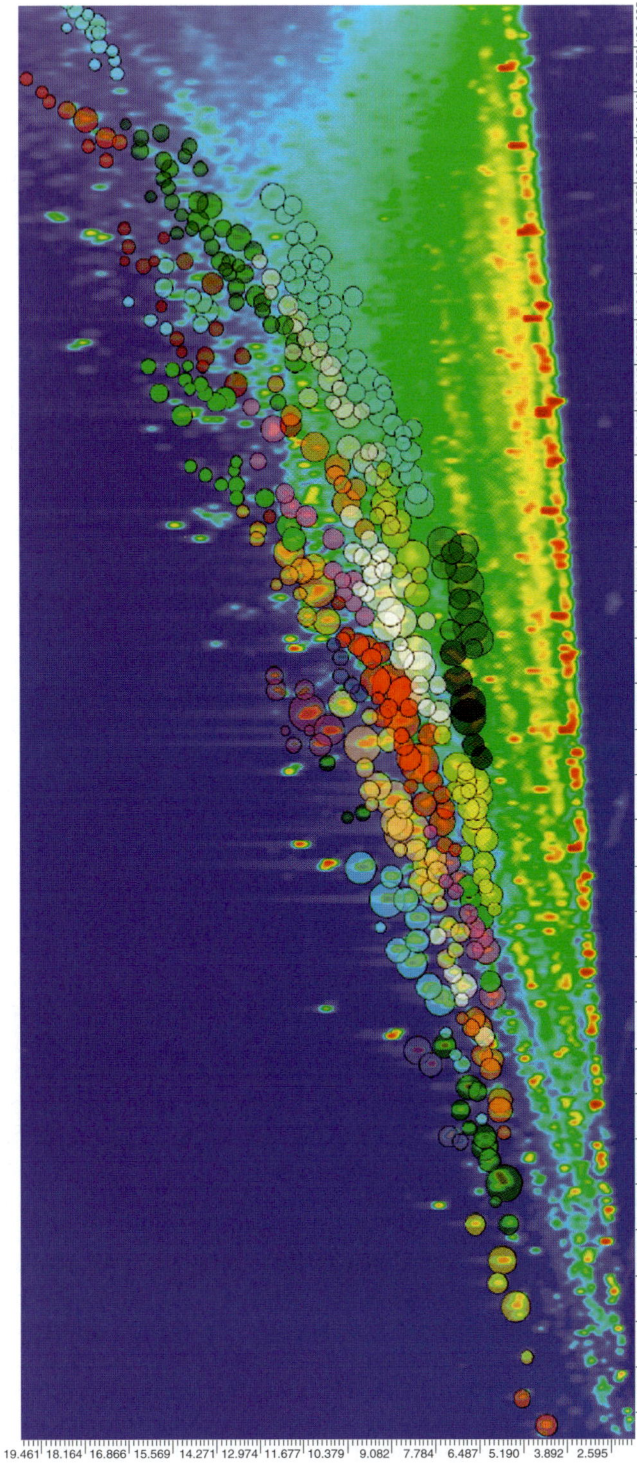

Figure 11 Group analysis of diesel aromatics by a Smart Template [47].

As of 2008, despite more than a decade of research demonstrating the increased selectivity and sensitivity of GC×GC, there are no standard GC×GC methods. One reason may be that GC×GC can be applied to standard GC methods to provide improved performance in a wide range of analyses, as described in the applications-oriented chapters of this book. Another possible reason is that prior to the availability of commercial GC×GC software in 2003, quantitative analysis was laborious and time consuming. Perhaps another reason is that GC×GC opens so many options for new method development that settling on specifics is more difficult and has required a period of research and development of technologies and methods for standardization.

6.2 Sample comparison, classification, and recognition

The first level of intersample comparison is qualitative visualization and tabular comparison of sample constituents. Hollingsworth et al. [55] described various approaches for such comparisons. The visualizations begin with registering (aligning) the chromatograms to minimize the mean-square difference between peak retention times and normalizing the intensities with respect to a standard peak or set of peaks. Methods for visualization include flicker between images (i.e., cycle from one image to another) and display combination images (subtraction, ratio, addition) with grayscale or pseudocolorization. A method for "fuzzy differences" adjusts the difference image for residual differences due to peak shape and/or misregistration. Tables can be used to report quantitative differences. Frysinger and Gaines [56] used flicker visualization to find differences between regular and super gasoline for forensic analysis of fire debris. To track an oil spill, Nelson et al. [57] used difference, ratio, and addition images to show chemical changes over time. In Figure 12, the upper visualization shows the arithmetic difference (after registration and normalization) for samples in May and November 2003, and the lower visualization shows the color addition with the May data in green and the November data in red. The color addition image shows not only the magnitudes of the peaks (with intensity), but also the degree of change — from near complete weathering (indicated by the color green) of the n-alkane peaks along the bottom and the more volatile aromatics in the left half of the image to almost no weathering (indicated by the color yellow) of the less volatile aromatics. Their qualitative and quantitative analyses of peak intensity differences showed the differing effects of evaporation, water washing, and biodegradation on different compounds over time.

The classification of samples is another important analytical problem. For example, the search for biomarkers in metabolomic and proteomic research has the goal of finding sample characteristics indicative of a disease state or other biological condition. When samples are reduced to peak sets, the GC×GC classification problem is not significantly different from classification with GC, but the selectivity of GC×GC can be critical for classification accuracy. Frysinger and Gaines [40] demonstrated the utility of GC×GC for separating known biomarkers in crude oil. Shellie et al. [58] analyzed derivatized tissue samples from two classes of mice, obese and lean, and identified the 10 most likely biomarkers in the data

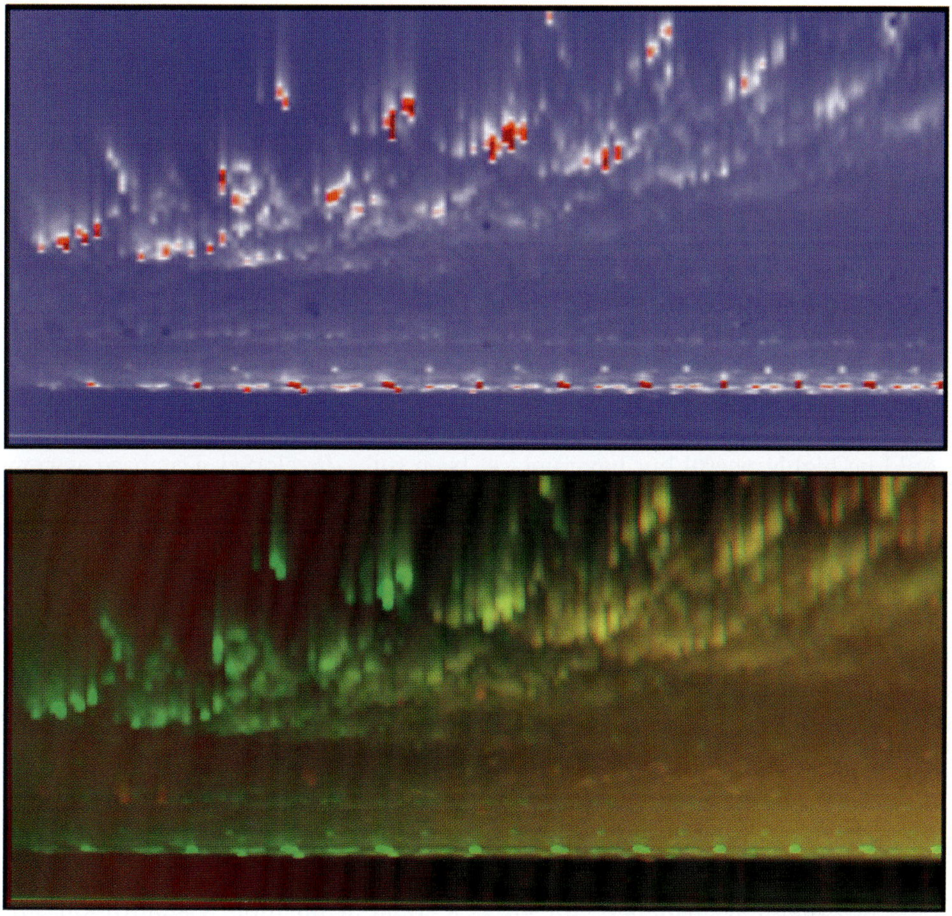

Figure 12 Comparison of oil spill samples in a difference image (between samples in May and November 2003) and a color addition image with the May sample in green and the November sample in red [57].

using *t*-test values for the spectra of deconvolved peaks. To classify yeast samples grown under either fermenting or respiring conditions, Mohler et al. [59,60] used multivariate methods to identify chromatographic regions with significant interclass differences prior to peak detection. In [59], principal component analysis (PCA) was applied to normalized selected-ion chromatograms to identify regions for peak detection with deconvolution. In [60], they identified chromatographic regions of interest by totaling the mean-signal-weighted Fisher ratio at each point in each spectral channel. Regions of interest were deconvolved, and the detected peaks were evaluated by the *t*-test. Others have used analysis of variance (ANOVA) methods to select chromatographic features for GC×GC classification [61,62]. These methods are discussed in detail in Chapter 5.

Fingerprinting focuses the classification problem to recognize one of multiple individuals (i.e., classes of size one). Gaines et al. [63] used GC×GC FID fingerprints to identify which of two potential sources was responsible for an oil spill. They used fingerprint features from four chemical groups: naphthalenes, anthracenes/phenanthrenes, alkenes and cycloalkenes, and aliphatics. In each group, they utilized three minutiae (points of interest), each computed as the response ratio of a specific analyte peak within the chemical group to a fourth peak in the chemical group. The fingerprints provided convincing evidence for source identification. Investigating the problem of classifying crude oils by their source reservoir, van Mispelaar et al. [64] did not find any individual chemical markers sufficient for classification, but successfully classified samples based on small differences in many peaks using principal-component discriminant analysis (PCDA).

6.3 Databases and information systems

Software for higher-level database and information queries for GC×GC datasets would be highly useful but have not yet been fully developed. Database systems could support content-based data and information retrieval, for example, list the datasets for which the ratio of Chemical A to Chemical B is greater than x. Such queries could support fingerprint identification [63] on large databases. Information systems could support higher-level queries, for example, to support automated classification based on statistically significant peak-to-peak variations between two groups of datasets. Such queries could support the type of classification Shellie et al. [58] used to chemically distinguish obese and lean mice from tissue samples. Such systems would be useful not only for applications but for quality control, for example, finding differences in datasets of standard runs over time.

7. CONCLUSION

Many of the initial challenges for GC×GC data acquisition, visualization, and analysis have been surmounted, and solutions are available in commercial GC×GC software. Available software supports the following basic operations:

- Reading data from file formats produced by chromatographic systems.
- Displaying data in various modes, for example, as two-dimensional images, as projections of three-dimensional surfaces, as one-dimensional profiles, and so on.
- Preprocessing data to remove acquisition artifacts, such as modulation phase shift and signal baseline.
- Peak detection, including deconvolution/unmixing co-eluted peaks.
- Chemical identification using both retention-time and spectral data.
- Chemical quantification using the same approaches as for GC analysis.
- Multi-dataset analyses such as qualitative and quantitative comparisons.

Some problems require further research and development, notably:

- A standard file format for GC×GC data.
- More effective tools for chromatographic-spectral visualizations and multi-dataset visualization.

- Deconvolution/unmixing of difficult co-elutions.
- Multi-dataset analyses for classification and fingerprinting.
- Advanced queries for GC×GC databases.

These and other challenges are the subjects of ongoing research and development.

ACKNOWLEDGMENT

This chapter is based upon work supported by the National Science Foundation Division of Information and Intelligent Systems under Grant No. IIS-0431119. Zoex Corporation supported this work with example data and GC Image supported this work with GC×GC software.

REFERENCES

1 GC Image, LLC, GC Image® software, http://www.gcimage.com (2008).
2 J.V. Seeley, *J. Chromatogr. A*, 962 (2002) 21.
3 P.C. Kelly and G. Horlick, *Anal. Chem.*, 45(3) (1973) 518.
4 K.A. Duell, J.P. Avery, K.L. Rowlen and J.W. Birks, *Anal. Chem.*, 63(1) (1991) 73.
5 R.E. Murphy, M.R. Schure and J.P. Foley, *Anal. Chem.*, 70(8) (1998) 1585.
6 Agilent, Mini-IRS for: IQ Data File Format, Agilent Technologies, revised July 7, 1995.
7 B. Allen, M. Allen, R. Bingham, G. Carpanese, D. Gedcke, S. Haywood, G. Jackson, and J. Peck, FASTFLIGHT™, a Digital Signal Averager for Continuous High-Speed Data Acquisition with Electrospray Time-of-Flight Mass Spectrometers Coupled to Chromatographs, ORTEC, 1998.
8 ASTM International, Standard specification for analytical data interchange protocol for chromatographic data, Tech. Rep. E1947-98, ASTM, West Conshohocken, PA (1998).
9 ASTM International, Standard specification for analytical data interchange protocol for mass spectrometric data, Tech. Rep. E2077-00, ASTM, West Conshohocken, PA (2000).
10 Unidata Program Center, University Corporation for Atmospheric Research, NetCDF (network Common Data Form), http://www.unidata.ucar.edu/software/netcdf/ (2008).
11 P. J. Linstrom, A proposed data model for ASTM E13.15, http://animl.sourceforge.net/Linstrom-Data-Model-old.pdf (2003).
12 M. Fiege, T. Davies, T. Fröhlich, and P. Lampen, The AnIML core, sample, and technique shells: Proposal for an AnIML schema, http://animl.sourceforge.net/CLC Waters AnIML Proposal.pdf (2004).
13 T. Bray, J. Paoli, C. Sperberg-McQueen, and E. Maler, Extensible Markup Language (XML) 1.0, World Wide Web Consortium (2000).
14 R.C. Gonzalez and R.E. Woods, *Digital Image Processing*, Prentice-Hall, Englewood, Cliffs, NJ, 2008.
15 A. Visvanathan, S.E. Reichenbach and Q. Tao, *J. Electron. Imaging*, 16(3) (2007) 033004.
16 ASTM, Standard test method for boiling range distribution of petroleum fractions by gas chromatography, Tech. Rep. D5580-02, ASTM, West Conshohocken, PA (2007).
17 S.E. Reichenbach, M. Ni, D. Zhang and E.B. Ledford, Jr., *J. Chromatogr. A*, 985(1) (2003) 47.
18 J. Beens, H. Boelens, R. Tijssen and J. Blomberg, *J. High Resolut. Chromatogr.*, 21(1) (1998) 47.
19 S. Peters, G. Vivó-Truyols, P. Marriott and P. Schoenmakers, *J. Chromatogr. A*, 1156(1–2) (2007) 14.
20 S.E. Reichenbach, M. Ni, V. Kottapalli and A. Visvanathan, *Chemom. Intell. Lab. Syst.*, 71(2) (2004) 107.
21 S. Beucher and C. Lantuejoul, In: *International Workshop on Image Processing, Real-Time Edge and Motion Detection/Estimation*, 1979, pp. 17–21.
22 GC Image, LLC, GC Image™ Users' Guide, http://www.gcimage.com/usersguide (2008).
23 J.L. Snyder, In: R.L. Grob and E.F. Barry (Eds.), *Modern Practice of Gas Chromatography*, John Wiley and Sons, New York, 2004, pp. 769–882.
24 IUPAC, Compendium of Chemical Terminology, http://goldbook.iupac.org (2007).
25 R. Shellie, L.-L. Xie and P. Marriott, *J. Chromatogr. A*, 968 (2002) 161.

26 M. Ni, S.E. Reichenbach, A. Visvanathan, J.R. TerMaat and E.B. Ledford, Jr., *J. Chromatogr. A*, 1086(1–2) (2005) 165.
27 R.J. Western and P.J. Marriott, *J. Sep. Sci.*, 25(13) (2002) 832.
28 R.J. Western and P.J. Marriott, *J. Chromatogr. A*, 1019(1–2) (2003) 3.
29 J. Arey, R. Nelson, L. Xu and C. Reddy, *Anal. Chem.*, 77(3) (2005) 7172.
30 S. Bieri and P.J. Marriott, *Anal. Chem.*, 78(23) (2006) 8089.
31 J.V. Seeley and S.K. Seeley, *J. Chromatogr. A*, 1172 (2007) 72.
32 M. Ni, Point pattern matching and its application in GC×GC, Ph.D. thesis, University of Nebraska (2004).
33 M. Ni and S.E. Reichenbach, In: *IEEE Workshop on Statistical Signal Processing*, 2003, pp. 369–372.
34 M. Ni, Q. Tao and S.E. Reichenbach, In: *IEEE Workshop on Statistical Signal Processing*, 2003, pp. 497–500.
35 M. Ni and S. E. Reichenbach, In: Automatic Target Recognition XIV, Proc. SPIE 5426, 2004, pp. 155–163.
36 M. Ni and S. E. Reichenbach, In: Visual Information Processing, Proc. SPIE 5438, 2004, pp. 101–110.
37 M. Ni and S. E. Reichenbach, In: *Proceedings of the International Conference on Pattern Recognition*, Vol. 2, IAPR/IEEECS, 2004, pp. 145–148.
38 NIST/EPA/NIH Mass Spectral Library with Search Program, NIST Standard Reference Database 1A (2005).
39 F.W. McLafferty, *Interpretation of Mass Spectra*, 4 th Edition, University Science Books, Herndon, VA, 1996.
40 G.S. Frysinger and R.B. Gaines, *J. Sep. Sci.*, 24(2) (2001) 87.
41 W. Welthagen, J. Schnelle-Kreis and R. Zimmermann, *J. Chromatogr. A*, 1019 (2003) 233.
42 S.E. Reichenbach, V. Kottapalli, M. Ni and A. Visvanathan, *J. Chromatogr. A*, 1071(1–2) (2004) 263.
43 H. Lohninger and K. Varmuza, *Anal. Chem.*, 59(2) (1987) 236.
44 K. Varmuza and W. Werther, *J. Chem. Inf. Comput. Sci.*, 36(2) (1996) 323.
45 L. Vogt, T. Gröger and R. Zimmermann, *J. Chromatogr. A*, 1150(1–2) (2007) 2.
46 S.E. Reichenbach, P. Carr, D. Stoll and Q. Tao, *J. Chromatogr. A*, 1216(16) (2009) 3458.
47 S.E. Reichenbach, S.B. Cabanban, E.B. Ledford, H.A. Pham, W.E. Rathbun, Q. Tao, and H. Wang, In: *Pittcon Conference and Expo*, Chicago, IL, 2009, p. 540.
48 J. Dallüge, J. Beens and U.A.Th. Brinkman, *J. Chromatogr. A*, 1000 (2003) 69.
49 M. Adahchour, J. Beens, R. Vreuls and U.A.Th. Brinkman, *Trends Anal. Chem.*, 25(6) (2006) 540.
50 O. Amador-Muñoz and P.J. Marriott, *J. Chromatogr. A*, 1184(1–2) (2007) 323.
51 R.B. Gaines, E.B. Ledford, Jr. and J.D. Stuart, *J. Microcol. Sep.*, 10(7) (1998) 597.
52 A.L. Lee, K.D. Bartle and A.C. Lewis, *Anal. Chem.*, 73(6) (2001) 1330.
53 J. Dallüge, R. Vreuls, J. Beens and U.A.Th. Brinkman, *J. Sep. Sci.*, 25(4) (2002) 201.
54 P. Korytár, P. Leonards, J. de Boer and U.A.Th. Brinkman, *J. Chromatogr. A*, 958(1–2) (2002) 203.
55 B.V. Hollingsworth, S.E. Reichenbach and Q. Tao, *J. Chromatogr. A*, 1105(1–2) (2006) 51.
56 G. Frysinger and R. Gaines, *J. Forensic Sci.*, 47(3) (2002) 471.
57 R.K. Nelson, B.S. Kile, D.L. Plata, S.P. Sylva, L. Xu, C.M. Reddy, R.B. Gaines, G.S. Frysinger and S.E. Reichenbach, *Environ. Forensics*, 7(1) (2005) 33.
58 R.A. Shellie, W. Welthagen, J. Zrostliková, J. Spranger, M. Ristowe, O. Fiehn and R. Zimmermann, *J. Chromatogr. A*, 1086(1–2) (2005) 83.
59 R.E. Mohler, K.M. Dombek, J.C. Hoggard, E.T. Young and R.E. Synovec, *Anal. Chem.*, 78(8) (2006) 2700.
60 R.E. Mohler, K.M. Dombek, J.C. Hoggard, K.M. Pierce, E.T. Young and R.E. Synovec, *Analyst*, 132 (2007) 756.
61 K.J. Johnson and R.E. Synovec, *Chemom. Intell. Lab. Syst.*, 60(1–2) (2002) 225.
62 M. Kallio, T. Hyötyläinen, M. Lehtonen, M. Jussila, K. Hartonen, M. Shimmo and M. Riekkola, *J. Chromatogr. A*, 1019 (2003) 251.
63 R.B. Gaines, G.S. Frysinger, C.M. Reddy and R.K. Nelson, Identification. In: S.S.Z. Wang (Ed.), *Oil Spill Environmental Forensics: Fingerprinting and Source Identification*, Academic Press, Burlington, MA, 2007, p. 169.
64 V. van Mispelaar, A. Smilde, J. Blomberg and P. Schoenmakers, *J. Chromatogr. A*, 1096 (2005) 156.

CHAPTER 5

Chemometric Approaches

Jamin C. Hoggard and **Robert E. Synovec**

Contents

1. INTRODUCTION

Within a gas chromatography (GC) instrument, the column is commonly coupled with a single-channel (univariate) detector that provides a single measured value at a given time. In terms of data produced, the result of a single separation is a vector containing the detected property sampled throughout the length of the run, that is, a first-order data structure. A vector thus obtained in comprehensive two-dimensional (2D) gas chromatography (GC×GC) takes on additional meaning because of the structure imposed on the data by modulation of the first-column effluent onto the second column for subsequent separation under different conditions. The vector of GC×GC data can be viewed as consisting of repeated separations of the first-column effluent in the second chromatographic dimension (i.e., on the second column). Assuming that the modulation period is constant (which is normally the case in GC×GC), the data vector can be reshaped or folded into an $m \times n$ or $n \times m$ array or matrix consisting of m second-column separations each of n data points, producing a second-order data structure. Since the data can be meaningfully shaped into a matrix, this allows the application of a wide variety of mathematical techniques (chemometrics) for

Comprehensive Analytical Chemistry, Volume 55
ISSN: 0166-526X, DOI 10.1016/S0166-526X(09)05505-6

gleaning useful information from the data and affords certain advantages over implementing "more traditional" first-order data analysis approaches. Herein, we define chemometrics as the use of multivariate data analysis methods to *optimally* glean useful information from data obtained from chemical measurements such as GC×GC. In the context of chemical separations, chemometrics essentially provides mathematical resolution to enhance (augment) the chromatographic resolution provided by the instrumentation and methodology. When properly applied, chemometrics significantly broadens the scope of GC×GC. Indeed, although almost any mathematical technique for gleaning information from chemical data can be considered chemometrics, in this chapter we will primarily examine chemometric techniques where either necessity or applicability arises from the special structure of GC×GC data [1], and hence these techniques have opened doors to important applications to GC×GC data.

Generally speaking, the application areas for chemometrics with GC×GC are the following [2–4]. First, there is the need to identify and quantify target analytes. This is essentially an extension of the traditional application of chromatographic data (often limited to a baseline resolution requirement), extended to much lower resolution with advanced chemometric methods. Second, there is the area of pattern recognition and classification, where the GC×GC chromatogram is essentially treated as a complex image or fingerprint. For this purpose, class membership may be unknown (using, e.g., principal component analysis, PCA) or known (using Fisher ratio analysis). Then, class-distinguishing analytes (biomarkers, contaminants, etc., located using Fisher ratio feature selection, for example) can be identified and quantified. In a related application, groups of compounds may be quantified in concert, for example, using partial least squares (PLS), and applied in method transfer, thus developing a fast on-line method to replace a tedious off-line lab method. Another area of broad interest is retention calculation and prediction for separation optimization. In order to apply the powerful chemometric methods, however, the analyst must be aware of, and deal with, various issues: data visualization (including comparative visualization) and data preprocessing (e.g., baseline correction, retention time alignment, etc.), both of which were treated in detail in Chapter 4, as well as other issues such as peak finding that we introduce in this chapter. We begin with a focus on those issues that have not already been presented.

2. RETENTION CALCULATION AND PREDICTION, AND SEPARATION OPTIMIZATION

In GC, retention times (or retention indices) correlate to chemical information and are used to provide supporting evidence for peak identification. GC×GC can comprehensively separate a sample using two different stationary phases, yielding a retention time in each separation dimension for each peak, and with retention information from both separation mechanisms, more confident identification and more chemical information can be obtained than from 1D GC.

Retention information can be obtained from various chemometric methods for peak finding, deconvolution, and feature selection, or by manual inspection. A variety of data processing and chemometric methods have been explored to interpret or predict retention time or index information. Autocovariance functions (ACVF) have been used in cases where there are linear relations between retention times in GC×GC to find and distinguish different retention structures that correlate to different classes of compounds (where a class of compounds means compounds sharing structural similarities) [5]. Retention correlation maps can be created from two or more classes of compounds for particular separation conditions [6]. Retention indices for compounds eluting in a GC×GC separation can be interpolated or extrapolated based on a retention correlation map using, for example, polynomial fits to the retention times of the known classes of compounds [7], as already explained in Chapter 3. GC×GC retention indices can also be predicted from retention indices from 1D GC separations using knowledge of instrumental parameters and extrapolation of retention factors from 1D GC van't Hoff plots. Predicted retention information can also be used to aid in optimizing GC×GC separation conditions [8]. Instrument parameter selection for GC×GC is somewhat more complicated than for 1D GC, partly because there are more parameters and partly because some of the parameters specific to GC×GC (such as modulation period, which affects resolution on both dimensions) can be critical in obtaining satisfactory separations. Aside from using predicted retention information, automated techniques have been investigated to optimize GC×GC separations with respect to the number of peaks detectable by peak finding and deconvolution software, which typically corresponds to optimizing for better resolution and limit of detection (LOD). One such technique involves the use of windows event scripting and global optimization algorithms to automatically adjust many instrument parameters over numerous runs to achieve optimized separation conditions [9]. Achieving more optimal separations is a key facet of GC×GC data analysis with chemometrics; while chemometrics can often partially make up for suboptimal separation conditions, it is only through the combination of optimized separation conditions with chemometrics that GC×GC can achieve its full potential.

3. PEAK FINDING

Before beginning a discussion of peak finding techniques, it is important to make the distinction between peak finding and feature selection. In feature selection, the goal is to find features of interest (differences or sometimes similarities) between samples (with the samples often known to belong to specific classes via the experimental design of a hypothesis-driven study). These features usually correspond to peaks, but the fact that these features are peaks does not mean that all peaks are features, and the identity of the analytes giving rise to these peaks is generally either not known or is not taken into account. On the other hand, the goal of peak finding is simply to find peaks, that is, signals in a GC×GC chromatogram arising from detection of an analyte (which may be a specific analyte or any analyte). That said, peak finding is often used as a preprocessing step for other chemometric techniques such as deconvolution (mathematical resolution), classification, feature selection, or prediction. This may be done to filter signals not arising from analyte detection from a dataset or to reduce the size of a dataset that must be processed while retaining important information. However, peak finding in itself is an important chemometric technique because it can provide useful information (such as retention times).

Because many techniques for peak finding are adversely affected by baseline and noise, preprocessing to remove baseline and noise is common before applying peak finding techniques. Aligning data across multiple chromatograms may also be performed if accurate retention time information is desired.

Peak finding methods for 1D GC data can often be applied to GC×GC data (which, as we will see, is often the case with the other chemometrics techniques as well). However, these will find multiple 1D peaks for each 2D (GC×GC) peak; that is, many of the modulations of the first separation dimension peak onto the second separation dimension will be found as separate peaks. This may be good enough for some purposes, but it is not ideal for many purposes, especially when as close to a one-to-one correspondence between peaks found in the chromatogram with analyte species is desired. One way to deal with this is to try and combine the hits (locations where peaks were detected) from the many 1D GC peaks back into a single GC×GC peak hit [10], which of course requires further processing and can be a point of difficulty, if not failure. The other primary approach is to find peaks treating the data as a matrix rather than as a vector. One such approach that has been used as a data-reducing preprocessing step in a feature selection method involves finding local maxima in GC×GC data that are above a threshold value, the idea being that large enough local maxima typically correspond to locations of peak maxima [11].

Finding peaks that correspond to particular analytes of interest in GC×GC data when using single channel detection can be difficult. Some types of detectors, such as electron-capture detection (ECD), may be able to reduce the number of candidate peaks in certain situations, but more definitive identification is usually desired. Short of the very reliable but time-consuming task of spiking the sample with the compounds of interest to verify identities (i.e., standard addition method), retention time or retention index information can be used to find and identify peaks corresponding to the compounds of interest with greater confidence (see Chapter 3 and Section 2 for useful methods). When using mass spectral detection, finding peaks corresponding to particular analytes may be much easier. Methods such as DotMap that scan the three-way array of the GC×GC–MS data (in its entirety or a just a subsection, depending on whether retention information is known) using a mass spectral similarity algorithm can find peaks corresponding to specific analytes [12], with some consideration to the mass spectral selectivity for the analytes of interest [13]. Finally, if an analyte identity for a peak is known in one sample from a set of similar samples, it may be possible to find or identify the same analyte peak (or lack thereof) in the other samples for which the identity is initially not known by using alignment. This, of course, relies on the local continuity in alignment between similar peak patterns around the known analyte, providing a reliable shift vector or transformation from the location of the known peak to the location of the same peak in the other samples (or vice versa), a condition that could reasonably be expected as long as the samples are sufficiently similar.

4. DECONVOLUTION (MATHEMATICAL RESOLUTION)

Although GC×GC can provide increased peak capacity and more selectivity compared with 1D GC, peak overlap is still predicted to occur [14] and indeed does occur in practice. Fortunately, the bilinear structure of GC×GC data allows for application of advanced (linear algebra based) chemometrics techniques for mathematically separating signal contributions in overlapping peaks. Many of these techniques differ greatly from those employed in 1D GC or even GC–MS deconvolution [15], which rely much more on peak shape assumptions and the presence of selective ions. As is usually the case with the other chemometric techniques, many 1D GC and GC–MS data analysis techniques can be employed on GC×GC data [16]. However, these techniques usually result in identification of multiple 1D peaks (one for each second-column separation) for a single GC×GC peak (i.e., per one analyte), and additional logic is needed to recombine the 1D peaks into a single GC×GC peak, which turns out to be a common point of difficulty (uncertainty and failure) in automated routines that deconvolute GC×GC data in a 1D GC context. There are other problems with using 1D deconvolution techniques (such as a number of critical parameters that can greatly influence the deconvolution results, each 1D peak having a somewhat different spectrum than the other 1D peaks that make up a given GC×GC–MS peak, and missed 1D peaks). In practice, chemometric techniques that take advantage of the special "bilinear" data structure for the 2D separation space, such as the generalized rank annihilation method (GRAM) and parallel factor analysis (PARAFAC), have shown very good results for GC×GC and higher order data [4,11,17–23].

GRAM and PARAFAC are essentially trilinear models such that the data, **R**, is computationally modeled as the sum of the outer product of an integer number, N (the number of factors in the model), of three vectors, **x**, **y**, and **z**, as given in the following equation.

$$\mathbf{R} = \sum_{n=1}^{N} \mathbf{x}_n \otimes \mathbf{y}_n \otimes \mathbf{z}_n + \mathbf{E}$$

Any part of the data that cannot be modeled in this manner falls into the residual or error matrix, **E**. There are other ways to write the same model, but the given expression seems the most intuitive in relation to application to GC×GC data. It may be important to note that the ordering of the dimensions in the data does not matter; that is, the second separation dimension may correspond to the first, second, or third dimension of **R**, as long as the other model conditions outlined below are met. For the purpose of clarity in this discussion, the first dimension of **R** (having the same length as the **x** vectors) will correspond to the first separation dimension, and the second dimension of **R** (having the same length as the **y** vectors) will correspond to the second separation dimension. In application to GC×GC data, the data consisted of aligned subsections of GC×GC chromatograms (of dimensions equal to the lengths of the **x** and **y** vectors). The model consists of *N* factors, each factor having a first and second separation dimension profile represented, respectively, by an **x** and **y** vector followed by scaling or weighing values between the chromatograms in the **z** vector. Hence, the 2D bilinear structure of the 2D separation space is defined by the **x** and **y** vectors.

One primary difference between GRAM and PARAFAC is the allowed length of the **z** vectors; whereas GRAM requires that each of these vectors have a length of two, PARAFAC operates with any vector length of two or greater. Because the third dimension of **R** must have the same length as the **z** vectors, it follows that GRAM can only model a given subsection from two GC×GC chromatograms in a single model. Another major difference is that PARAFAC can actually model data (**R**) having more than three dimensions by inclusion of more sets of vectors like the **x**, **y**, and **z** vectors in the outer products that are summed, so that there are *d* sets of vectors for modeling *d*-dimensional data [24,25]. This turns out to be useful for modeling higher dimensional data such as multiple aligned subsections of GC×GC–MS chromatograms. Also, there are usually differences in how these models are fit or created, but those details are beyond the scope of this chapter.

GRAM and PARAFAC share a number of important properties and characteristics. Both have the ability to, and indeed routinely do, model baseline as one or more factors of a model, thereby removing baseline from the factors representing peaks and improving quantitative results. Preprocessing for baseline is unnecessary and may even be counterproductive. Applied as described above, both methods can directly provide quantitative information for a peak if a chromatogram containing the peak from the analyte at a known concentration is included in the data to be modeled (and the signal is linear with concentration). Indeed, both techniques also provide an analysis advantage called the second-order advantage, which is that deconvolution (and accurate quantification) can be performed in situations in which unknown interferences or peaks overlap an analyte peak. This advantage does not exist for comparable standard 1D GC data analysis methods (same number of runs, same type of detection, etc.) [18].

Because GRAM and PARAFAC are trilinear models, elements of the signal that are not trilinear or even bilinear, such as most of the noise, cannot be modeled (at least not unless a very large number of factors are used) and are essentially filtered out of the signal (and end up in **E**). This effectively provides an signal-to-noise (S/N) enhancement for GC×GC peaks, which are generally bilinear in each chromatogram [4,19,20]. Both methods may allow the use of some modeling constraints that can be useful for chromatographic data, such as unimodality or nonnegativity constraints, on all or a selected set of vectors in the model, depending on the implementation of the method.

Another important property of GRAM and PARAFAC is that the models are unique (except for permutation of the order of the factors and scaling, which is handled by a type of normalization) and are not nested, which means that a $N + 1$ factor model is not just an N factor model with one additional factor tacked on, but is instead a model in which all the factors may be different than those of an N factor model (but in practice usually exhibit some similarity). That being the case, the selection of an appropriate number of factors to incorporate into a GRAM or PARAFAC model is by far the most important parameter that needs to be determined in order to create meaningful models. The selection of an appropriate number of factors (N) to use in the model is not necessarily straightforward, especially with an unknown number of analytes and interferences in the **R** data subsection, although there is often a narrow range of values of N that will yield good results. S/N is also an important consideration in determining an appropriate value for N. A poor selection of N, however, can result in a model that does not describe the data well in a number of ways: baseline may not be accounted for, signal from several peaks or sources may be lumped into one factor, peaks (especially small peaks) may not be considered, or signal from a single peak may be split across multiple factors of the model. In most reports in the literature, an analyst will manually select and evaluate the results for different values of N on a given set of data. A few metrics can be useful in choosing an appropriate number of factors, such as peak finding (to determine the number of peaks in a **R** data subsection), core consistency [26], and possibly the sum of the squares of the residuals, but these approaches do not always work well (e.g., at low analyte S/N, or when N is large).

For analysis of GC×GC–MS data (or another multichannel detector or separation dimension), GRAM cannot be used, but this higher order data provides a number of advantages. First, deconvolution, even in the presence of unknown interferences and overlapping peaks, can be carried out using a single GC×GC–MS chromatogram using PARAFAC [21], as demonstrated in Figures 1–4. This is referred to as the third-order advantage. Quantitative information, in the form of a peak sum (integrating the peak across all data dimensions), can also be

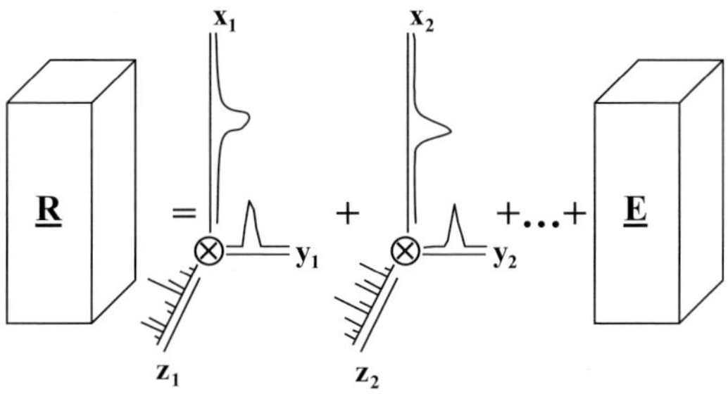

Figure 1 A graphical representation of the PARAFAC model as applied to a subsection of GC×GC–MS data from a single run. The **x** and **y** vectors represent the profiles of the factors in the two chromatographic separation dimensions, as with GC×GC data, but now the **z** vectors each represent the mass spectrum of the corresponding factor instead of relative intensity between multiple chromatograms.

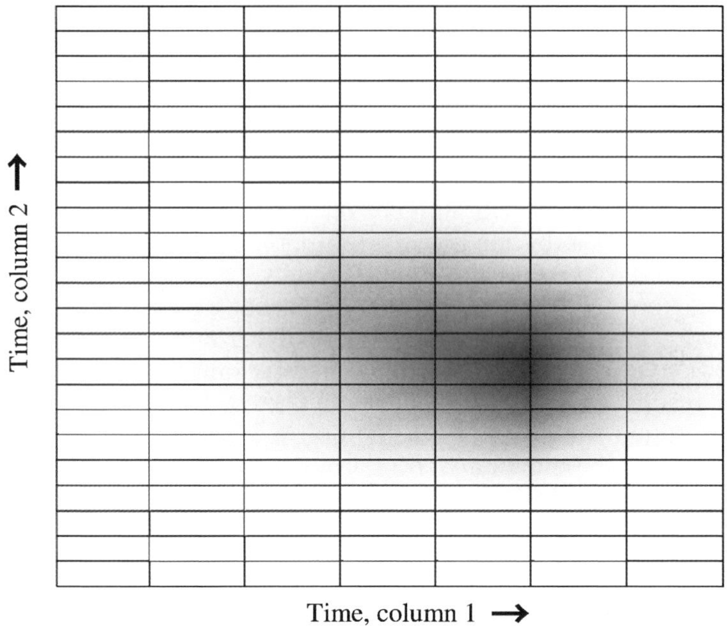

Figure 2 A subsection of GC×GC–MS data taken from an actual diesel sample dataset. The data has been summed in the mass spectral dimension to produce an image of the TIC.

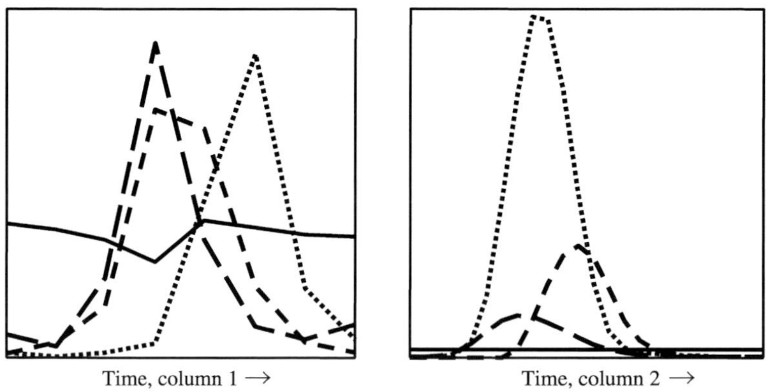

Figure 3 First and second separation dimension peak profiles obtained from PARAFAC of the subsection depicted in Figure 2, corresponding to the **x** (left) and **y** (right) vectors of the model. What appears to be a single peak in the TIC (or at least a highly overlapped group of peaks) can be seen to actually consist of several smaller peaks that are mathematically resolved by PARAFAC. Baseline is modeled by one factor (denoted by the solid line) and thus removed from the other peak profiles.

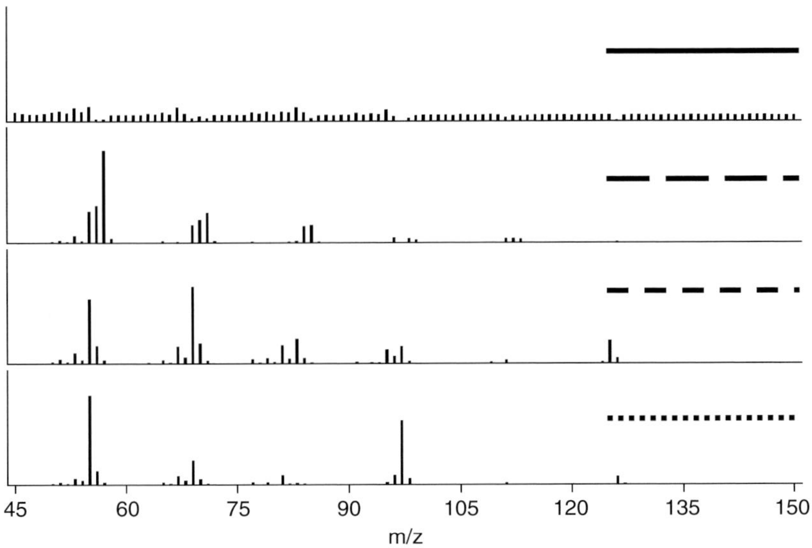

Figure 4 Mass spectra corresponding to the profiles shown in Figure 3, and the **z** vectors of the same PARAFAC model of the data visualized in Figure 2. Some similarity between the bottom three mass spectra is apparent; this is expected due to the structural similarity of the compounds that produced these signals.

obtained from application to a single GC×GC–MS chromatogram, and this value can be compared with values from other GC×GC–MS chromatograms for relative or absolute quantification (depending on whether or not absolute information, such as a concentration, is known for one or more of the chromatograms). In this type of analysis, alignment between chromatograms or different peak shapes are not pressing concerns because each chromatogram is analyzed one at a time. Alternatively, multiple GC×GC–MS chromatograms can be analyzed using PARAFAC by stacking the chromatograms into a four-way array, in which case run-to-run alignment between chromatograms must be adequate in the original data (or alignment preprocessing performed), and peak shapes for a given analyte across the 2D chromatograms must be sufficiently similar (as well as adhere to the bilinearity requirement). Another benefit specific to GC×GC–MS data is the use of the mass spectral mode to automate the selection of a model having an appropriate number of factors for an analyte by mass spectral matching of the deconvoluted mass spectra of the models [22], presuming the MS employs standard electron impact ionization technology.

Another chemometric technique that has been applied to GC×GC–MS data is PARAFAC2, which differs from PARAFAC in that one dimension need not be strictly trilinear [27,28]. This allows for analysis of data where there is shifting of peak profiles or changes in retention time in one separation dimension due to, for example, temperature programming, which affects the second-column retention times, or misalignment across samples [4,29]. PARAFAC2 can be used in most cases where PARAFAC is used, but is computationally more complex and expensive.

5. CLASSIFICATION, FEATURE SELECTION, DATA MINING, AND PREDICTION

Probably the most familiar chemometric techniques in application to GC×GC data are those that deal with classifying samples by sample type and concurrently finding features (analytes) that distinguish (or sometimes are common) between different types of samples. Chemometric techniques that deal with predicting (and quantifying) sample properties are closely related and also important.

Principal component analysis (PCA) is very well known and has been frequently applied to GC×GC data for classification, as illustrated in Figure 5. With PCA, the data matrix, **R**, is decomposed into two matrices **S** and **L**, which are called the scores and loadings, respectively, plus any remaining unaccounted signal, **E**, as given by the following equation.

$$\mathbf{R} = \mathbf{SL} + \mathbf{E} = \sum_{p=1}^{c} \mathbf{s}_p \otimes \mathbf{l}_p + \mathbf{E}$$

Given that **R** is of size $m \times n$ and c components are used in the mathematical decomposition, **S** will be of size $m \times c$ and **L** will be $c \times n$. Note that the term *"component"* in this context does not refer to a specific chemical species, but rather to a fundamental part or basis of the matrix **R**, as further explained below. If the number of components used in the decomposition, c, is less than the rank of the data matrix, then **E** will be a non-zero matrix of size $m \times n$. In a mathematical sense, rank can be defined as the number of linearly independent rows or columns, whichever is smaller, of a matrix. Without showing any derivations here, it follows that a matrix can be completely decomposed (leaving no error) into the sum of the outer product of a number vector pairs equivalent to the rank of the matrix; PCA is such a decomposition. In reality, almost all matrices of chromatographic data are, strictly speaking, of full rank (having rank which is the lesser of m or n) because of imperfections in the data. Fortunately, however, a

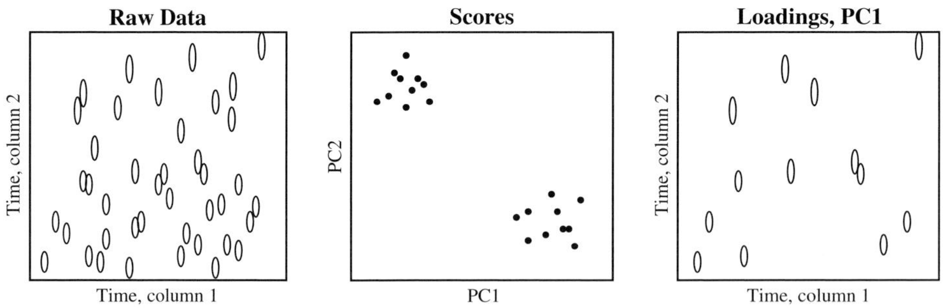

Figure 5 Illustration of PCA on simulated GC×GC data, one sample (from one sample type) of which is shown at left. Relative scores on the first two principal components (PCs) are given in the center pane for 20 chromatograms that fall into two sample types (or classes), indicated by the two clusters of 10 points each. Refolded loadings on the first PC are shown at right, indicating chromatographic locations that most affect (either positively or negatively) the scores for the first PC.

data matrix can almost always be estimated with little error using only a few vectors (such as c of the **s** and **l** vectors for PCA) because, ignoring the small imperfections, the primary sources of variance in the matrix are signals from chemical species. Actually, many matrix decompositions could fit the given equation, but the constraint that defines PCA is that the vectors given by the outer product of the columns of **S** with the rows of **L** are mutually orthogonal and ordered by magnitude so that the outer product of the first column from **S** ($\mathbf{s}_1$) and first row of **L** ($\mathbf{l}_1$), or the first component, gives the largest contribution to (or captures the largest variance in) **R**, followed by smaller and smaller contributions in the subsequent components. It has been mathematically proven that such constraints give a decomposition that captures the greatest variance in **R** using the fewest components.

As a consequence of the orthogonality constraints and the greedy nature of the decomposition (capturing as much variance as possible in as few components as possible), the individual principal components may not be as chemically meaningful as was the case with GRAM or PARAFAC, where each component or factor ideally models variance resulting from a distinct signal source (i.e., analytes, interferences, baseline, etc). This seems to especially be the case with the later principal components, partly because they capture less variance, but also because they are more constrained by all the previous principal components to which they must be orthogonal. Unlike PARAFAC, the selection of an appropriate number of components to use in the decomposition, though still important, is not as critical because PCA decompositions are nested. That is, a $c-1$ component decomposition is the same as a c component decomposition without the last component that captures the least variance. Therefore, a decomposition using a large number of principal components can be carried out (PCA is quite fast, even with many components); then any smaller number of relevant principal components can be used in the subsequent data analysis and interpretation. Determining which principal components are relevant in PCA can be somewhat challenging, but many automated methods have been developed to aid in or perform this task.

In relation to GC×GC data, for PCA, **R** is usually formed by concatenating a number of unfolded GC×GC chromatograms into a matrix. The loadings for a component (a row from **L**) show the distribution of the variance captured by that component in the unfolded chromatogram. The loadings can be refolded to show the same in a 2D visualization (as in the right pane of Figure 5). Scores for a component on the various principal components (columns of **S**) can be thought of as locations in the principal component space (see Figure 5 center pane). For classification purposes, scores of samples from the same class would ideally lie together in a tight cluster close to a particular location in principal component space that is far from the locations and clusters of the scores of other classes. Useful chemical information can often be gleaned from the relative positions of clusters in a scores plot. Sample classes that are near each other are more chemically similar than sample classes that are further apart. Specific changes in chemical composition from one class to the next can correlate along one principal component or another. Hence, "variance," and thus the position of a cluster of samples with respect to a given principal component in a scores plot, goes hand in hand with "chemical information." In addition, features (peaks) in the loadings plots often provide the locations of the chemical compounds that are responsible for the class separations visualized in the scores plots. These locations can be "mined" to identify the chemical compounds.

In order to obtain good clustering and separation in principal component space, it is important to remove extraneous sources of variance from the data (unwanted variance is essentially experimental imprecision), which is usually accomplished through preprocessing. Baseline correction and normalization (including mean centering) are routinely performed to remove some trivial sources of variance. PCA is fairly sensitive to alignment between samples, and it is often the case with misaligned chromatograms that the first few principal components capture mostly variance due to the misalignment. Misalignment is observed as "derivative of a gaussian-shaped" signal in the loadings plots (along a given column time axis). Thus, having aligned data (either well aligned from the instrument or aligned as part of preprocessing) is important for obtaining chemically meaningful information via PCA. Other preprocessing steps, such as variable or feature selection, are possible and can help to focus PCA on the meaningful attributes of the dataset [30]. PCA can be used with mass spectral detection by either analyzing selected m/z, [31,32] or performing a multi-way version of PCA (MPCA), which is actually the same as PARAFAC using orthogonality constraints on all dimensions and then multiplying the $\mathbf{y}$ and $\mathbf{z}$ vectors to obtain the multi-way equivalent of the $\mathbf{L}$ matrix. If PCA is applied in an unsupervised mode — that is, sample class membership is not known *a priori* and observance of class membership clustering is desired — then within-class variation can severely hamper PCA application. The within-class variation obscures the class-to-class variation of interest. Hence, for many studies a supervised mode is needed, in which sample class membership is known *a priori*, and other approaches such as a Fisher ratio method can be applied [17,30,33], as described below.

Several methods can be used for prediction of sample classes or characteristics. One method that has been used with GC×GC data is partial least squares (PLS, particularly the multi-way extension thereof (N-PLS)) [34]. The construction and fitting of these regression models is more complicated than PCA and will not be discussed in mathematical detail here. In short, PLS and N-PLS prediction relies on a calibration set of samples having known values of the properties of interest. A regression model is created using the calibration set and a specified number of latent variables, which are like factors in PARAFAC or components in PCA, and similar rules and methods apply in selecting a parsimonious number of these to include. For multi-way partial least squares (N-PLS), GC×GC chromatograms need not be unfolded and information in the structure of the chromatogram is preserved. Orthogonality is not enforced, so although it may take more latent variables to describe the variance (i.e., chemical information), the latent variables tend to be more meaningful than with principal component regression (PCR, the regression equivalent of PCA). PLS and N-PLS do not provide the second-order advantage; that is, it is not able to account for interferences not included in the calibration "training" set, so the calibration set must include any likely interferences or sources of variance not dealt with by preprocessing. With proper preprocessing, well-controlled instrumental parameters, and a wide enough training set, N-PLS of GC×GC data has demonstrated good results for content prediction in samples as complex as naptha and jet fuel [35,36]. Suggested applications include using N-PLS to *rapidly* provide GC×GC-based property values that are correlated to those from a "slow" benchmark method. Once this correlation has been determined, the GC×GC-based measurements can be used for benchmark prediction, which in turn can be correlated to some other property or parameter of interest (quality parameter, cost, product value, etc.).

In another class of techniques for prediction of sample properties using GC×GC data, the data is first processed to provide retention time or index information for relevant peaks. The retention information can then be analyzed using much simpler but effective regressions that relate the retention information to properties of interest, such as enthalpies of vaporization (ΔH_{vap}), liquid vapor pressures (p_L), and partition coefficients [37], or number of carbon atoms and double bonds [38]. This kind of approach represents a large reduction in the size of the dataset down to the most important features, that is, the location of peaks arising from compounds in the samples. As such, much noise and other sources of extraneous variance are removed from the data. Some properties can be calculated or estimated directly from retention information and knowledge of the separation conditions (based on chromatographic theory), but for other properties, corresponding property information for each compound is required as is the case with PLS. It is noteworthy that for prediction of some properties it may be necessary to inject unretained compounds, such as methane, in order to obtain dead time measurements in the two chromatographic dimensions (see also Chapter 3).

Building from the prior discussion of PCA, analysis of variance (ANOVA) and the Fisher ratio (*f*-ratio) parameter that is calculated in that method have been used for feature selection in GC×GC data. These approaches deal with within-class variation for studies in which class membership of the samples is known via the experimental design, and the analyst is generally interested in learning about the chemical compounds that distinguish the different analyte classes. Fisher ratios are calculated as follows. First, in application to GC×GC data, chromatograms are unfolded and concatenated into a matrix, **X**. Next, the class-to-class variance, σ_{cl}^2, is calculated as

$$\sigma_{cl}^2 = \frac{1}{k-1}\sum_{i=1}^{k}(\bar{x}_i - \bar{x})^2 n_i$$

where k is the number of classes, n_i is the number of chromatograms for the ith class, $\bar{x}_i$ is the mean across the chromatograms of the ith class at the data point of interest, and $\bar{x}$ is the mean across all of the chromatograms at the data point of interest. The within-class variance, σ_{err}^2, is given by

$$\sigma_{err}^2 = \frac{1}{(N-k)}\left(\left(\sum_{i=1}^{k}\sum_{j=1}^{n_i}(\bar{x}_{i,j} - \bar{x})^2\right) - \left(\sum_{i=1}^{k}(\bar{x}_i - \bar{x})^2 n_i\right)\right)$$

where N is the total number of chromatograms across all classes. The Fisher ratio, f, is then

$$f = \sigma_{cl}^2 / \sigma_{err}^2$$

using the class-to-class variance and within-class variance for the given data point in the unfolded chromatograms. The f-ratios calculated at each data point can be refolded back into a matrix for visualization, as shown in Figure 6. A threshold value is usually specified so that only chromatographic locations with an f-ratio above this threshold value, and thus a class-to-class variance that is large enough with respect to the within-class variance, are accepted as class-distinguishing features of importance. Methods based on f-ratio feature selection have been used as a preprocessing or feature selection step for PCA in GC× GC–FID data [30] and for finding peaks that distinguish between classes in GC×GC–ToF MS data [17,33]. For GC×GC–ToF MS data, features identified by f-ratio analysis have been used as a "template" for subsequent data mining using combined ChromaTOF and PARAFAC for deconvolution, refined mass spectral identification, and quantification [17]. For example, the sum of the 2D Fisher ratio plot in Figure 6 gives the locations of signals from yeast metabolites that are up or down regulated based on growth conditions, similar to a loadings plot from PCA (see Figure 5, right pane). One such location in the GC×GC–ToF MS data, shown boxed in Figure 6, was analyzed using both ChromaTOF and PARAFAC to provide identification of the metabolite (citrate) as well as relative quantitative information between growth conditions.

Another method for finding class-distinguishing features, but not necessarily between known classes, is the signal ratio method (S_{ratio} method). The details of the method will not be presented here, but it is designed to find the chromatographic (and mass spectral) locations of greatest difference, measured as a ratio, in signal intensity across all samples and mass channels in a set of

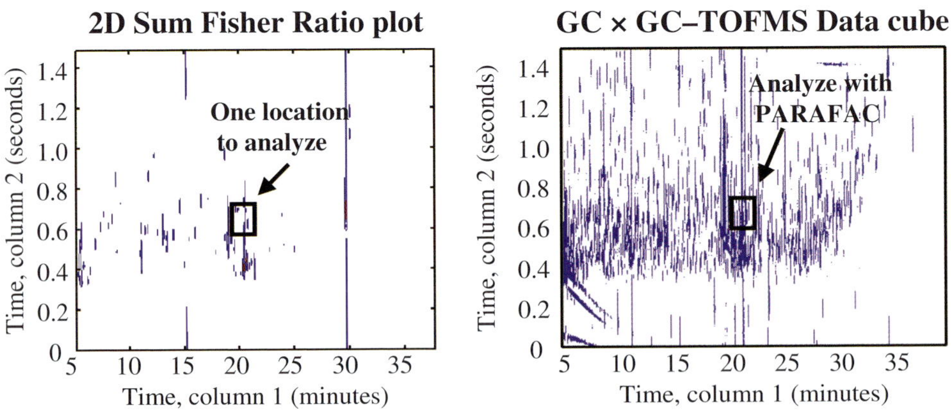

Figure 6 Refolded Fisher ratios are visualized in the left pane (summed over all m/z of interest). Locations identified in the Fisher ratios indicating features of interest can be analyzed in the full dataset, in the right pane, using PARAFAC or other methods.

GC×GC–ToF MS chromatograms. The method uses baseline correction, alignment, and then peak finding as preprocessing. Again, PARAFAC can be used for data mining involving deconvolution, identification, and quantification of the peaks of interest [11].

6. COMPUTATIONAL CONSIDERATIONS

Chemometric techniques vary greatly in computational complexity. Some are very simple, requiring only a few operations per data point (such as PCA), while others may require many iterations of complex calculation for all data points (such as PARAFAC and PARAFAC2). However, it is generally noted that chemometric techniques that require more preprocessing, such as PCA, are less computationally expensive than those that require little or no preprocessing, like PARAFAC2. In the end, the selection among chemometric techniques depends on the analysis situation and on what information is required to answer the analytical question(s) of interest. For some process analysis situations, for example, calculation simplicity and time may be an important factor, but for applications where comprehensive data analysis is the goal, much more involved methods and more time may be required and is usually available.

7. CONCLUSIONS

Chemometric techniques are emerging as essential tools to more fully analyze GC×GC data. This chapter has touched on many of the recent advances in this area. Better utilization of chemometrics by the chromatography community will require the continued development of user-friendly software that implements, for the most part, existing fundamentally sound chemometric algorithms. In addition, there is the challenge to the individual chromatographer to explore the use of these techniques in order for the field of GC×GC to more readily move forward. The rewards for this effort are indeed promising. Other chemometric techniques used in LC×LC and other two-dimensional separation methods could, in principle, be applied to GC×GC with further study and adaptation, but they are beyond the scope of this chapter. Chemometric and data processing techniques from more distant disciplines have crossed into separation science in the past and will certainly continue to do so in the future. Undoubtedly, many chemometric techniques are waiting to be discovered that could bring further benefits to the analysis of GC×GC data by exploiting its special characteristics and data structure.

REFERENCES

1 K.J. Johnson, B.J. Prazen, R.K. Olund and R.E. Synovec, *J. Sep. Sci.*, 25 (2002) 297.
2 R.E. Synovec, B.J. Prazen, K.J. Johnson, C.G. Fraga and C.A. Bruckner. In: P.R. Brown and E. Grushka (Eds.), *Advances in Chromatography*, Vol. 42, Marcel Dekker, New York, 2003, p. 1.

3 K.M. Pierce, J.C. Hoggard, R.E. Mohler and R.E. Synovec, *J. Chromatogr. A*, 1184 (2008) 341.
4 V.G. van Mispelaar, A.C. Tas, A.K. Smilde, P.J. Schoenmakers and A.C. van Asten, *J. Chromatogr. A*, 1019 (2003) 15.
5 N. Marchetti, A. Felinger, L. Pasti, M.C. Pietrogrande and F. Dondi, *Anal. Chem.*, 76 (2004) 3055.
6 R.J. Western and P.J. Marriott, *J. Sep. Sci.*, 25 (2002) 832.
7 T. Pang, S. Zhu, X. Lu and G. Xu, *J. Sep. Sci.*, 20 (2007) 868.
8 C. Vendeuvre, F. Bertoncini, D. Thiebaut, M. Martin and M.C. Hennion, *J. Sep. Sci.*, 28 (2005) 1129.
9 S. O'Hagan, W.B. Dunn, J.D. Knowles, D. Broadhurst, R. Williams, J.J. Ashworth, M. Cameron and D.B. Kell, *Anal. Chem.*, 79 (2007) 464.
10 S. Peters, G. Vivo-Truyols, P.J. Marriott and P.J. Schoenmakers, *J. Chromatogr. A*, 1156 (2007) 14.
11 R.E. Mohler, B.P. Tu, K.M. Dombek, J.C. Hoggard, E.T. Young and R.E. Synovec, *J. Chromatogr. A*, 1186 (2008) 401.
12 A.E. Sinha, J.L. Hope, B.J. Prazen, E.J. Nilsson, R.M. Jack and R.E. Synovec, *J. Chromatogr. A*, 1058 (2004) 209.
13 J.L. Hope, A.E. Sinha, B.J. Prazen and R.E. Synovec, *J. Chromatogr. A*, 1086 (2005) 185.
14 J.M. Davis, *J. Sep. Sci.*, 28 (2005) 347.
15 S.E. Stein, *J. Am. Soc. Mass Spectrom.*, 10 (1999) 770.
16 H. Kong, F. Ye, X. Lu, L. Guo, J. Tian and G. Xu, *J. Chromatogr. A*, 1086 (2005) 160.
17 R.E. Mohler, K.M. Dombek, J.C. Hoggard, K.M. Pierce, E.T. Young and R.E. Synovec, *Analyst*, 132 (2007) 756.
18 C.G. Fraga, B.J. Prazen and R.E. Synovec, *J. High Resolut. Chromatogr.*, 23 (1999) 215.
19 C.G. Fraga, C.A. Bruckner and R.E. Synovec, *Anal. Chem.*, 73 (2001) 675.
20 L. Xie, P.J. Marriott and M. Adams, *Anal. Chim. Acta*, 500 (2003) 211.
21 A.E. Sinha, C.G. Fraga, B.J. Prazen and R.E. Synovec, *J. Chromatogr. A*, 1027 (2004) 269.
22 J.C. Hoggard and R.E. Synovec, *Anal. Chem.*, 79 (2007) 1611.
23 X. Guo and M.E. Lidstrom, *Biotechnol. Bioeng.*, 99 (2007) 929.
24 R.A. Harshman, *UCLA Working Papers in Phonetics*, 22 (1972) 111.
25 R. Bro, *Chemom. Intell. Lab. Syst.*, 38 (1997) 149.
26 R. Bro and H.A.L. Kiers, *J. Chemom.*, 17 (2003) 274.
27 R.A. Harshman, *UCLA Working Papers in Phonetics*, 22 (1972) 30.
28 H.A.L. Kiers, J.M.F. ten Berge and R. Bro, *J. Chemom.*, 13 (1999) 275.
29 R. Bro, C.A. Andersson and H.A.L. Kiers, *J. Chemom.*, 13 (1999) 295.
30 K.J. Johnson and R.E. Synovec, *Chemom. Intell. Lab. Syst.*, 60 (2002) 225.
31 K.M. Pierce, J.L. Hope, J.C. Hoggard and R.E. Synovec, *Talanta*, 70 (2006) 797.
32 R.E. Mohler, K.M. Dombek, J.C. Hoggard, E.T. Young and R.E. Synovec, *Anal. Chem.*, 78 (2006) 2700.
33 K.M. Pierce, J.C. Hoggard, J.L. Hope, P.M. Rainey, A.N. Hoofnagle, R.M. Jack, B.W. Wright and R.E. Synovec, *Anal. Chem.*, 78 (2006) 5068.
34 R. Bro, *J. Chemom.*, 10 (1996) 47.
35 K.J. Johnson, B.J. Prazen, D.C. Young and R.E. Synovec, *J. Sep. Sci.*, 27 (2004) 410.
36 B.J. Prazen, K.J. Johnson, A. Weber and R.E. Synovec, *Anal. Chem.*, 73 (2001) 5677.
37 J.S. Arey, R.K. Nelson, L. Xu and C.M. Reddy, *Anal. Chem.*, 77 (2005) 7172.
38 B. Vlaeminck, J. Harynuk, V. Fievez and P.J. Marriott, *European J. Lipid Sci. Tech.*, 109 (2007) 757.

CHAPTER 6

Comprehensive Multidimensional Systems Incorporating GC×GC

Hans-Gerd Janssen, Erwin Kaal and **Sjaak de Koning**

Contents

1. INTRODUCTION

Comprehensive two-dimensional (2D) chromatography is a very powerful tool for the separation of complex samples because it offers a significant increase of peak capacity as a result of the expansion of the available separation space. Co-eluting peaks in one-dimensional (1D) chromatography might be separated in the second dimension, meaning that more peaks can be individually detected in comprehensively interfaced two-dimensional chromatography.

In comprehensive two-dimensional chromatography, time fractions of the first-dimension separation are transferred onto the second-dimension column for a further separation, as explained in previous chapters. Of course, there is no

Comprehensive Analytical Chemistry, Volume 55
ISSN: 0166-526X, DOI 10.1016/S0166-526X(09)05506-8

reason why this comprehensive interfacing should remain restricted to just two dimensions. Clearly, higher order comprehensively interfaced systems could also be envisaged, such as comprehensive three-dimensional (3D) systems like GC×GC×GC or LC×GC×GC. Not only would such systems result in an even further enhanced peak capacity as compared to comprehensive two-dimensional systems, but they would also allow researchers to obtain a better match of sample dimensionality and system dimensionality. A good match between sample and system dimensionality is a prerequisite for obtaining detailed and complete structured information on complex samples as illustrated by Giddings [1]. Sample dimensionality as defined by Giddings refers to the number of individual parameters that must be specified to characterise the components in a sample. System dimensionality pertains to the number of different separation mechanisms in a multidimensional separation approach. The so-called complex sample, such as mineral oil samples, often contain a close to infinite number of compounds, yet the number of dimensions of interest is far more limited.

The keyword when dealing with complex samples is selectivity, that is, the ability to distinguish the compounds or compound groups of interest from those that are not relevant — in other words, the ability to extract the relevant information from the "noisy" chromatogram. Here the word "noise" does not refer to random fluctuations in the background signal but to noninformative regions of the chromatogram. In a recent publication, it is argued that there are only three routes to convert chromatograms into information: (1) the target compound route, (2) group-type analysis, and (3) profiling approaches [2]. In *target compound analysis* one looks for a limited number of known compounds. Clearly, the chances of the system to resolve these from the numerous other compounds present in a complex sample is higher if the peak capacity of the separation system is higher. Comprehensive systems here offer a clear advantage over one-dimensional systems, while comprehensive three- or four-dimensional systems in turn would perform better than comprehensive two-dimensional setups.

Most of the initial applications of GC×GC did not deal with target compound analysis but focused on *group-type separations*, in particular those of oil products. In group-type separations, there is no need to separate all components in a sample individually. Rather, the aim is to get all compounds belonging to the same class in one band. Separation between the groups should be maximised, whereas separation between compounds of the same group should be minimized. Comprehensively interfaced systems are ideal for group-type separation because they allow independent separation according to different sample dimensions. Since this property of comprehensive multidimensional chromatography is vital, we will elaborate on this a bit further. For instance, GC×GC allows decoupling the volatility and polarity contributions to GC retention [3]. All compounds that co-elute from a non-polar first-dimension GC column have the same volatility. If now these compounds are introduced onto a more polar second-dimension column, they can be separated more or less on the basis of polarity alone. For samples that have molecules that differ only in two basic properties, here volatility and polarity, such a comprehensive 2D system provides a matching system dimensionality; that is, it separates according to the two independent factors deemed relevant. Group-type separation of higher-dimensionality samples requires comprehensive systems of a higher dimensionality. A typical example is the separation of fatty acid methyl esters (FAMEs) derived from vegetable oils or fats. The FAMEs derived from such oils and fats differ according to at least three independent parameters: chain length, number of aromatic rings, and the position of the aromatic rings on the fatty acid chain. The matched-dimensionality system would be a comprehensive 3D system where first the compounds are separated based on chain length, and then each chain-length fraction is sampled as several fractions onto a second dimension for a number-of-double-bonds separation, with each fraction again being transferred as several fractions to a column that separates the compounds according to the position of the double bonds.

A final generic application type of chromatography is chromatographic *fingerprinting or profiling*, an approach that is rapidly becoming relevant as a result of the increased popularity of metabolomics. In chromatographic profiling, the chromatogram of a sample is seen as a fingerprint. Advanced chemometrical methods are then used to correlate fingerprints to certain specific properties of the sample. Using such approaches samples can be rapidly tested, or compounds responsible for a certain effect can be discovered and identified.

Comprehensive 3D chromatographic systems offer two main advantages: better match of sample dimensionality and system dimensionality, which is relevant for group-type separations; and higher overall peak capacity. In this chapter, we will present a short overview of the principles, difficulties, and applications of multidimensional systems incorporating a GC×GC separation procedure. We will first assess, from a theoretical viewpoint, the possibilities of incorporating GC×GC in a truly three-dimensional comprehensive setup with either GC or LC in the first dimension, here abbreviated as XC×GC×GC. Next, we will summarize the practical aspects of GC×GC×GC, LC×GC×GC and LC–GC×GC. From the limitations imposed by the theory and the problems of the implementation, practical progress in the area of comprehensive and heart-cut multidimensional chromatography involving GC×GC is slow, and the number of applications is still rather limited. Anyhow, the continuous need to obtain more detailed information on complex samples will continue to call for more selective and efficient separation methods with higher peak capacities and more order in the multicomponent chaos.

2. THEORETICAL ASPECTS OF USING GC×GC IN COMPREHENSIVE 3D SYSTEMS

2.1 Separations in space and in time

Two distinctly different methods to perform comprehensive chromatography can be distinguished based on how the chromatography is performed: *in space* or *in time*. This classification was recently proposed by Guiochon and co-workers [4] and is highly relevant when considering the best options for multidimensional and comprehensive chromatography. The first type of separation, separation in space, sorts out the sample components in a physical separation plane. The analysis time is the same for all compounds, but their migration distance is different. Thin-layer chromatography (TLC) is an example of a separation in space. In the second type of separation, separation in time, compounds travel the same distance and are separated in the time space. Classical LC and GC are examples of separations in time. To express the idea that TLC is a separation that separates in space or distance, Guiochon used the superscript^{x} to yield TLC^{x}. For separations in time, the superscript^{t} was used: hence, classical LC and GC are denoted as LC^{t} and GC^{t}, respectively. Unlike in the case of LC, for which one could imagine it can be done in the "place mode", with TLC actually being LC^{x}, it is hard to imagine how GC should be done in the GC^{x} mode. GC currently can only be GC^{t}, and comprehensive 3D GC will therefore necessarily be GC^{t}×GC^{t}×GC^{t}. Although Ledford et al. [5] and Synovec and co-workers [6] have demonstrated successful GC^{t}×GC^{t}×GC^{t} separations, work by Schoenmakers et al. [7] and the results of Moore and Jorgenson in SEC^{t}×LC^{t}×CZE^{t} [8], clearly show three-dimensional comprehensive coupling of separations in time as GC^{t}×GC^{t}×GC^{t} to be impractical and provide only mediocre improvements over GC^{t}×GC^{t}. The impractical nature of GC^{t}×GC^{t}×GC^{t} approaches is largely a consequence of the very long analysis times, as will be shortly addressed in the section below. In addition, there are experimental difficulties, the most important one being the limited sensitivity arising from the low sample transfer from injector to detector if split transfer is used between the first dimension (^{1}D) and the second dimension (^{2}D) and/or between the ^{2}D and the third dimension (^{3}D) as is done by Synovec in his pioneering work [6].

The differences between comprehensive couplings in time and in space can be illustrated with a comparative example. Standard GC×GC as we know it is comprehensive 2D in time: Second-dimension chromatograms are recorded one after the other. The total runtime equals *NR* times the runtime of a second dimension run, where *NR* is the number of second-dimensional runs performed. All comprehensive couplings of separations *in time* are slow. Comprehensive couplings of separation methods that separate *in space* can be much faster. Comprehensive 2D *in space* is what is done in two-dimensional TLC: After having developed the TLC plate in one direction, it is turned 90° and is again developed, yet now in the second dimension. The total runtime now is only the sum of the runtimes of the ^{1}D and ^{2}D separation. Comprehensive 2D chromatography *in space* can therefore be much faster than comprehensive 2D chromatography *in time*.

2.2 Considerations on analysis time

The higher overall peak capacity of comprehensive 3D chromatography is mainly attractive for target compound analysis and profiling applications of chromatography, as mentioned in the Introduction. The total peak capacity of a comprehensive 3D system is given by

$$n_{\mathrm{total}} = {}^{1}n * {}^{2}n * {}^{3}n \qquad (1)$$

Here, ^1n, ^2n and ^3n are the peak capacities of the first, second, and third dimension respectively. In the nomenclature for comprehensive chromatography as proposed by Schoenmakers et al. [9], which will be followed in this and other chapters in this book, "kth dimension" is denoted as a superscript k, before the actual property (see also Chapter 3). Since GC×GC×GC will always be GC^t×GC^t×GC^t, the total analysis time can be calculated by realising that the number of third-dimension runs that has to be performed determines the total time required. If we assume that we want four reinjections into the next dimension across every lower dimension peak, we can write the total number of ^{3}D runs (^3NR) as:

$$^3NR = (4 * {}^2n) * (4 * {}^1n) \tag{2}$$

The total analysis time t_{total} is then given by

$$t_{\text{total}} = {}^3NR * {}^3t_r = 16 * {}^2n * {}^1n * {}^3t_r \tag{3}$$

Here, ^3t_r is the runtime of a single third-dimension run. Using Equations (1) to (3), we can compare different scenarios with regard to column selection in the three dimensions based on their consequences for the total peak capacity and analysis time. Here, we will consider three column sets for GC^t×GC^t×GC^t. Since all GC separations are GC^t separations, we will from now on refrain from using the superscript^t in GC^t. In the first scenario, the second- and third-dimension columns can be selected to be the standard column set currently used in GC×GC, connected to a standard GC column of, for example, 30 meters, 320 μm internal diameter in the ^{1}D (we call this the *maximum peak capacity* scenario). Clearly, the first dimension should be operated at a very low speed. This can be done either by operating it at a very low linear velocity or by implementing the stop-and-go method as discussed by the Yates group [10] and Bedani et al. [11]. In the stop-and-go method, chromatography in the first-dimension column is stopped after transfer of a fraction to the next dimension (in this case, the ^{2}D×^{3}D) and is only resumed after completion of the comprehensive ^{2}D×^{3}D analysis. A second scenario would be to connect a standard GC×GC column set to a very short, yet very selective precolumn for very selectively separating some 10 fractions (the *selective detailed-separation scenario*). A third scenario would be the *all peak capacities are equal* option, having three columns each with a peak capacity of, for example, 15. The consequences of these three options for the total peak capacity and total analysis time are summarized in Table 1. In the calculations, it is assumed that a standard column set for GC×GC generates a peak capacity of approximately 250 in the first dimension and 15 in the second dimension.

The calculation results in Table 1 indicate what is called an impractical solution by Schoenmakers et al. [7]. Analysis times of over 1100 hours, or over 45 days, can undoubtedly be called impractical. The second scenario, the *selective detailed-separation,* would result in a sample throughput of about three samples per week, assuming the instrument to be running 24 hours a day, 7 days a week. Although slightly less impractical than the *maximum peak capacity* solution, this is also hardly practical. There is a partial solution to this problem if we deny the

Table 1 Comparison of various options for comprehensive 3D XC[a]. For a definition of the scenarios, see text. Assumptions: $^{3}t_r = 4$ s

Scenario	^{1}n	^{2}n	^{3}n	Peak capacity (-)	Time (h)
Maximum peak capacity	250	250	15	937500	1111
Selective detailed-separation	10	250	15	37500	44.4
All peak capacities are equal	15	15	15	3375	4

[a]XC is used to indicate that this can either be LC or GC.

self-imposed constraint that we want four reinjections across every $^{n-1}$D peak onto the ^{n}D column and operate the system in the multiple heart-cut mode. This is not illogical if the first-dimension separation in this scenario is optimised for maximum selectivity. Compounds belonging to a group elute in perfectly separated bands, and hence no resolution loss would occur by sampling a peak as just one fraction. If we decide to operate the system as an XC–XC×XC system with just one reinjection across a ^{1}D peak, this would require the XC×XC analysis of only 10 fractions (at a ^{1}D peak capacity of 10), or a total analysis time of "only" some 11 h. This is the mode that we and other authors selected in the attempt to combine the excellent selectivity of LC with the very high separation power and group-type selectivity of GC×GC [12,13].

The *all peak capacities are equal* scenario is not very realistic from the perspective of the timing of the separate dimensions. In truly comprehensive operations, the time available to perform the separation of a fraction from the $^{n-1}$D on the ^{n}D is only one quarter of the baseline width of the peak in $^{n-1}$D, as follows from the desire to have four reinjections across an $^{n-1}$D peak. In the case of a three-dimensional GC×GC×GC system, with each of the columns having a peak capacity of 15, the ^{2}D column should therefore be 60 times faster than the ^{1}D column. The ^{3}D column analogously should be 60 times faster than the ^{2}D column and hence 3600 times faster than the 1D column. In GC, there are several ways to obtain faster separations [14,15]. The most logical route here would be the use of columns of progressively lower internal diameter going from ^{1}D to ^{2}D to ^{3}D. Since in GC the analysis speed is inversely proportional to the square of the inner diameter (in case of negligible pressure drop and assuming the columns are operated at their respective optimum velocities), the ratio of the column diameters $^{1}d_c : {}^{2}d_c : {}^{3}d_c$ should be 60 : 7.75 : 1. If we accept the minimum practical column diameter in GC to be 50 μm, this would result in a column assembly of columns of 3000 μm, 387 μm, and 50 μm in the ^{1}D, ^{2}D, and ^{3}D, respectively. The lengths of the columns can be estimated from the well-known equation for the peak capacity in isothermal GC:

$$n = \frac{\sqrt{N}}{4R_s} \ln\left(\frac{1+k_\omega}{1+k_\alpha}\right) + 1 \tag{4}$$

Here, N is the plate number, R_s is the desired resolution, and k_α and k_ω are the retention factor of the first and the last eluting peaks, respectively. With $k_\alpha = 0.5$ and $k_\omega = 10$, 15 peaks can be separated with $R_s = 1.5$ on a column with some

1800 theoretical plates. To a rough approximation, N equals L, the column length, divided by the column diameter d_c. This would mean that our "all peak capacities are equal" column set would consist of the following columns:

$$^{1}\mathrm{D}: L = 5.4\ \mathrm{m},\ d_c = 3\ \mathrm{mm}$$
$$^{2}\mathrm{D}: L = 0.7\ \mathrm{m},\ d_c = 387\ \mu\mathrm{m}$$
$$^{3}\mathrm{D}: L = 0.09\ \mathrm{m},\ d_c = 50\ \mu\mathrm{m}$$

Flow splitting would be needed between subsequent dimensions to make sure every column is operated at its optimum velocity.

Apart from difficulties in finding the right stationary-phase chemistries to create a truly orthogonal system, it is clear from the above that operation of the above comprehensive system is very difficult at best. It is for this reason that in literature, to the best of our knowledge, only two groups have performed a GC×GC×GC separation [5,6]. Ledford et al. [5] used three 100-μm columns with lengths of 20, 3, and 0.035 m, respectively, in the three dimensions. Modulation between the first and second, and between the second and third column could be achieved with just one sweeper thermal-modulator. The separation shown was not very impressive: the three compounds could easily have been separated on just one column, but the experiments clearly demonstrated that it was possible to perform GC×GC×GC separations. Synovec's group [6] used columns with inner diameters of 530, 250, and 100 μm at lengths of 25, 5, and 0.55 m in the three dimensions. Flow modulation was provided by two diaphragm values. The authors clearly demonstrated the proof-of-principle of GC×GC×GC, yet also concluded that the system needs to be improved for practical use.

Because of the before-mentioned problems and of its scarce current use, GC×GC×GC is not further discussed in the present chapter. On the other hand, GC×GC×GC appears to be more realistic than a standard 1D separation, giving a peak capacity of 3375. Obtaining such a peak capacity on a standard 1D column would require a staggering 110 million plates! Clearly, GC×GC×GC is interesting, but many improvements are needed to make the technique truly meet its potential in practice.

3. PRACTICAL ASPECTS OF HIGHER-DIMENSION GC×GC SYSTEMS

3.1 Operational approaches towards LC×GC×GC and LC–GC×GC

As illustrated above, GC×GC×GC is currently not yet practical. LC×GC×GC appears to be more feasible. In the development of a system for the latter technique, the question of how to interface the various dimensions holds a key position. To answer this question, it is wise to first look in more detail at how subsequent dimensions in a comprehensive system can be interfaced. In GC×GC, just as in $\mathrm{LC^t}\times\mathrm{LC^t}$, the ^{2}D analyses are usually performed on-the-fly. That is, the separation on the first column continues while the previous time fraction is analysed on the second column. Two other operational approaches could, however, also be envisaged. This gives a total of three possible routes of operating a comprehensive XC×XC system:

- *Real-time operation*: Time fractions from the ^{1}D column are reinjected onto the ^{2}D column. While the time fraction *n* is separated on the ^{2}D column, the eluate from the ^{1}D column is temporarily stored in a valve (LC and GC) or in a cooled modulator (GC only). Peak width of the ^{1}D peak and the allowed ^{2}D separation time are clearly interrelated:

$$^1\sigma_1 = {}^2t_r \tag{5}$$

  with $^1\sigma_1$ being the standard deviation of the peaks eluting from the first column and ^2t_r the analysis time on the second-dimension column.
- *Stop-flow or "stop-and-go" operation of the first dimension*: Flow in the first dimension is halted while the analysis of the transferred fraction is carried out on the second-dimension column. This mode eliminates any time constraints on the second-dimension separation, but still avoids the storage of large numbers of fractions. We successfully deployed this mode of operation in LC×GC of triglycerides and mineral oil compounds [16,17], in SEC×LC of peptides [11] and in SEC–pyrolysis GC–MS of natural and synthetic polymers [18]. Band-broadening during the repeated stop-flow periods is not a source of concern, as diffusion of larger molecules in liquids is very slow. For GC in the first dimension, this mode is less obvious. Here, a very significant zone broadening due to longitudinal diffusion would occur.
- *Intermediate collection of fractions with storage in vials or multiple sample loops for subsequent reinjection onto the second column* [19]: Again, this mode is more applicable to a liquid-phase first-dimensional separation than in case of a GC ^{1}D analysis. Once a liquid fraction is collected, various operations can be performed with it in the vials prior to reinjection onto the ^{2}D column. Preconcentration and solvent exchange are logical operations, but also chemical or enzymatic conversion reactions can be performed. We successfully applied an intermediate hydrolysis and methylation in the fatty acid analysis of triacylglycerides separated intact on a silver-phase LC column [19]. From the instrumental perspective, this mode is very simple. Moreover, it allows analysts to evaluate the merits of LC×GC for their laboratory without basically any investment. A standard LC and standard GC are all that is needed.

In LC×GC×GC there are two interfacing steps: the comprehensive interfacing between the ^{1}D LC and the ^{2}D GC, and that between the ^{2}D and ^{3}D GC. Looking to the comprehensive interfacing of the ^{2}D and ^{3}D GC dimension, we find that real-time operation is basically the only realistic operational mode. Stop-flow operation very likely results in too much band broadening, while intermediate collection of large numbers of fractions containing highly volatile materials is difficult at best. Fortunately, several modulators for GC×GC are now commercially available.

For LC×GC, all three modes of operation identified above are available, but unfortunately none of them is straightforward. Real-time operation imposes the stringent demand of completing a ^{2}D GC run within a quarter of the peak width of a ^{1}D LC peak. Knowing that GC is much faster than LC, one might easily be led to believe that real time on-the-fly LC×GC is therefore straightforward. Unfortunately, this is not the case. In our experience, the applicability of real-time operation in LC×GC is very limited. For most samples studied by LC×GC so far, it turns out that LC and GC are highly orthogonal, with very little correlation between retention in LC and GC. LC×GC is clearly much more orthogonal than GC×GC. The much higher degree of orthogonality of LC×GC as compared to GC×GC has important consequences for the way LC×GC systems have to be operated. Below we will elaborate on this issue a bit further.

In GC×GC, there is a very strong correlation between retention in the two dimensions. Compounds that are strongly retained on a given stationary phase will also have a high retention factor on any other stationary phase. This is because retention is largely determined by temperature. As a consequence, only a small part of the two-dimensional retention plane is used, as is schematically illustrated in Figure 1 for the separation of a mixture of alkanes, alcohols, and fatty acids on a non-polar×polar GC×GC column set. The limited orthogonality of GC×GC means that compounds co-eluting from the ^{1}D column can easily be separated isothermally on any other ^{2}D stationary phase. In (Normal Phase) LC×GC, there is little or no correlation between retention in the two dimensions. The NPLC ^{1}D separates the compounds in groups according to polarity, with little or no contribution of size. Thereby, all alkanes co-elute in one peak. In mineral oil products, this peak can easily contain alkanes covering 20 carbon atoms in distillation fractions, or up to 80 carbon atoms in raw feeds! It is needless to say that the ^{2}D GC separation of such a wide range of compounds requires a temperature-programmed run covering a wide temperature range. Even with the fastest heating and cooling systems, it is hard to perform such a GC analysis in less than 1 min. This value contrasts with that of the ^{2}D GC runs in GC×GC that typically can be done in less than 6 to 8 s.

Figure 2 shows a comparison of an LC×GC and a GC×GC separation of a real sample. Group separation is much better in the LC×GC system. An interesting question is which technique is faster. Because of the excellent class separation in the LC dimension, in fact only three fractions have to be transferred to the GC dimension resulting in a total analysis time of about 30 min. This is probably very

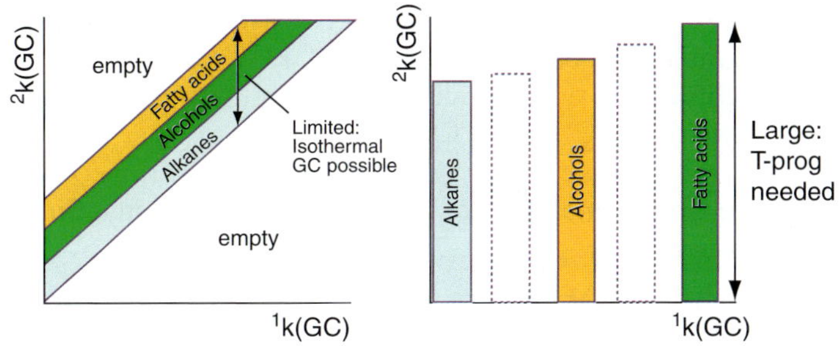

Figure 1 Schematic representation of the differences in orthogonality between GC×GC (left) and LC×GC (right) and the consequences for the operation of the ^{2}D GC separation.

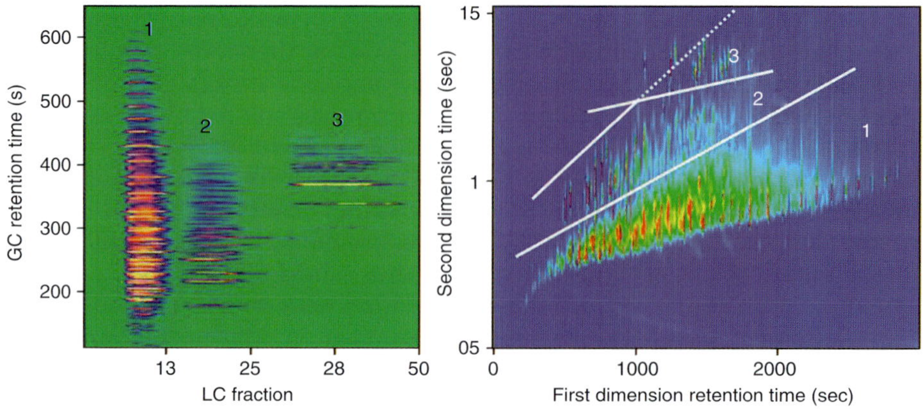

Figure 2 Comparison of LC×GC (left) and GC×GC (right) in group-type analysis of mineral oil product.

similar to the analysis time in GC×GC. Admittedly, the LC×GC separation is in fact a multiple heart-cut LC–GC method. But if group-type information is the target information desired, it will deliver that in a very short time.

True LC×GC is slow, regardless of which operational mode is selected. Real-time operation is feasible, but in that case the flow rate in the ^{1}D column should be kept very low to meet the requirement of four reinjections across a peak. Since diffusion coefficients in liquids are low, this will not result in measurable band broadening [16]. Stop-flow operation is equally slow, but again will not result in additional diffusional band-broadening. Collection in vials is feasible as well.

In the above discussion, the differences between LC^{x} and LC^{t} have been neglected. Whereas GC can realistically be performed only as GC^{t}, for LC this choice is less obvious. In the present manuscript, however, we discard the possibility of LC separations in space. TLC of course is available, but the separation quality of TLC is much poorer than that of modern column LC$^{(t)}$. Moreover, instrumentation for automated LC^{x}×GC obviously is not commercially available, and it is even difficult to imagine what such instrumentation would look like. Most importantly, however, whereas in LC^{x}×LC^{x} all ^{2}D separations could be developed in parallel giving a tremendous time benefit, LC^{x}×GC would necessarily be LC^{x}×GC^{t}, thereby nullifying the advantage of the ^{1}D LC^{x} separation, unless multiple GC columns are operated in series. LC×GC can only be performed as LC^{t}×GC^{t}, which, combined with previous statements, means that LC×GC×GC can only be performed as LC^{t}×GC^{t}×GC^{t}.

LC×GC is slow because the GC dimension is time consuming as a consequence of the orthogonal nature of the two dimensions. Since GC×GC is even (much) slower than regular 1D GC, LC×GC×GC is even slower. The fastest GC×GC separation ever performed requires approximately 5 min [20]. At four reinjections across every ^{1}D LC peak and a ^{1}D LC peak capacity of 50, this means a total runtime of some 17 h. At a more realistic GC×GC runtime of 30 min, this would be about 100 h, or 4 days! Clearly, fully comprehensive

LC×GC×GC is not really practical. The more likely reality is that only a few fractions, or a few peaks of interest, are transferred for GC×GC analysis, basically converting the fully comprehensive LC×GC×GC into multiple heart-cut LC–GC×GC. From the hardware point of view, LC×GC×GC or LC–GC×GC is rather straightforward. Although there are no fundamental problems with real-time operation, the very low flow rates required for the LC separation will make this method of interfacing unattractive. Much more practical is stop-flow operation or intermediate collection of fractions in sample vials. Additional information on the hardware aspects of LC×GC×GC and LC–GC×GC is given in the next section.

3.2 Hardware solutions for LC×GC×GC and LC–GC×GC

As discussed in previous chapters, the hardware for GC×GC is nowadays well developed (e.g., see Chapter 4), and reliable modulators are commercially available from various manufacturers (Chapter 2). For automated transfer of fractions from the LC to the GC×GC, we have recently described two approaches, one based on a six-port switching valve and a second one based on a syringe [16]. The two interfaces are shown schematically in Figure 3.

The configuration of the valve system is basically identical to that used by numerous authors in the past for on-line LC–GC [e.g., 21]. The effluent from the LC column is transferred to one of the ports of a six-port valve. From there, it flows through a deactivated fused-silica column into the heated GC injector. The transfer capillary is simply inserted into the heated injector liner through the septum. In the transfer position (solid line), the LC effluent flows to the GC(×GC) system. Once transfer of the fraction is complete, the valve is turned.

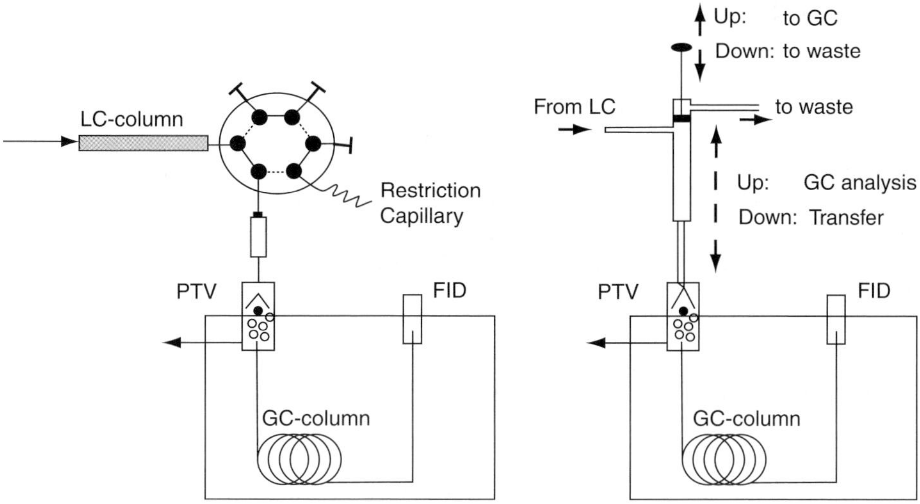

Figure 3 Schematic representation of the valve-based (left) and syringe-based (right) interface for interfacing LC to GC or GC×GC [16].

The LC is now isolated from the GC×GC. The LC flow is stopped, and the GC(×GC) separation can now take place. A small leak flow of carrier gas is created by installing a restriction capillary in one of the valve ports. This serves the purpose of allowing an exit for discharge of the residual liquid from the transfer capillary to waste, in that way preventing diffusion of solvent vapour into the injector. Although this interfacing method performs well, the small leak flow required is a disadvantage because part of the sample is lost via this exit.

The second interface developed, the syringe-based interface, uses commercially available side-port syringes. Two side-ports are needed, one slightly higher than the other. These side-ports can be positioned either at the upper end of the syringe or closer to the needle exit. With the ports in the top section of the syringe, the syringe barrel can be used for temporary storage of an LC fraction. If the ports are situated at the bottom side of the syringe, close to the needle, the syringe basically just acts as a transfer capillary with no storage capacity. Which option to select depends on whether one wishes to be able to independently vary the LC flow rate and the injection rate into the GC. This is only feasible if the syringe barrel is used for storage. A drawback of using the syringe barrel for storage is the slightly increased risk of carryover. Regardless of which option is selected, the lower of the two connections is used as the LC effluent entrance, while the upper port is connected to waste. The syringe plunger is used to direct the flow either to waste or to the GC inlet. With the plunger stamp below, the bottom entrance flow is sent to waste. With the plunger stamp between the two side-ports, the LC effluent flows down through the barrel and eventually into the GC. In addition to the plunger position, the syringe position can also be selected. With the syringe out of the injector, the physical connection between the LC and the GC is interrupted. This is the default position. With the syringe down, transfer of a fraction into the GC can take place. Once the transfer of a fraction is completed, the syringe is withdrawn from the injector and the GC separation can be started. With this interface no sample loss occurs because no leak flow is needed to remove solvent vapours from the system.

An important step in the transfer of an LC fraction to the GC×GC separation is the elimination of the solvent. Typical peak widths at baseline in LC are around 40 s. At four transfers across a peak and an LC column diameter of 2.1 mm, viz. a flow rate of 200 μL/min, the corresponding fraction volume is approximately 35 μL. Clearly, special precautions are needed for introducing such a volume into a capillary GC(×GC) system. Since the LC separation mode used in LC×GC×GC is very likely to be a Normal Phase LC separation, the introduction of such a volume is feasible. It actually is rather straightforward, especially if the compounds of interest are not too volatile. A detailed discussion of methods for large-volume injection in GC is beyond the scope of the present chapter. For an excellent overview of such methods, see Mol et al.'s work published several years ago [22].

In many of the LC×GC applications reported so far, as well as in the limited number of LC×GC×GC separations published, the actual analytical problem is more the characterisation of the sample itself rather than the analysis of specific

Table 2 Protocol for automated comprehensive LC×GC×GC using a syringe-based interface with fraction collection in the syringe barrel

	GC×GC	Syringe	Syringe plunger	Injector	LC pump	Action/Status
Initialisation						
1	ready	up	down	ready	running	Sample injected in LC LC void volume send to waste
2	ready	up	up	ready	running	Collection of fraction in barrel
Repeated cycle						
3	ready	down	moving	ready	hold	Transfer of fraction into injector
4	ready	up	up	running	hold	Elimination of solvent
5	running	up	up	running	hold	GC×GC analysis
6	ready	up	up	ready	running	Collection of next fraction in barrel
Repeat from step 3						

Source: Adapted from [16]

compounds in the sample. The typical application is group-type analysis of a complete mineral oil or an edible oil. Hence, the concentrations are very high, and simple sample splitting in a hot injector is all that is needed. More sophisticated large-volume injection methods are needed only if low levels of compounds have to be measured as, for instance, will be the case in target compound analysis of pesticides or flavour and taste compounds in food products.

Automation of comprehensive LC×GC×GC is straightforward, if at least the GC×GC dimensions are operated in the stop-flow mode. In addition to the valve or syringe interface and the large-volume injection method for GC, the only extra item needed is a control box controlling the timing of the many events in a run. The full procedure for an LC×GC×GC analysis is described in Table 2. Recently, commercial instrumentation for (stop-flow) LC×GC has been introduced by two instrument manufacturers. This equipment would also allow LC×GC×GC. An aspect that is neglected here is data processing and quantification. These operations are of course highly relevant for all practical users. Good solutions are available for comprehensive two-dimensional chromatography. For higher dimensional comprehensive systems, far more manual processing is still generally needed. In-depth discussions of signal processing and data handling are presented in other chapters in this book (Chapters 4 and 5).

4. APPLICATIONS OF LC×GC×GC AND LC–GC×GC

4.1 Additional dimensions prior to GC×GC

GC×GC has by now been adopted by a large number of laboratories for a wide range of applications. For an overview of the current GC×GC applications, the reader is referred to recent review articles [23,24] and the following chapters in this book. The significantly enhanced peak capacity of the method has resulted in strong acceptance of the technique in areas where samples containing numerous peaks have to be separated. Typical areas in which this is the case are flavour analysis (Chapter 9) and environmental analysis (Chapter 11). In both cases, the availability of mass-spectrometric detection contributed to an even faster growth of the methodology. Group-type analysis remains a second important generic application area of GC×GC in the mineral oil industry (Chapter 7) as well as in edible-oil characterisation (Chapter 10). More recently, the applicability of GC×GC as an unsurpassed method for sample profiling was realised by researchers in the area of metabolomics, where the technique is now also being successfully implemented [25]. In addition to providing researchers in all areas with strongly enhanced separation capabilities, the success of comprehensive 2D separations also led to the awareness that further gain could be achieved by adding yet another dimension.

4.2 Coarse prefractionation methods

The simplest form of LC–GC×GC is the prefractionation of specific fractions from a sample using solid-phase extraction (SPE) or LC. An example of an even more basic prefractionation prior to GC×GC was published by Lu et al. [26,27]. In tandem in these two papers, the authors describe the use of liquid–liquid extraction at different pH values to fractionate smoke condensates from cigarette smoke in a basic and an acidic fraction. Despite the huge peak capacity of GC×GC, the authors found the resolving power of this technique still insufficient for identification of the many thousands of peaks in cigarette smoke condensates if no prefractionation was applied. This limitation of "standard" GC×GC had previously already been reported by Dallüge et al. [28], who further improved the resolving power of the technique by combining it with rapid scanning time-of-flight mass spectrometry (ToF MS) with automated deconvolution. Using this method, Dallüge et al. could detect 30,000 peaks in the large series of ^{2}D chromatograms. Since many compounds were detected in several ^{2}D runs, this "only" represented some 7500 compound names. After applying a filter based on spectral similarity, 520 compounds could be identified unambiguously. The final number of confidently identified peaks was rather low, mainly because of the severe overlapping of the numerous compounds. A suitable prefractionation of the sample was needed to reduce the complexity of the sample and obtain cleaner spectra. In their work, Lu et al. [26,27] elaborated on this by applying a simple acidic/basic prefractionation, that is, a prefractionation with a peak capacity of only around two. In the acidic fraction, over 1000 compounds were found with a signal-to-noise ratio better than 100. From these, 139 were identified to be organic acids. In addition, over 150 compounds were identified as phenolic compounds. In the basic fraction, 377 nitrogen-containing compounds, including 155 pyridine derivatives, 104 quinoline/isoquinoline derivatives, and 56 pyrazine derivatives could tentatively be identified. The prefractionation, albeit simple, clearly helped identify the compounds from the very complex mixture.

Several methods for sample prefractionation can be used to divide complex samples into a set of slightly less complex subsamples. In addition to liquid–liquid extraction at different pH values, other methods could be extraction with solvents of increasing polarities or, for example, SPE, multistep thermal desorption and, of course, LC prefractionation. In a recent study, we applied automated multistep, direct thermal desorption (DTD) and pyrolysis with GC×GC–ToF MS to yield a technique with the acronym DTD–GC×GC–ToF MS [29]. Fresh and aged olive oils were treated at four stepwise increasing desorption and pyrolysis temperatures (i.e., 70, 180, 250, and 600°C) to allow a prefractionation based on vapour pressure prior to the GC×GC–ToF MS analysis of the very complex mixture of volatiles. The number of compounds detected in the fractions ranged from about 90 when desorbing the sample at 70°C, to some 850 compounds when desorbing at 250°C. Not only did the multistep thermal treatment help bring the number of compounds down to a more realistic level for GC×GC analysis, but it also allowed us to draw conclusions on the release and formation of specific compounds during baking and cooking processes. The separation power of the combined multistep DTD–GC×GC–ToF MS was further improved by using advanced mass spectral filtering processes [30], allowing ordered structures of specific subclasses to be rapidly retrieved from the datasets. Again this study, where only a coarse prefractionation method was applied, confirmed the findings of the previously quoted authors: Despite the huge peak capacity of GC×GC, and even if combined with MS deconvolution methods, a simple prefractionation of the sample greatly simplifies the identification of unknowns.

4.3 LC dimensions for GC×GC

The above-mentioned prefractionation techniques are rather coarse, but high-resolution preseparation methods can also be used prior to GC×GC. The most logical method is LC. The first researchers to describe its use as a preseparation method were Edam et al. [12]. Starting from Giddings's principle of sample dimensionality, the authors show that for their samples, highly complex mineral oil fractions, the dimensionality of the samples is much higher than the number of separation dimensions that can possibly be achieved. Realizing this fact, the authors acknowledged that the sample cannot be fully and independently separated according to all dimensions, and so they selected a lower order separation system in a way that the most relevant sample dimensions would be matched with a few separation mechanisms. The three most important dimensions selected were aromaticity, polarity/ring structure, and volatility. Three separation dimensions were selected to separate the samples according to these three properties. In that way, a truly ordered separation according to these three different sample dimensions could be obtained. The first dimension used was an LC separation based on a standard method from the petrochemical area, method IP 391 [31]. In this separation dimension, a group-type separation of the sample into compound classes with equal numbers of aromatic rings is desired. The four desired classes are the compound groups containing zero, one, two, and more aromatic rings. A peak capacity of just four in the LC ^{1}D separation would suffice to obtain this separation. Indeed, the actual IP 391-method delivers a peak capacity just slightly above four, yet with a near perfect selectivity for the number of aromatic rings. The compound classes containing zero, one, and two double bonds elute just baseline separated in the forward-flush mode, whereas all compounds containing three or more double bonds are eluted as one fraction in back-flush. Systematic elution of ring-structure classes and separations based on boiling point (or volatility) could be achieved in subsequent GC×GC runs.

Because of the very high selectivity of the ^{1}D LC separation in the above example, with the compound classes being baseline separated, the LC×GC×GC setup could actually be operated in the multiple-fraction LC–(GC×GC) mode with just four fractions being transferred. When the authors realised that the LC separation was very selective, they also considered the use of LC×GC instead of multiple-fraction LC–(GC×GC). Unfortunately, the faster LC×GC route failed to resolve the tri-naphthenic compounds from the mono-aromatics, and LC–GC×GC was needed to obtain the required level of detail. This process of developing separation systems that match sample properties and separation modes to arrive at a system delivering the desired level of information is schematically shown in Figure 4. A final conclusion from the work was that LC–GC×GC allowed information to be provided on naphthenic and aromatic compounds, which remained partially convoluted in a GC×GC separation. The selectivity of the ^{1}D LC separation was the cause of the enhanced separation power of GC×GC. A peak capacity of only four appears to be sufficient to perform separations that cannot possibly be obtained even with a careful

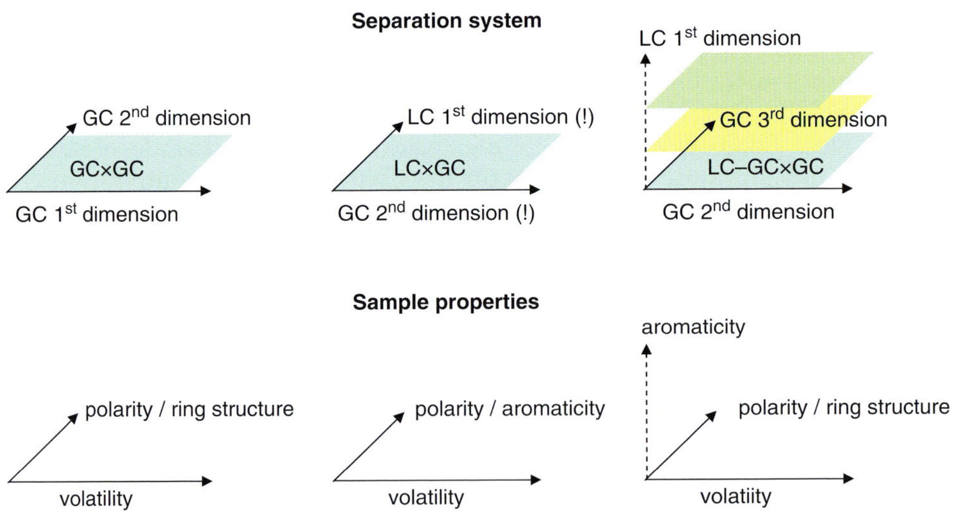

Figure 4 Separation dimensions in GC×GC, LC×GC, and LC–GC×GC of diesel fuels. Redrawn from [12].

selection of the GC×GC column set. But the price to pay is analysis time. Edam et al. had GC×GC runtimes of 100 min, which means that even the analysis of only four fractions takes 6 to 7 h!

The pioneering work by Edam et al. indicated one application area of LC–GC×GC: the characterisation of samples with a sample dimensionality exceeding the two dimensions that can be separated in GC×GC. The addition of an LC ^{1}D separation resolves the issue, but only at the expense of a very long total analysis time. As already discussed, Edam et al. also briefly considered the use of LC×GC. The technique demonstrated a marvelous selectivity but did not provide a sufficiently high system dimensionality. In fact, this was a confirmation of our earlier studies on LC class-type fractionation in mineral oil analysis [17]. Clearly, three dimensions are needed: LC and GC×GC. Fortunately, because of the excellent group selectivity of the ^{1}D LC dimension, the system can be operated in the LC–GC×GC mode rather than in the more demanding LC×GC×GC mode. An alternative route to ease the requirements to be imposed on the GC×GC separation is through the use of mass-spectrometric (MS) detection, as has been convincingly demonstrated by several authors. Particularly interesting is the work by Wang et al. [32] and Schoenmakers et al. [33] in which GC×GC is compared to GC–MS. Both author groups reach a similar conclusion: if a universal soft ionisation mode were to be made available for MS, GC–MS would offer much greater separation power for group-type analysis of mineral oil products than GC×GC. Compounds of different chemical groups that would co-elute in the GC dimension are likely to have different molecular weights. So if no fragmentation occurred and the signal intensity was proportional to the mass eluting, the MS information could be used to distinguish between groups and to quantify groups. However, because no ionisation mode meets these requirements, GC–MS is of little use when quantitative group-type information is needed. MS remains an important technique in mineral oil analysis, but only for its ability to identify compounds and compound classes.

In a series of studies, we incorporated an MS dimension in GC×GC and LC×GC separation studies [16,17,29,34]. The resulting systems were used for analysis of both edible and mineral oils. The main reason for incorporating MS as a detection step was the desire to make use of the selectivity offered by MS and to study how selectivity gained by selective-ion MS detection can be mutually traded against chromatographic selectivity. The conclusions from the work were rather clear. Specific compounds do show specific fragments, but if compounds are structurally similar, separation based solely on MS spectra is generally not perfect. The role of MS in such a system is to permit further identification of individual compounds or compound classes rather than to provide quantitative information on the compound groups. Thus, MS is a helpful route to study whether separation between the relevant compound groups in the system is acceptable. In that sense, MS helps in developing multidimensional separation systems that eventually will be using flame-ionisation detection for quantification. An example of the selectivity offered by MS is shown in Figure 5. This figure shows the GC×GC specific mass chromatograms of the saturates peaks from the LC–GC×GC–ToF MS separation of a diesel product. The characteristic mass

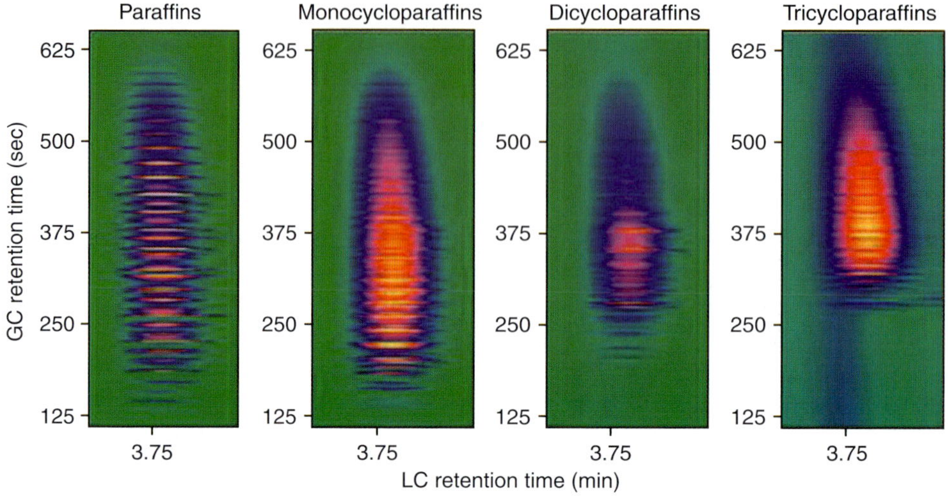

Figure 5 GC×GC mass chromatograms of various saturates classes after an LC fractionation [17].

Table 3 Characteristic mass fragments for middle-class distillates

Compound class	Mass fragments (*m/z*)
Group 1	
Paraffins	71 + 85
Monocycloparaffins	67 + 68 + 69 + 81 + 82 + 83 + 96 + 97
Dicycloparaffins	123/124 + 137/138 + ... + 249/250
Tricycloparaffins	149/150 + 163/164 + ... + 247/248
Group 2	
Alkylbenzenes	91/92 + 105/106 + ... + 175/176
Indanes and tetralines	103/104 + 117/118 + ... + 187/188
Indenes	115/116 + 129/130 + ... + 185/186
Groups 3	
Naphthalenes	141/142 + 155/156 + ... + 239/240
Acenaphthenes	153/154 + 167/168 + ... + 251/252
Acenaphthylenes	151/152 + 165/166 + ... + 249/250

fragments used to distinguish the specific compound groups are presented in Table 3.

In our comparative studies on GC×GC, LC×GC and LC×GC×GC, all with and without MS detection, we reached two interesting conclusions: First, if the proper LC columns and conditions are selected, LC can offer a remarkable and unsurpassed group selectivity. An example of this very strong selectivity is shown in Figure 6 where the group-type separations obtained with GC×GC and LC×GC×GC are compared. Defining the exact borders between the compound groups is difficult, if not impossible, from the GC×GC data only. Because of the

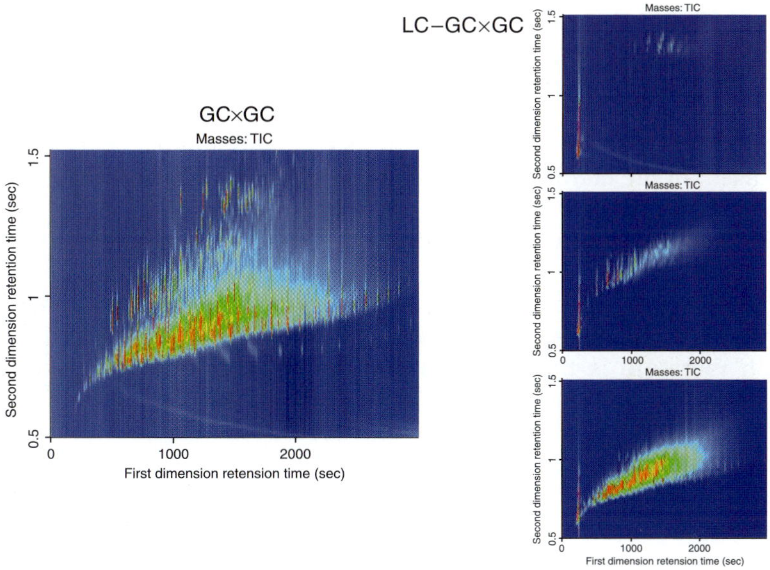

Figure 6 Comparison of GC×GC (left) and LC–GC×GC (right) in the separation of a mineral oil product sample. The LC–GC×GC chromatograms (bottom to top) represent saturates, mono-aromatics and di-aromatics.

much higher selectivity, separation of the three fractions in LC is easy. The peaks in LC were baseline separated. Indeed, GC×GC analysis of the three fractions collected from the LC shows no presence of compounds from the "wrong classes" in any of the fractions. Indeed, in the ^{1}D LC fraction, the peaks were baseline separated, allowing transfer of just three adjacent fractions with still no "wrong" compounds being present in any of the fractions. A second interesting remark that cannot be seen from Figure 6 but that was clear from the comprehensive LC×GC×GC experiments is that we obtain "reversed selectivities" for the ^{1}D LC and the ^{2}D/^{3}D GC×GC separations. Within a class of compounds, those molecules that elute relatively late in the ^{1}D LC run elute early in the GC×GC separation. In the LC system, the small molecules most likely have a slightly stronger polarity interaction with the stationary phase than the heavier molecules. For example, in the di-aromatics LC peak, this would mean that the nonsubstituted di-aromatics elute later than the heavily substituted di-aromatics. In the GC dimensions, where elution is largely based on size, the opposite occurs. This phenomenon had previously been reported in our work on LC×GC for triacylglycerides and resulted in very characteristic bands grouping the compounds [19].

The price to pay for the improved group-type fractionation of the mineral oil product in the LC–GC×GC run shown in Figure 6 is again time. The total time for the three GC×GC separations is evidently three times that of one GC×GC run. However, there are two ways to minimize the total analysis time. First, one could decide not to use LC–GC×GC for every sample, but instead use it just occasionally to determine the exact boundaries between the various classes in the GC×GC separation. Second, now that an LC prefractionation is performed, the requirements to be imposed on the GC×GC run are relieved. If the only purpose of the group-type separation is to determine the levels of the individual groups, a separation within a group is no longer necessary, and a much faster GC×GC run could be used. In the literature, GC×GC separations with a runtime of only 5 min have been shown [20]. Possibly even the GC×GC run can be replaced by a single GC analysis basically converting the LC–GC×GC method into a much simpler LC–GC run. It is interesting to think about the role of the GC dimension in the latter system. Basically, the GC run is performed only to allow quantification of the mass of material in the individual LC fractions. The final conclusion therefore is that what originally looked like a complex analytical problem in which the resolving power of GC×GC was insufficient in the end actually turned out to be more a quantification issue. For a simple group-type analysis of low-dimensionality samples, LC provides sufficient selectivity needed for the simple separation. The next GC dimension provides the sensitive and universal detector. The point here is that a serious risk exists that one will try to solve a selectivity issue by brute force and excessive dimensions. The analytical problem at hand in the analysis described in Figure 6 was a simple separation according to one sample dimension: the degree of aromaticity. The GC×GC system could not provide the required selectivity because the within-group separation obscured the between-group resolution.

In addition to mineral oil characterisation, analysis of edible oils is an interesting application area of comprehensive multidimensional systems incorporating GC×GC. A very important analytical method for characterising edible oils and fats is the so-called FAME analysis. These FAMEs are prepared from the actual ingredients of the fat or oil, the triacylglycerides (TAGs), by hydrolysis and methylation. The merits of GC×GC for FAME analysis have been discussed by numerous authors [35,36]. Less well known is the possibility of characterising the intact TAGs by comprehensive multidimensional chromatography. Because of the very high molecular weight of the TAGs, GC has only limited direct applicability, but LC offers very nice selectivities for TAG analysis. In a recent publication on the use of multidimensional comprehensive methods for TAG analysis, we identified four sample dimensions relevant for the detailed description of FAMEs, with three additional dimensions to fully characterise the TAGs from which the FAMEs are prepared. An overview of these sample dimensions is given in Table 4.

Using GC×GC, FAME analysis according to the first three dimensions is possible [35]. Again, there are several possibilities for adding the additional dimension. By using a very long, highly polar cyanopropyl ^{1}D column, information on the first four sample dimensions can be obtained using GC×GC only [37]. Including information on the sample dimensions five to seven, however, requires the addition of additional separation dimensions. Using a silver phase LC×GC×GC system with intermediate conversion, we could characterise TAGs according to six out of seven sample dimensions [13]. The silver phase LC (AgLC) dimension was used to separate TAGs according to the number of double

Table 4 Sample dimensions for edible oil and fat analysis (FAMEs and TAGs)

Fatty acid methyl esters
1 Chain length
2 Number of double bonds
3 Positions of the double bonds
4 *Cis* or *trans* orientation
Triacylglycerides
5 Total number of carbon atoms
6 Total number of double bonds
7 Stereospecific position on glycerol backbone

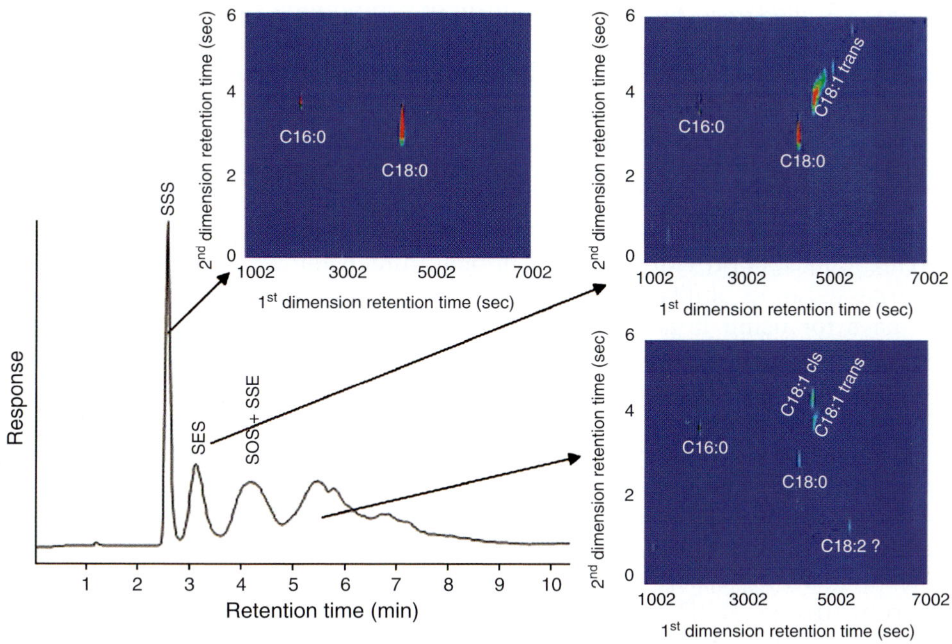

Figure 7 AgLC×GC×GC analysis of a hardened vegetable oil. For further details, see [13]. Fractions are converted from TAGs into FAMEs between the AgLC and GC×GC dimensions.

bonds present in the intact TAG molecule, with additional information on the presence of *trans* double bonds. Fractions collected from the AgLC separation were then converted into FAMEs using a rapid transesterification procedure [38]. For relatively simple vegetable oils, the separation power of the comprehensive 3D system was sufficient. However, for much more complex animal fats, for example, fish oils, it clearly was not. And of course there was one additional drawback: time. A small selection of the AgLC×GC×GC run is shown in Figure 7. Typically, some 30 fractions from the AgLC had to be analysed by

GC×GC. At a total GC×GC run time of about 1.5 to 2 h, this results in a total analysis time per sample of more than two days, which was not a very practical solution. Operating the system in the heart-cut AgLC–GC×GC mode is not possible. The selectivity of the AgLC is limited, which is not a problem when it is comprehensively interfaced to the GC×GC because the "reversed selectivity" of the AgLC and GC×GC again results in a good group-type separation. In heart-cut interfacing, however, group separation is not good enough.

5. CONCLUSIONS

In many of the samples that have to be analysed today, the number of compounds that should be determined significantly exceeds the peak capacity of a single GC analysis. Separation of all these compounds from one another and from interfering (matrix) analytes then becomes impossible. Comprehensive higher-dimensional systems can help to resolve this issue. In particular, systems incorporating comprehensive GC×GC dimensions are very powerful because of their very high peak capacity. Unfortunately, GC×GC×GC is not yet practically feasible, but LC×GC×GC is. For group-type separation, comprehensive LC×GC×GC or multiple heart-cut LC–GC×GC provide a significantly enhanced selectivity. In both approaches, mass spectrometry can be used as a specific detector providing an additional dimension of information. LC is unsurpassed in its ability to provide group-type selectivity, but time is the price one pays for using it as an additional dimension prior to GC×GC. Comprehensive 3D chromatography is very slow, as shown here for some theoretical settings, so care should be taken not to transfer more fractions than necessary. Ultimately, the comprehensive 3D systems can provide the strongly desired match between sample dimensionality and system dimensionality. Other objectives of comprehensive multidimensional chromatography are sample profiling and fingerprinting. The more detailed the characterization provided, the higher the chances of finding differences that correlate with properties of the sample. Unfortunately, the separation scientists here are ahead of the chemometricians, and comparisons of 3D or 4D datasets (if MS is included) are still far from easy (for more details on the use of chemometrics in comprehensive multidimensional chromatography, see Chapter 5). Anyhow, higher dimensionality systems are the road for the future. LC dimensions will most likely be involved to provide selectivity, GC×GC can add peak capacity, MS will add structural information, sensitivity and identification help. However, more work in different aspects of XC–GC×GC or XC×GC×GC is needed to fully reap the benefits of the higher-dimension comprehensive systems.

REFERENCES

1 J.C. Giddings, *J. Chromatogr. A*, 703 (1995) 3.
2 V. van Mispelaar, H.G. Janssen, A. Tas and P.J. Schoenmakers, *J. Chromatogr. A*, 1071 (2005) 229.
3 J. Blomberg, P.J. Schoenmakers and U.A.Th. Brinkman, *J. Chromatogr. A*, 972 (2002) 137.

4 G. Guiochon, N. Marchetti, K. Mriziq and R.A. Shalliker, *J. Chromatogr. A*, 1189 (2008) 109.
5 E.B. Ledford, Jr., C.A. Billesbach and Z. Zhu, *J. High Resolut. Chromatogr.*, 23 (2000) 205.
6 N.E. Watson, W.C. Siegler, J.C. Hoggard and R.E. Synovec, *Anal. Chem.*, 79 (2002) 8270.
7 P.J. Schoenmakers, G. Vivó-Truyols and W.M.C. Decrop, *J. Chromatogr. A*, 1120 (2006) 282.
8 A.W. Moore, Jr. and J.W. Jorgenson, *Anal. Chem.*, 67 (1995) 3456.
9 P.J. Schoenmakers, P. Marriott and J. Beens, *LC-GC Europe*, 16 (2003) 335.
10 D.A. Wolters, M.P. Washburn and J.R. Yates, *Anal. Chem.*, 73 (2001) 5683.
11 F. Bedani, W. Kok and H.G. Janssen, *J. Chromatogr. A*, 1133 (2006) 126.
12 R. Edam, J. Blomberg, H.G. Janssen and P.J. Schoenmakers, *J. Chromatogr. A*, 1086 (2005) 12.
13 S. de Koning, H.G. Janssen and U.A.Th. Brinkman, *LC-GC Europe*, 19 (2006) 590.
14 P. Korytár, H.G. Janssen, E. Matisová and U.A.Th. Brinkman, *Trends Anal. Chem.*, 21 (2002) 558.
15 E. Matisová, M. Šimeková, S. Hrouzková, P. Korytár and M. Dőmőtőrova, *J. Sep. Sci.*, 25 (2002) 1325.
16 S. de Koning, H.G. Janssen, M. van Deursen and U.A.Th. Brinkman, *J. Sep. Sci.*, 27 (2004) 397.
17 S. de Koning, H.G. Janssen and U.A.Th. Brinkman, *J. Chromatogr. A*, 1058 (2004) 217.
18 E.R. Kaal, G. Alkema, M. Kurano, M. Geissler and H.G. Janssen, *J. Chromatogr. A*, 1143 (2007) 182.
19 H.G. Janssen, W. Boers, H. Steenbergen, R. Horsten and E. Flöter, *J. Chromatogr. A*, 1000 (2003) 385.
20 M. Junge, S. Bieri, H. Huegel and P.J. Marriott, *Anal. Chem.*, 79 (2007) 4448.
21 K.K. Verma, A.J.H. Louter, A. Jain, E. Pocurull, J.J. Vreuls and U.A.Th. Brinkman, *Chromatographia*, 44 (1997) 372.
22 H.G. Mol, H.G. Janssen, C.A. Cramers and U.A.Th. Brinkman, *Trends Anal. Chem.*, 15 (1996) 206.
23 M. Adahchour, J. Beens, R.J.J. Vreuls and U.A.Th. Brinkman, *Trends Anal. Chem.*, 25 (2006) 438.
24 M. Adahchour, J. Beens, R.J.J. Vreuls and U.A.Th. Brinkman, *Trends Anal. Chem.*, 25 (2006) 821.
25 M.M. Koek, B. Muilwijk, M.J. van der Werf and T. Hankemeier, *Anal. Chem.*, 78 (2006) 1272.
26 X. Lu, J. Cai, H. Kong, M. Wu, R. Hua, M. Zhao, J. Liu and G. Xu, *Anal. Chem.*, 75 (2003) 4441.
27 X. Lu, M. Zhao, H. Kong, J. Cai, J. Wu, M. Wu, R. Hua, J. Liu and G. Xu, *J. Sep. Sci.*, 27 (2004) 101.
28 J. Dallüge, L.L.P. van Stee, X. Xu, J. Williams, J. Beens, R.J.J. Vreuls and U.A.Th. Brinkman, *J. Chromatogr. A*, 974 (2002) 169.
29 S. de Koning, E. Kaal, H.G. Janssen, C. van Platerink and U.A.Th. Brinkman, *J. Chromatogr. A*, 1186 (2007) 228.
30 W. Welthagen, J. Schnelle-Kreis and R. Zimmermann, *J. Chromatogr. A*, 1019 (2003) 233.
31 IP391, *Standard Methods for Analyzing and Testing of Petroleum Related Products*, Aromatic Hydrocarbon Types in Diesel Fuels and Distillates, The Institute of Petroleum, 1997, London, UK.
32 F.C.Y. Wang, K. Qian and L.A. Green, *Anal. Chem.*, 77 (2005) 2777.
33 P.J. Schoenmakers, J.L.M.M. Oomen, J. Blomberg, W. Genuit and G. van Velzen, *J. Chromatogr. A*, 892 (2000) 29.
34 H.G. Janssen, S. de Koning and U.A.Th. Brinkman, *Anal. Bioanal. Chem.*, 378 (2004) 1944.
35 L. Mondello, A. Casilli, P.Q. Tranchida, P. Dugo and G. Dugo, *J. Chromatogr. A*, 1019 (2003) 187.
36 H.J. de Geus, I. Aidos, J. de Boer, J.B. Luten and U.A.Th. Brinkman, *J. Chromatogr. A*, 910 (2001) 95.
37 H.M. Steenbergen. S. de Koning, and H.-G. Janssen, Poster presented at 10th International Symposium on Hyphenated Techniques (HTC-10), Bruges, Belgium, February 2008.
38 B.S.J. Jeffrey, *J. Am. Oil Chem. Soc.*, 68 (1991) 289.

PART II
Application Areas

CHAPTER 7

Petrochemistry

Jan Beens and **Jan Blomberg**

Contents

1. INTRODUCTION

Unlike most other samples that have to be analysed, petrochemical samples generally contain a limited number of classes of compounds, notably, alkanes, alkenes, cyclic alkanes, and aromatics with a different number of fused rings. The sample dimensionality therefore is rather limited (see below). The absolute number of compounds, however, is not only tremendous, but they all need to be analysed. In other words, the matrix itself is the goal of the analysis. So instead of looking for the needle in the haystack, the haystack itself needs to be characterised. In that respect, it is not too surprising that GC×GC from the conception of the technique on, has been developed using petrochemical samples as a subject [1,2].

Comprehensive Analytical Chemistry, Volume 55
ISSN: 0166-526X, DOI 10.1016/S0166-526X(09)05507-X

2. SAMPLE DIMENSIONALITY

In his paper on multidimensional analyses [3], Giddings explains sample dimensionality (s) as

- Intrinsic property of samples (other than the number m of components) that determines their amenability by multidimensional techniques.
- Number of independent variables requiring specification in order to identify the components of a sample.

In view of the large number of components that have to be separated in petrochemical samples, the use of multidimensional analyses is inevitable. But in the same paper Giddings also states:

> While the use of multiple dimensions in separation systems can create very high peak capacities, the effectiveness of the enhanced peak capacity in resolving large numbers of components depends strongly on whether the distribution of component peaks is ordered or disordered. Peak overlap is common in disordered distributions, even with a very high peak capacity. It is therefore of great importance to understand the origin of peak order/disorder in multidimensional separations and to address the question of whether any control can be exerted over observed levels of order and disorder and thus separation efficacy.

To understand the origin of peak order and disorder in multidimensional analyses, let us extend the term *sample dimensionality* in the petrochemical sample.

Sample dimensionality	s
Sample parameter	$p_1, p_2, \ldots, p_s$
Separator (or analysis) dimensionality	n

Since the set of s values of p uniquely establishes molecular identity, it also specifies its position (displacement) X_i along each of the separate axes x_i in the multidimensional chromatogram:

$$\text{Expression of a sample variable:} \quad \Delta X_i/\Delta p_i$$

If the behaviour of a sample with a weakly expressed variable closely resembles that of a sample of lower dimensionality, this may be thought of as an *apparent sample dimensionality* s', equal to the number of variables strongly enough to produce suitable resolution.

Required sample dimensionality s'': number of sample variables that must be determined for the purpose of the analysis (see also discussion in Chapter 6).

For the relationship between sample dimensionality and ordered or disordered chromatograms, see Table 1.

An example of an ordered separation in which the required sample dimensionality fits the separation dimensionality is shown in Figure 1. And since the combination of the individual (orthogonal) selectivities toward the

Table 1 Relationship between sample dimensionality and chromatogram separation pattern

s, s', n relationship	Separation pattern
$s > s' > n$	Disordered[a]
$s > s' = n$	Pseudo-ordered
$s = s' = n$	Ordered

[a]May be pseudo-ordered or ordered with secondary pattern when $s' = n+1$.

boiling-point (volatility or vapour pressure) in the first-dimension (^{1}D) and polarity in the second-dimension (^{2}D) results in highly ordered separations as far as aromaticity is concerned ($s'' = n = 2$); the four groups are nicely separated.

The target groups of the saturated, mono-, di-, and tri-aromatic species are separated by most medium-polar stationary phases applied in the second-dimension. However, within the group of saturates, no information on the (important) subclass of saturated cyclic hydrocarbons is obtained. Thus far, only phases that exhibit phenylic interactions have shown selectivity toward the saturated cyclic hydrocarbons (naphthenics).

An example of a separation with a phenylic interacting phase in the second-dimension BPX-50 is shown in Figure 2, which illustrates a separation of a purely saturated sample, where all three groups — alkanes (paraffins), mono-cyclic alkanes (mono-naphthenes), and di-cyclic alkanes (di-naphthenes) — are not only separated from each other, but also exhibit a nice separation within the groups according to the number of C-atoms in the molecules. Note that even within the roof-tiled subclasses of mono-naphthenes there is a separation between five- and six-membered rings (e.g., n-heptyl-cyclohexane apart from n-octyl-cyclopentane). Within all of these subclasses, the compounds with the highest degree of branching have the lowest retention time, in both dimensions. The degree of branching decreases with higher retention times up to the n-alkyl branched compounds. This phenomenon is a valuable tool for identifying and characterizing the sample.

A selected overview of reported papers of GC×GC on petrochemical products are summarised in Table 2. This table indicates the column combination, modulation system, and detection system used in the reported work.

Von Mühlen et al. presented a more comprehensive overview of the use of GC×GC for petrochemical samples and derived products and discussed the applications [4].

3. GROUP-TYPE SEPARATION

Because the majority of petrochemical analyses involve a group-type separation rather than a target analysis, a number of examples of this type of separation will be discussed.

Frysinger and Gaines nicely demonstrated the feasibility of GC×GC for forensic analyses by comparing the analysis of biomarkers in the remainder of

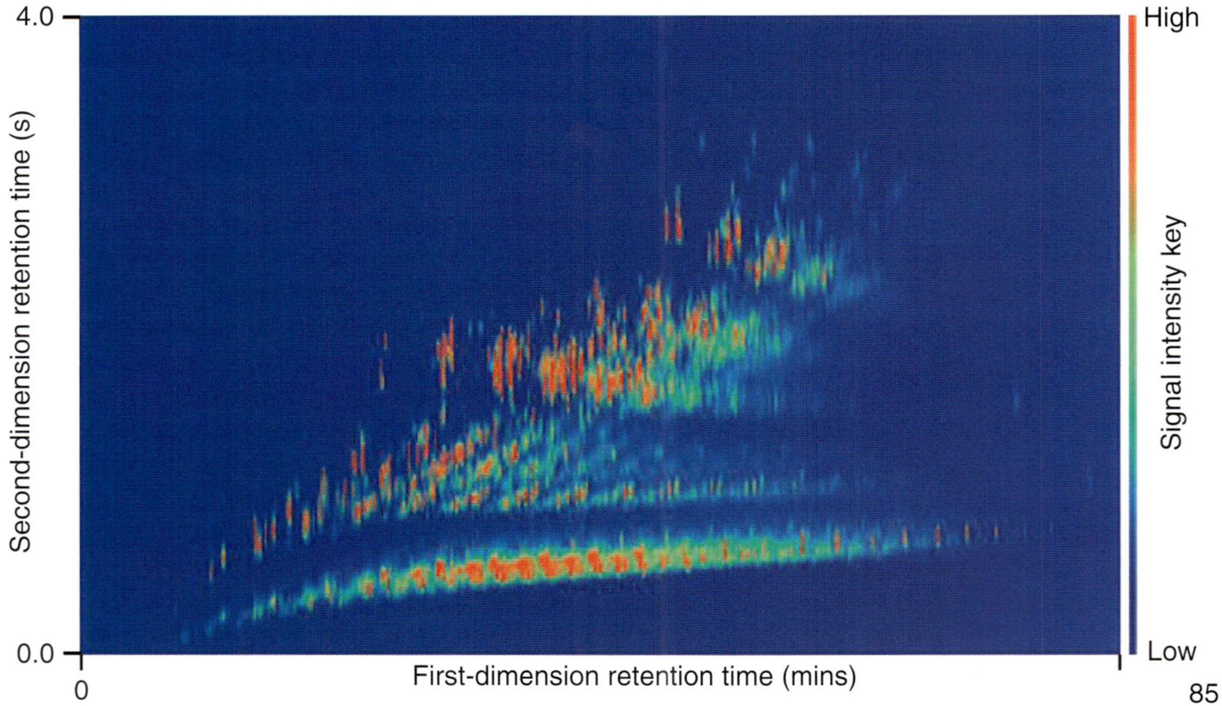

Figure 1 Example of an ordered GC×GC chromatogram of a light cycle oil. A nonpolar polysiloxane column and a medium-polar cyano-propyl polysiloxane column were used as first and second-dimension, respectively [5].

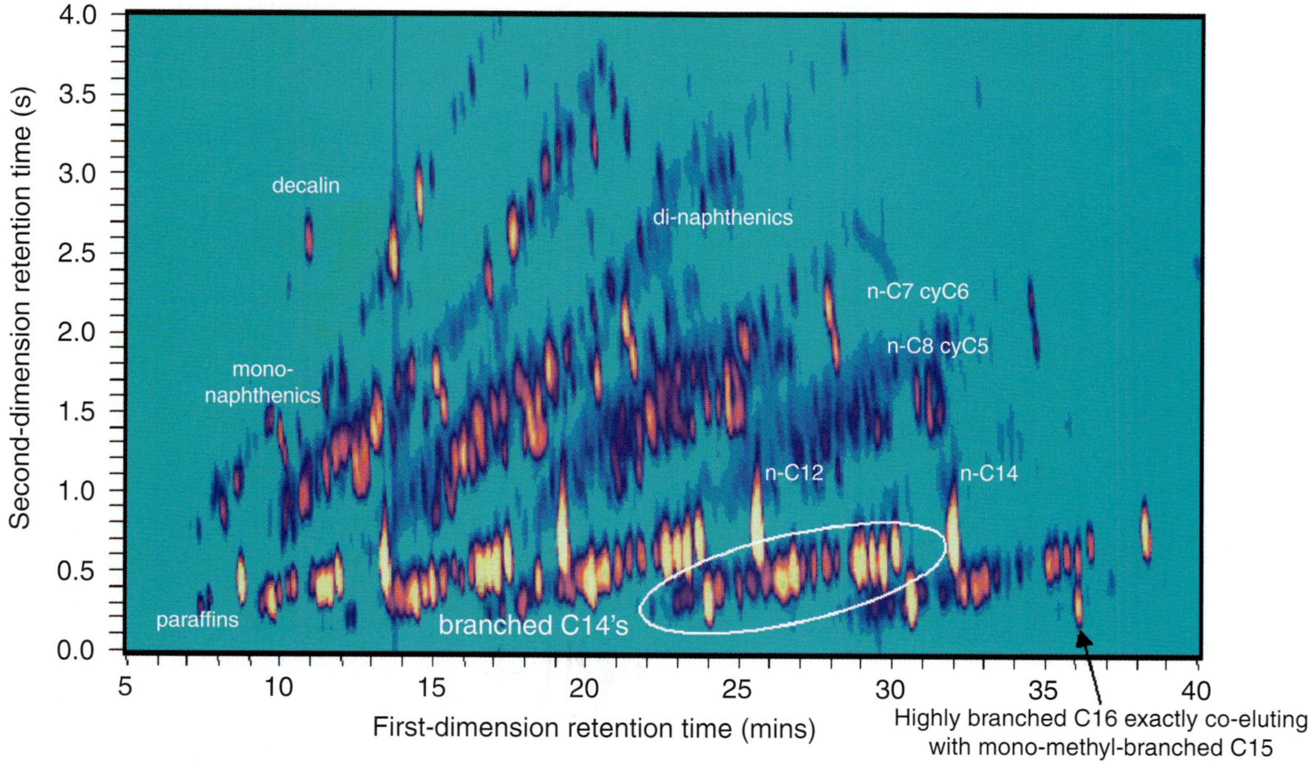

Figure 2 Separation of a purely saturated sample in its different groups [5].

Table 2 A selected overview of reported papers of GC×GC on petrochemical products

Subject matter	Modulator	Column combination (m×mm ID×μm d_f)	Detector	Reference
Hydrocarbons				
Quantitative comparison of 1D GC and GC×GC for analysis of petrochemical samples	Sweeper	SE-54 (2×0.1×3)× OV-1701 (0.5×0.1×0.14) DB-5MS (10×0.25×0.25)× OV-1701 (0.5×0.1×0.14)	FID	[5]
Qualitative comparison of 1D GC and GC×GC for analysis of petrochemical samples	2-jet CO_2	DB-1 (20×0.2×0.5)× BPX-50 (1.1×0.1×0.1) DB-1 (20×0.2×0.5)× BPX-70 (1.1×0.1×0.1) DB-1 (20×0.2×0.5)× CPWax (1.1×0.1×0.1)	SCD	[6]
Image background removal and computer language for analyte identification	4-jet LN_2	DB-1×CPWax (ND[a])	FID	[7,8]
Group-type analysis of middle distillates	Sweeper	DB-1 (20×0.2×0.5)× BPX-50 (1×0.1×0.1)	ToF MS	[9]
Quantification of naphthalenes in jet fuel with Tri-PLS and windowed rank-minimization retention-time alignment	Valve-based	AT-50 (10×0.32×0.25)× AT-Wax (2×0.18×0.2)	FID	[10]
Separation of saturated ring-type structures	2-jet CO_2	DB-1 (10×0.25×0.25)× BPX-50 (2×0.1×0.1)	FID	[5]
Optimization of analysis of complex volatile mixtures	2-jet LN_2	DB-5×DB-1701	ToF MS	[11]
PIONA analysis of heavy naphtha	2-jet CO_2	PIONA (10×0.2×0.5)× BPX-50 (0.8×0.1×0.1)	FID	[12]
Comparison of GC×GC and stop-flow GC×GC	2-jet LN_2	VF-1MS (30×0.25×1)× BP-20 (1.55×0.15×0.25)	FID	[13]

Characterization of petrochemical and related samples	Review	Review	FID	[4]
Group-type analysis of oxygenated compounds	2-jet LN_2	HP-5 (10×0.18×0.18)× DB-17 (1.9×0.1×0.1)	ToF MS	[14]
Tracking the weathering of an oil spill	2-jet LN_2	RTX-1 (7×0.1×0.4)× BPX-50 (0.82×0.1×0.1)	FID	[15]
Crude oil and bitumen (comparison of column phase configurations)	LMCS	SolgelWax (30×0.25×0.25)× BP-1 (1×0.1×0.1) BPX-5 (30×0.25×0.25)× BP-20 (0.8×0.1×0.1)	FID	[16]
Aliphatics/aromatics ratios in hydrocarbons	LMCS	HP-1 (25×0.2×0.33)× CP-Sil 19 CB (1.15×0.1×0.2)	PDD[b]	[17]
Alkene-based drilling fluids in crude oils	2-jet LN_2	RTX-1 (7.5×0.1×0.1)× BPX-50 (2×0.1×0.1)	FID	[18]
Coal liquids from a coal liquefaction process	2-jet LN_2	HP-5 (10×0.18×0.18)× DB-17 (1.9×0.1×0.1)	ToF MS	[19]
Characterization of middle distillates	2-jet CO_2	PIONA (10×0.2×0.5)× BPX-50 (0.8×0.1×0.1) BPX-50×(10×0.25×0.25)× DB-1 (0.8×0.1×0.1)	FID	[20]
S-containing compounds				
Correlation of GC×GC–AED and GC×GC–ToF MS data, with application in petrochemical analysis	2-jet CO_2	DB-1 (15×0.25×0.25)× BPX-50 (0.6×0.1×0.1)	ToF MS, AED	[21]
S-containing compounds in diesel oils	4-jet LN_2	VB-5 (6×0.18×3.5)× 00-17 (2×0.1×0.1) DB-petro (50×0.2×0.5)× OV-17 (1×0.1×0.1) 007-1 (4×0.1×3.5)× DB-17ht (2×0.1×0.1) 007-5MS (10×0.25×0.25)× DB-17ht (0.8×0.1×0.1) 007-1 (4×0.1×3.5)× DB-17ht (2×0.1×0.1)	SCD	[22]

Table 2 *(Continued)*

Subject matter	Modulator	Column combination ($m\times mm\ ID\times \mu m\ d_f$)	Detector	Reference
S-containing compound speciation in diesel oils	4-jet LN_2	SPB-5×BPX-50	SCD	[23]
S-containing compounds in crude oil	2-jet LN_2	VB-5 (6×0.18×3.5)× 007-17 (2×0.1×0.1)	SCD	[24]
Comparison of GC×GC–SCD to standard methods for speciation of S-containing compounds in middle distillates	2-jet CO_2	PIONA (10×0.2×0.5)× BPX-50 (0.8×0.1×0.1)	SCD	[25]
Classifying chromatographic applications, exemplified by GC×GC and multivariate analysis	ND[a]	ND[a]	SCD	[26]
Miscellaneous				
Nitrogen compounds in middle distillates	2-jet CO_2	SPB-5 (30×0.25×1)× BPX-50 (1×0.1×0.1) SPB-5 (30×0.25×1)× BPX-70 (1×0.1×0.2) SPB-5 (30×0.25×1)× Solgel Wax (1×0.1×0.2) SPB-5 (30×0.25×1)× Solgel Wax (1.6×0.1×0.2)	NCD	[12]
Large-volume injection for trace analysis of PAHs in diesel fuel	2-jet CO_2	RTX-5 (30×0.32×0.25)× BPX-50 (0.8×0.1×0.1)	FID	[27]

[a]ND, not defined.
[b]miniature PDD.

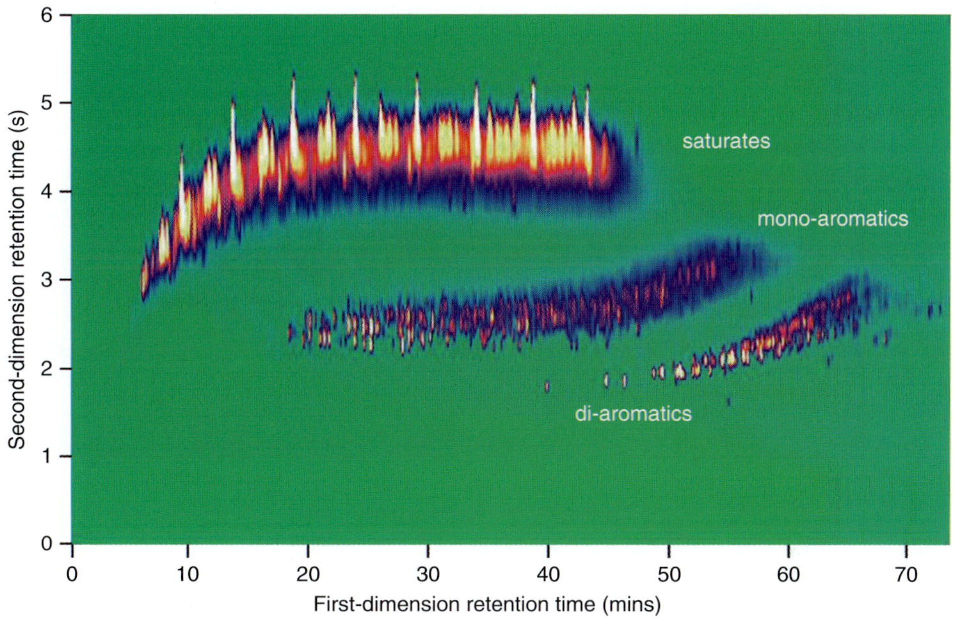

Figure 3 Colour plot of a "reversed-phase" GC×GC separation of a diesel oil [30]. Although the column combination (first polar, second nonpolar) is decidedly nonorthogonal, the plot still exhibits the well-known clustering of related chemical groups. But now compounds of the least polar group, the paraffins, have the longest second-dimension retention times. The more polarizable groups — the mono- and di-aromatics — now have lower second-dimension retention times.

a crude oil from a beach in Alaska with that of an oil tanker wreckage [28]. They also compared fire debris material with the group-type analyses of ignitable fluids [29] for arson analyses.

As regards the ultimate group-type separation in terms of no overlap at all between the different nonsaturated groups, Adahchour et al. [30] showed that a so-called reversed-phase column set is the solution (Figure 3). An excellent solution for the separation in the nonaromatics is given in the paper of Mao et al., where a silver-loaded HPLC column has been used to separate the saturates from nonsaturates prior to the GC×GC separations of the fractions [31]. In general, in view of the complexity of petrochemical samples, as already discussed in the Sample dimensionality section (i.e., a few dimensions in the sample that, however, do not differ too much in polarity or "polarizability"), a preseparation of some samples prior to the GC×GC separations of these fractions could greatly improve the applicability of the technique. For a more in-depth discussion of this matter, the reader is referred to Chapter 6.

The GC×GC separation of a mixture of C_{15} and C_{16} alkenes, which provides one unidentifiable hump of more or less separated peaks in 1D GC, where no distinction can be made between the compounds with 15 or 16 carbon atoms, is depicted in Figure 4 showing the group-type separation [5].

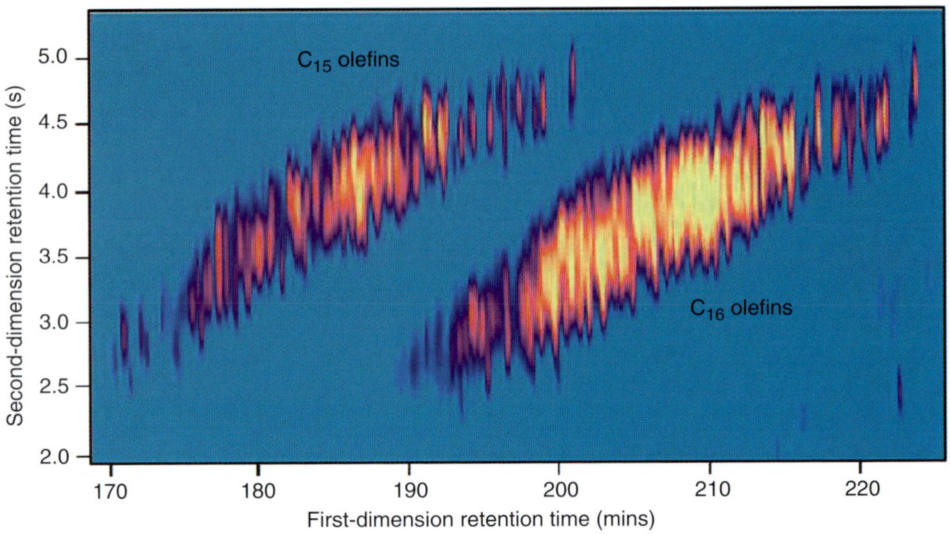

Figure 4 Colour plot of the GC×GC separation of C_{15}-C_{16} olefins [5].

An unusual GC×GC "high-resolution" separation has been demonstrated by Blomberg et al. in the six-hour separation of a diesel sample shown in Figure 5 [32]. The columns used for this separation were of regular length (^{1}D column, 10 m and ^{2}D column, 2 m), but the authors used a very slow temperature program of 0.5 °C/min. Note the modulation time of 20 s! Although the separation is impressive, the added selectivity for naphthenic species now causes some classes (di-naphthenes and mono-aromatics) to overlap. In this case, we are demanding too much of the separation system, which has only two separate dimensions: the apparent sample dimensionality now clearly exceeds the separation dimensionality. Edam et al. reported on adding another separation dimension, that is, an aromatic group-type separation by NPLC before the GC×GC analysis [33]. Three fractions were obtained from the NPLC: saturates, mono-aromatics, and di- and higher aromatics. These three fractions were subsequently separated by GC×GC, yielding chromatograms that did not contain any overlap (see also Figure 4 of Chapter 6).

Slater et al. used GC×GC–ToF MS to follow the biodegradation of a petroleum spill [34]. Natural abundance, molecular-level ^{14}C analysis was combined with GC×GC to investigate, in situ, the role of intrinsic biodegradation in the loss of petroleum hydrocarbons from the rocky, intertidal zone impacted by the Bouchard 120 oil spill. The analysis presented in Figure 6 indicated accelerated losses of n-alkane components of the residual hydrocarbons between day 40 and day 50 after the spill. ^{14}C analysis of bacterial phospholipid fatty acids from the impacted zone on day 44 showed that the polyunsaturated fatty acids attributed to the photoautotrophic component of the microbial community had the same dissolved ^{14}C (D^{14}C) as the local dissolved inorganic carbon (DIC), indicating that this DIC was their carbon source. (See also [9].)

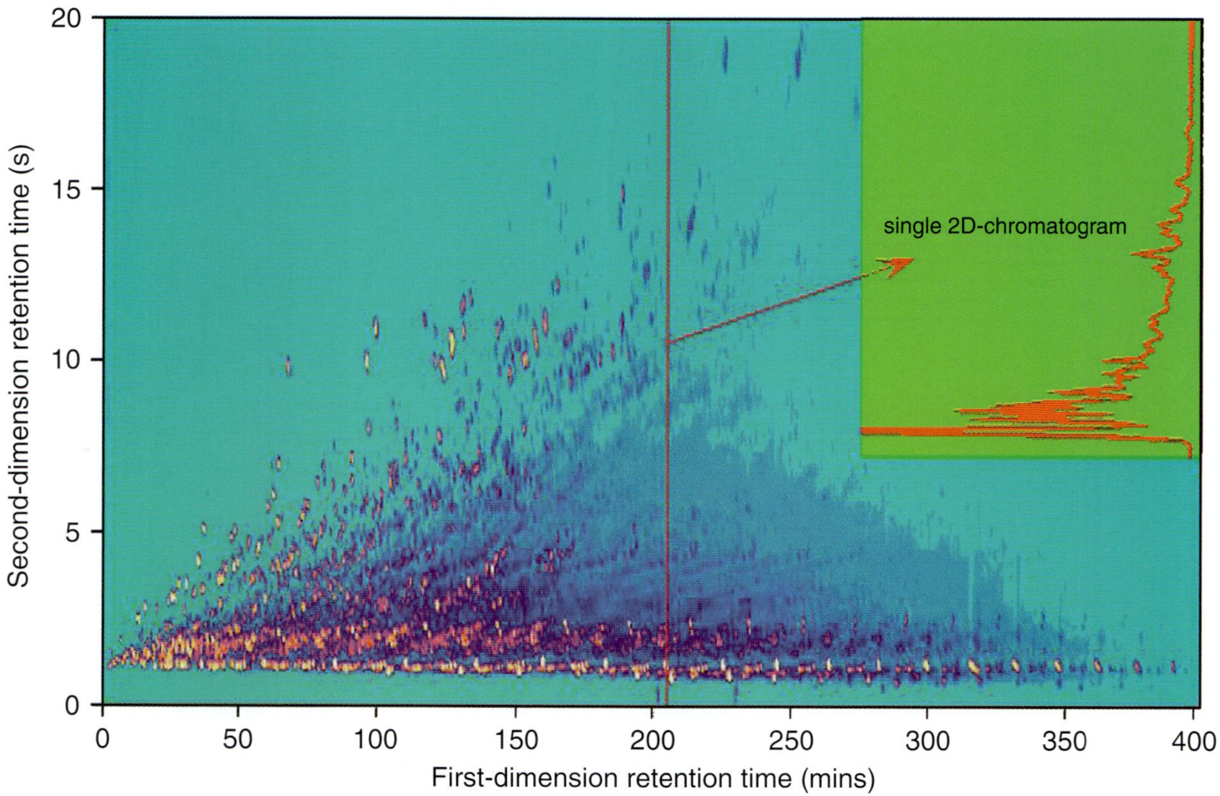

Figure 5 Colour plot of the GC×GC "high-resolution" separation of a diesel [32]. Because of the clustering, all the hydrocarbon classes can be identified, from n-C_7 through n-C_{28} and the branched alkanes in between, up to toluene through C_{20}-monoaromatics. From naphthalenes (second-dimension retention times 10 s) through the triaromatics in the top of the plot. The insert depicts one single second-dimension chromatogram, showing that in a single one-dimensional peak at least thirty compounds may co-elute.

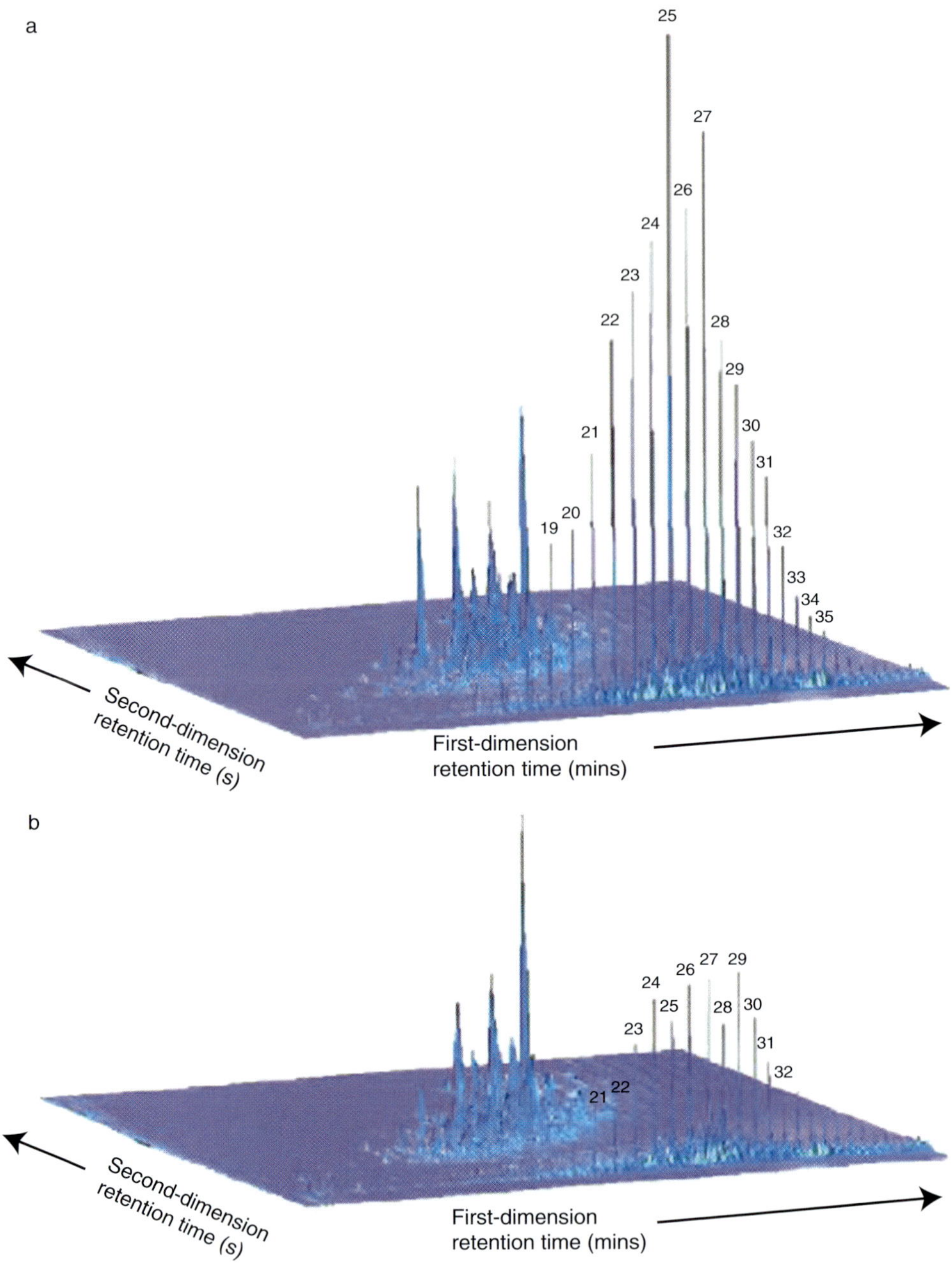

Figure 6 GC×GC 3 D-colour plots of the petroleum hydrocarbons at the site on day 40 (a) and day 50 (b) after the spill illustrating the extent of loss of the n-alkane envelope (n-alkanes are denoted by the carbon number at the top of the peaks) relative to the aromatic components of the petroleum hydrocarbons that appear behind the alkanes (i.e., at higher second-dimension retention times) [11].

4. TARGET ANALYSIS

Most of the target analyses performed in petrochemistry are analyses of heterocompounds, that is, analyses of oxygenates or sulphur- or nitrogen-containing compounds. In view of the extra complexity of these types of samples, 1D GC seldom provides a sufficient separation result.

Frysinger and Gaines modified the "Sweeper" heated thick-film modulator to enable the analysis of gasolines for oxygenates [35]. The sample consisted of a reformulated gasoline with added oxygenate standards (alcohols and ethers). A mixed ^{2}D phase, Wax/β-cyclodextrin was used to improve the separation of ether-containing oxygenates (see Figure 7). Quantitative applications with results directly comparable to those obtained with standard ASTM methods were derived.

Blomberg et al. separated the different types of sulphur compounds in hydro desulfurization (HDS) feed and product to follow the hydrogenation of the different types and species of sulphur-containing compounds (Figure 8) [36]. In order to enable maximum detector selectivity and linearity, the authors modified (the electronics of) an existing sulphur chemiluminescence detector, so that it could adequately follow the narrow peaks.

A year earlier Dallüge et al. had reported more or less the same separation [37]. In their study, these authors used a ToF MS to detect and identify all the different sulphur compounds, including 4,6-dimethyl-dibenzothiophene, one of the compounds that is very refractory to hydrogenation.

Similarly, Wang et al. [38] used a nitrogen chemiluminescence detector coupled to GC×GC for the analyses of nitrogen compounds (20 μg/g) in diesel. The indoles and carbazoles showed up as distinct bands, with a further roof-tile subdivision of $\tilde{C}_0$-C_6 alkyl-substituted indoles and $\tilde{C}_0$-C_5 alkyl-substituted carbazoles within these bands The basic reacting nitrogen compounds are particularly interesting because they are toxic to most hydrocracking catalysts.

5. QUANTIFICATION

GC×GC–FID can also be used for quantification [39]. In an early stage of the development of GC×GC, data for group-type analyses of heavy gas oils obtained using this technique were found to agree very well with those obtained with NPLC–GC, but the comprehensive technique provided more detailed results in a much shorter time. Also, two test mixtures containing different types of hydrocarbons (22 compounds each) exhibited a relative standard deviation of 0.9%. Since the FID response of hydrocarbons depends primarily on the mass of carbon in a molecule, and the gross formula of each group or subgroup is known, quantification can be performed even when no individual standards are available. The relative FID response to a hydrocarbon relates to the effective number of carbon atoms divided by the molecular weight. The slight deviation that occurs because of structural differences (i.e., aromatic ring or double bonds) can be corrected for by applying the so-called effective carbon-number concept [39].

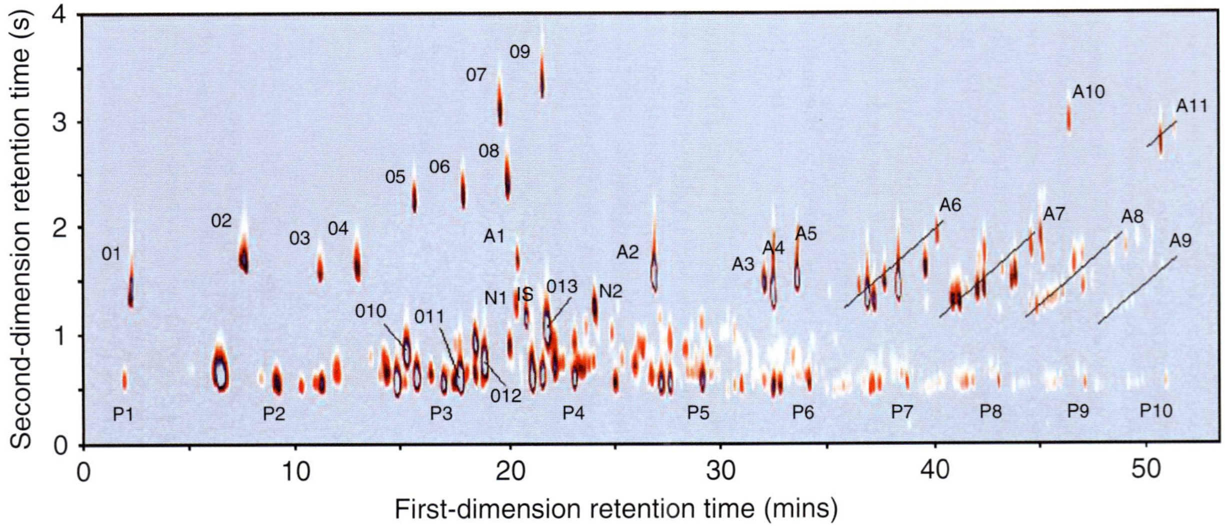

Figure 7 Colour plot of a GC×GC separation of reformulated gasoline [35]. O1. methanol, O2. ethanol, O3. isoproyl alcohol, O4. *tert*-butanol, O5. n-propanol, O6. *sec*-butanol, O7. iso butanol, O8. *tert*-pentanol, O9. n-butanol, O10. methyl *tert*-butyl ether, O11. diisopropyl ether, O12. ethyl *tert*-butyl ether, O13. *tert*-amyl methyl ether, N1. cyclo hexane, N2. methylcyclo hexane, IS. 1,2-dimethoxy ethane as internal standard. P1. n-C_4, P2. n-C_5, P3. n-C_6, P4. n-C_7, P5. n-C_8, P6. n-C_9, P7. n-C_{10}, P8. n-C_{11}, P9. n-C_{12}, P10. n-C_{13}. A1. benzene, A2. toluene, A3. ethyl benzene, A4. *para* + *meta* xylene, A5. *ortho* xylene, A6. C_9 *mono*-aromatics, A7. C_{10} *mono*-aromatics, A8. C_{11} *mono*-aromatics, A9. C_{12} *mono*-aromatics, A10. naphthalene, A11. methylnaphthalenes.

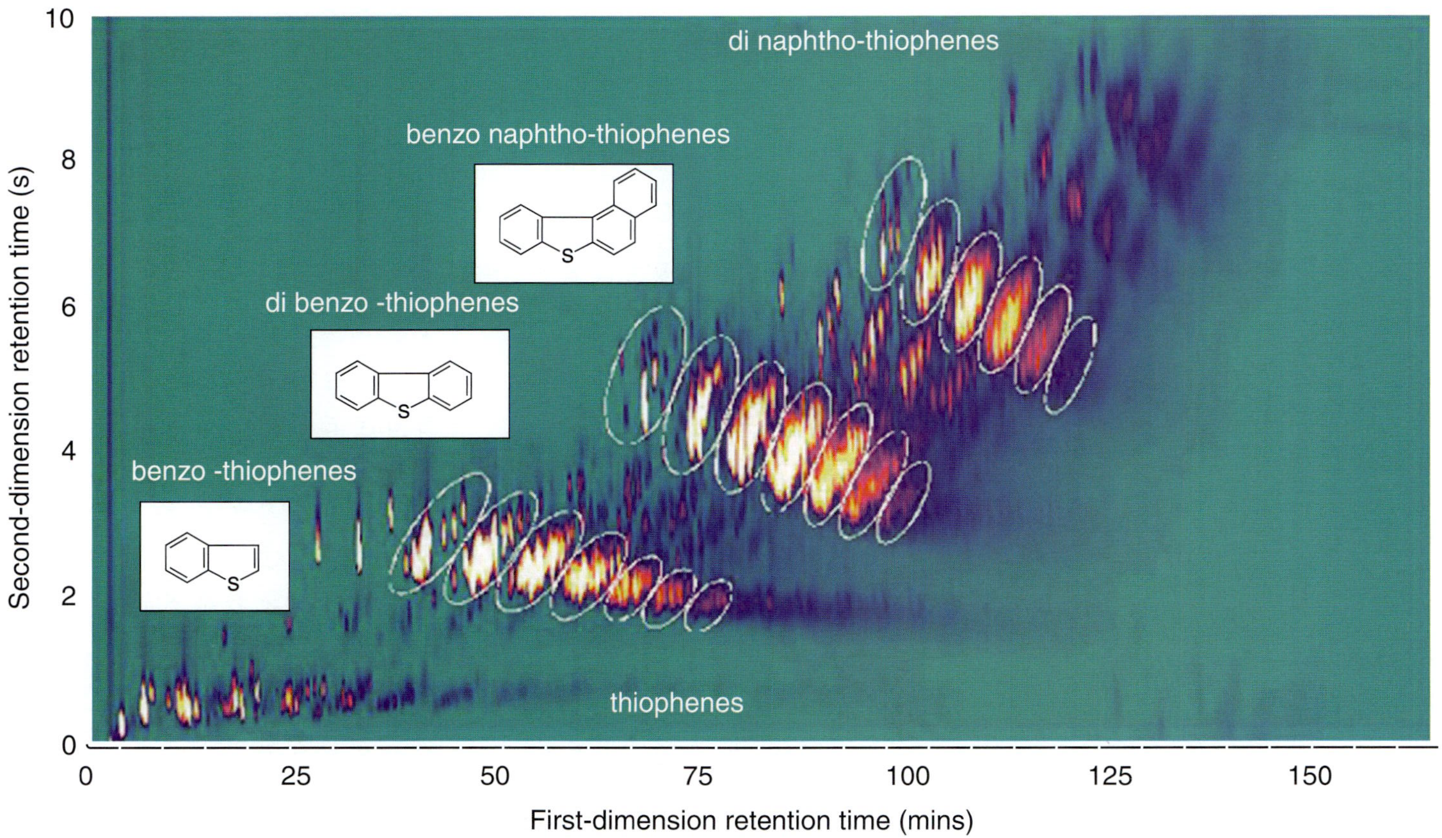

Figure 8 Colour plot of the GC×GC separation of a hydrodesulfurization feed [36]. The different groups indicated are the major sulphur compounds. In between these groups are the groups that also contain a saturated (hydrocarbon) ring in the molecule.

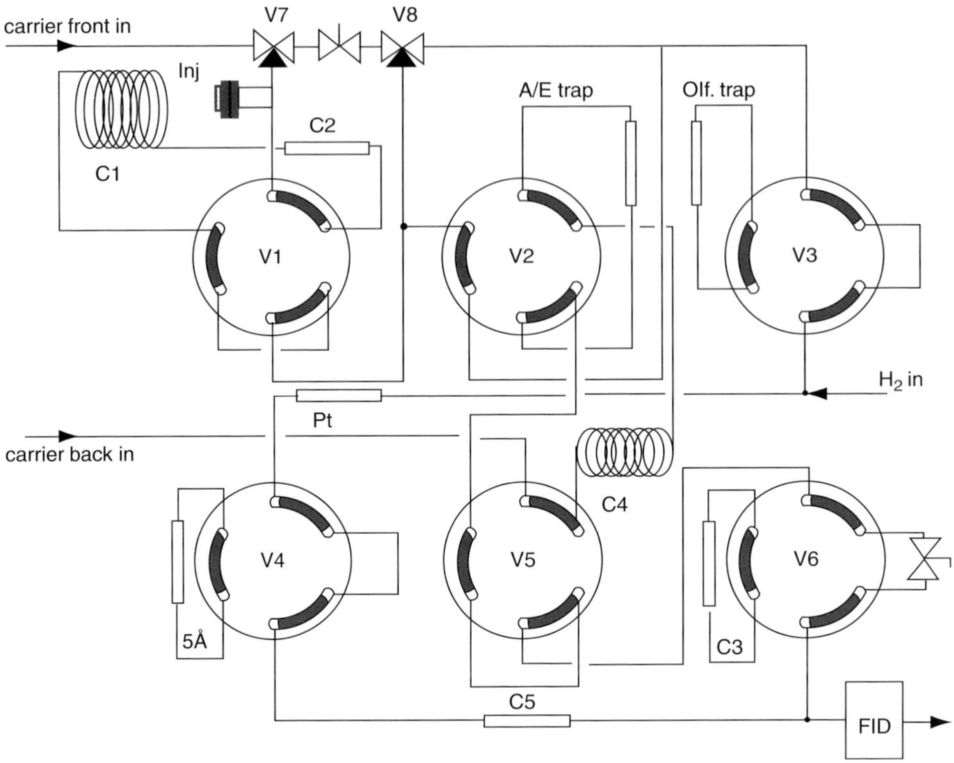

Figure 9 Schematic diagram of the Reformulyzer. Inj – split injector, C1 – polar capillary column, C2 – packed column to retain the alcohols, C3 – packed Porapak column for the separation of the oxygenates, C4 – capillary nonpolar column, C5 – packed 13X column, A/E trap – Tenax trap to retain the aromatics, Olf. trap – trap to retain the olefins, Pt – olefins hydrogenator, 5 Å – trap to retain the n-alkanes.

The collected data from detectors in GC×GC has to be reorganized in order to produce meaningful results, and several manufacturers produce software dedicated to this technique (see Chapter 4). Nevertheless, in view of the tremendous amount of data that is produced by a single analysis, some analysis results request a large effort of the user, but may also contain extended information. Several investigators have reported on extracting this extra information by using chemometrics as Multivariate Analyses (MVA), Generalised Rank Annihilation Method (GRAM), PARAFAC, and trilinear Partial Least Squares (tri-PLS) to improve or extend the results [40–43]. An overview of the use of these chemometric techniques is presented in [44], and a deeper discussion on the topic can be found in Chapter 5.

6. PIONA ANALYSES

Although it is not a GC×GC separation, PIONA analysers can be considered to produce comprehensive results of gasoline and naphtha-type samples with a

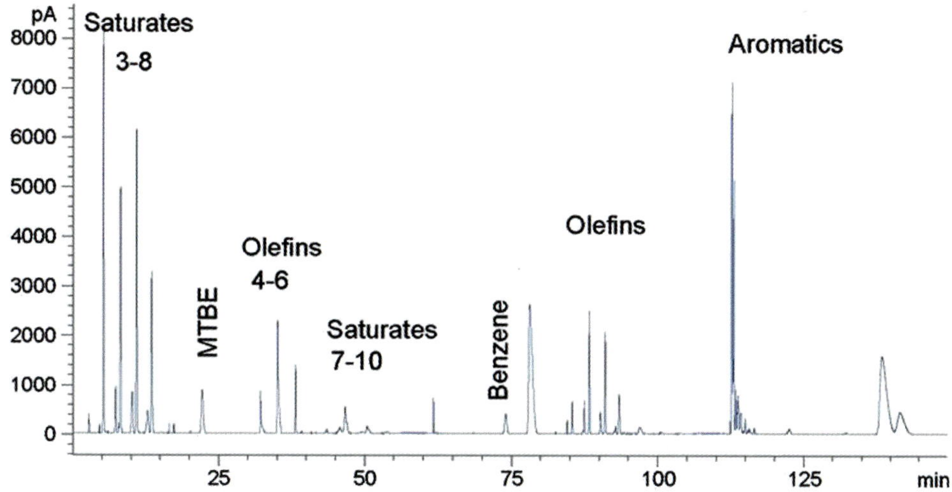

Figure 10 Chromatogram of the Reformulyzer.

multidimensional system. The system can be defined merely as a multiple heart-cut system and be designated as GC–GC–GC. Because these analysers are used in the majority of refinery laboratories, an explanation of their functions may be interesting for the reader. The system was developed in the early 1970s [45] as a PIONA-analyser (standing for Paraffins, Iso-paraffins, Olefins, Aromatics) and has subsequently been extended [46–48] for the analysis of reformulated gasolines, containing also blends of alkylates and oxygenates, to the "Reformulyzer" analyser as it exists now. As can be seen in Figure 9, the system contains four six-port switching valves, two separation columns, and four traps. The eventual separation produced is between branched, cyclic, and n-alkanes; branched, cyclic and n-alkenes, aromatics and oxygenates. All of these are separated by carbon number up to including C_{12}. The system is fully automated and, as can be seen from the chromatogram in Figure 10, the analysis takes about 150 min.

For a PIONA-type GC×GC analysis of middle distillates, Vendeuvre et al. developed an analysis system containing a column to retain olefinic compounds [9]. The eluents of this column were subsequently released from this column and separated further just after the separation of the rest of the sample. The final result produced by the system consists of saturates, olefins, and aromatics separated by carbon number.

REFERENCES

1 J. Blomberg, P.J. Schoenmakers and U.A.Th. Brinkman, *J. Chromatogr. A*, 972 (2002) 137.
2 G.S. Frysinger and R.B. Gaines, *J. Sep. Sci.*, 24 (2001) 87.
3 J.C. Giddings, *J. Chromatogr. A*, 703 (1995) 3.
4 C. von Mühlen, C.A. Zini, E.B. Caramão and P.J. Marriott, *J. Chromatogr. A*, 1105 (2006) 39.
5 J. Beens, J. Blomberg and P.J. Schoenmakers, *J. High Resolut. Chromatogr.*, 23 (2000) 182.
6 C. Vendeuvre, F. Bertoncini, L. Duval, J.L. Duplan, D. Thiébaut and M.C. Hennion, *J. Chromatogr. A*, 1056 (2004) 155.

7 S.E. Reichenbach, M.T. Ni, D.M. Zhang and E.B. Ledford, *J. Chromatogr. A*, 985 (2003) 47.
8 S.E. Reichenbach, V. Kottapalli, M. Ni and A. Visvanathan, *J. Chromatogr. A*, 1071 (2005) 263.
9 S. Penet, C. Vendeuvre, F. Bertoncini, R. Marchal and F. Monot, *Biodegradation*, 17 (2006) 577.
10 K.J. Johnson, B.J. Prazen, D.C. Young and R.E. Synovec, *J. Sep. Sci.*, 27 (2004) 410.
11 J.M.D. Dimandja, *Amer. Lab.*, 35 (2003) 42.
12 C. Vendeuvre, F. Bertoncini, D. Espinat, D. Thiébaut and M.C. Hennion, *J. Chromatogr. A*, 1090 (2005) 116.
13 J. Harynuk and T. Górecki, *J. Chromatogr. A*, 1105 (2006) 159.
14 J.F. Hamilton, A.C. Lewis, M. Millan, K.D. Bartle, A.A. Herod and R. Kandiyoti, *Energ. & Fuels*, 21 (2007) 286.
15 R.K. Nelson, B.M. Kile, D.L. Plata, S.P. Sylva, L. Xu, C.M. Reddy, R.B. Gaines, G.S. Frysinger and S.E. Reichenbach, *Environ. Forens.*, 7 (2006) 33.
16 T.C. Tran, G.A. Logan, E. Grosjean, J. Harynuk, D. Ryan and P. Marriott, *Org. Geochem.*, 37 (2006) 1190.
17 B.L. Winniford, K. Sun, J.F. Griffith and J.C. Luong, *J. Sep. Sci.*, 29 (2006) 2664.
18 Ch.M. Reddy, R.K. Nelson, S.P. Sylva, L. Xu, E.E. Peacock, B. Raghuraman and O.C. Mullins, *J. Chromatogr. A*, 1148 (2007) 100.
19 J.T. Scanlon and D.E. Willis, *J. Chromatogr. Sci.*, 23 (1985) 333.
20 C. Vendeuvre, R. Ruiz-Guerrero, F. Bertoncini, L. Duval, D. Thiébaut and M.C. Hennion, *J. Chromatogr. A*, 1086 (2005) 21.
21 L.L.P. van Stee, J. Beens, R.J.J. Vreuls and U.A.Th. Brinkman, *J. Chromatogr. A*, 1019 (2003) 89.
22 R. Hua, Y. Li, W. Liu, J. Zheng, H. Wei, J. Wang, X. Lu, H. Kong and G. Xu, *J. Chromatogr. A*, 1019 (2003) 101.
23 F.C.Y. Wang, W.K. Robbins, F.P. Di Sanzo and F.C. McElroy, *J. Chromatogr. Sci.*, 41 (2003) 519.
24 R. Hua, J. Wang, H. Kong, J. Liu, X. Lu and G. Xu, *J. Sep. Sci.*, 27 (2004) 691.
25 R. Ruiz-Guerrero, V. Colombe, Th. Didier, B. Fabrice and E. Didier, *J. Chromatogr. Sci.*, 44 (2006) 566.
26 V.G. van Mispelaar, H.G. Janssen, A.C. Tas and P.J. Schoenmakers, *J. Chromatogr. A*, 1071 (2005) 229.
27 D. Cavagnino, P. Magni, G. Zilioli and S. Trestianu, *J. Chromatogr. A*, 1019 (2003) 211.
28 G.S. Frysinger and R.B. Gaines, *J. Sep. Sci.*, 24 (2001) 87.
29 G.S. Frysinger and R.B. Gaines, *J. Forens. Sci.*, 47 (2002) 471.
30 M. Adahchour, J. Beens, R.J.J. Vreuls, A.M. Batenburg and U.A.Th. Brinkman, *J. Chromatogr. A*, 1054 (2004) 47.
31 D. Mao, H. van de Weghe, L. Diels, N. de Brucker, R. Lookman and G. Vanermen, *J. Chromatogr. A*, 1795 (2007) 919.
32 J. Blomberg, P.J. Schoenmakers and U.A.Th. Brinkman, *J. Chromatogr. A*, 972 (2002) 137.
33 R. Edam, J. Blomberg, H.G. Janssen and P.J. Schoenmakers, *J. Chromatogr. A*, 1086 (2005) 12.
34 G.F. Slater, R.K. Nelson, B.M. Kile and C.M. Reddy, *Org. Geochem.*, 37 (2006) 981.
35 G.S. Frysinger and R.B. Gaines, *J. High Resolut. Chrom.*, 23 (2000) 197.
36 J. Blomberg, T. Riemersma, M. van Zuijlen and H. Chaabani, *J. Chromatogr. A*, 1050 (2004) 77.
37 J. Dallüge, J. Beens and U.A.Th. Brinkman, *J. Chromatogr. A*, 1000 (2003) 69.
38 F.C.Y. Wang, W.K. Robinson, F.P. Di Sanzo and F.C. McElroy, *J. Chromatogr. Sci.*, 41 (2003) 519.
39 J. Beens, H. Boelens, R. Tijssen and J. Blomberg, *J. High Resolut. Chromatogr.*, 21 (1998) 47.
40 A.E. Sinha, B.J. Prazen, C.G. Fraga and R.E. Synovec, *J. Chromatogr. A*, 1019 (2003) 79.
41 B.J. Prazen, C.A. Bruckner, R.E. Synovec and B.R. Kowalski, *J. Microcol. Sep.*, 11 (1999) 97.
42 V.G. van Mispelaar, A.C. Tas, A.K. Smilde, P.J. Schoenmakers and A.C. van Asten, *J. Chromatogr. A*, 1019 (2003) 15.
43 B.J. Prazen, K.J. Johnson, A. Weber and R.E. Synovec, *Anal. Chem.*, 73 (2001) 5677.
44 A.E. Sinha, B.J. Prazen and R.E. Synovec, *Anal. Bioanal. Chem.*, 378 (2004) 1948.
45 H. Boer and P. van Arkel, *Chromatographia*, 4 (1971) 300.
46 H. Boer, P. van Arkel and W.J. Boersma, *Chromatographia*, 13 (1980) 500.
47 P. van Arkel, J. Beens, H. Spaans, D. Grutterink and R. Verbeek, *J. Chromatogr. Sci.*, 25 (1988) 141.
48 J. Beens, H.T. Feuerhelm, J.C. Fröhling, J. Watt and G. Schaatsbergen, *J. Chromatogr. Sci.*, 41 (2003) 564.

CHAPTER 8

Air and Aerosols

Tuulia Hyötyläinen and **Minna Kallio**

Contents

1. INTRODUCTION

Air is a complex mixture containing both inorganic and organic compounds, which are present in both the gas and particle phase. Typical atmospheric trace constituents are ozone, nitrogen oxides, sulphur oxides, carbon monoxide, volatile organic compounds (VOCs), and aerosol particles. Some atmospheric organic compounds occur entirely in the gas phase, whereas others occur as liquids or solids in aerosols. Partitioning between the gas and aerosol phases in the atmosphere depends on the liquid- or solid-phase vapour pressure of the compound, whether it occurs as a pure substance or as a mixture in the aerosol phase, and its water solubility when the aerosol phase is primarily aqueous. Most emissions of organic compounds to the atmosphere are gaseous.

VOCs, which are present mainly in the gas phase, play important roles in a range of environmental issues. These important roles involve their global greenhouse effect, their toxic or carcinogenic human health effects, their accumulation and persistence in the environment, and their role in secondary aerosol formation. The definition of VOC is often very diffuse. In physicochemical terms, the term *VOC* refers to those organic compounds that are present in the atmosphere

Comprehensive Analytical Chemistry, Volume 55
ISSN: 0166-526X, DOI 10.1016/S0166-526X(09)05508-1

as gases but that under normal conditions of temperature and pressure would be liquids or solids. By definition, a VOC has vapour pressure at ambient conditions (ca. 20°C) less than 101.3 kPa and greater than 0.13 kPa [1]. Generally, compounds in the atmosphere will be gas phase if their vapour pressure at ambient temperature is higher than 1.01×10^{-3} kPa and semivolatile if vapour pressures are between 1.01×10^{-3} kPa and 1.01×10^{-11} kPa and aerosols at vapour pressures below this [1,2]. Semivolatile organic compounds (SVOCs) are distributed between the vapour and the particle phase. As temperature decreases (or altitude increases), vapour pressure decreases and more compounds become aerosols.

The organic compounds emitted directly to atmosphere typically have high vapour pressures, but owing to chemical reactions in the atmosphere, secondary organic aerosol (SOA) is formed, and the vapour pressures of these typically more polar reaction products are lower. The vapour pressure decreases with increasing polarity, as oxidation proceeds first to carbonyls and esters. With further oxidation to alcohols and carboxylic acids, the vapour pressure decreases further because of hydrogen bonding. The partitioning favours the aerosol phase for alkanes with more than 20 carbon atoms and for dicarboxylic acids with three or more carbon atoms [2]. Figure 1 gives an overview of the distribution of organic compounds in atmosphere in relation to their carbon number.

The particle phase, alternatively referred to as particulate matter (PM), aerosols, or fine particles, plays an important role by scattering or absorbing solar radiation [3,4] and by acting as cloud condensation nuclei [5]. In addition, aerosols are also associated with damaging effects on human health [6–8]. It has recently been established that significant amounts of particulate matter are organic in nature, deriving from both primary and secondary sources. VOCs emitted to the atmosphere have the ability to react rapidly with NO_x in the presence of OH radicals to form ozone and photochemical smog. SOA is formed

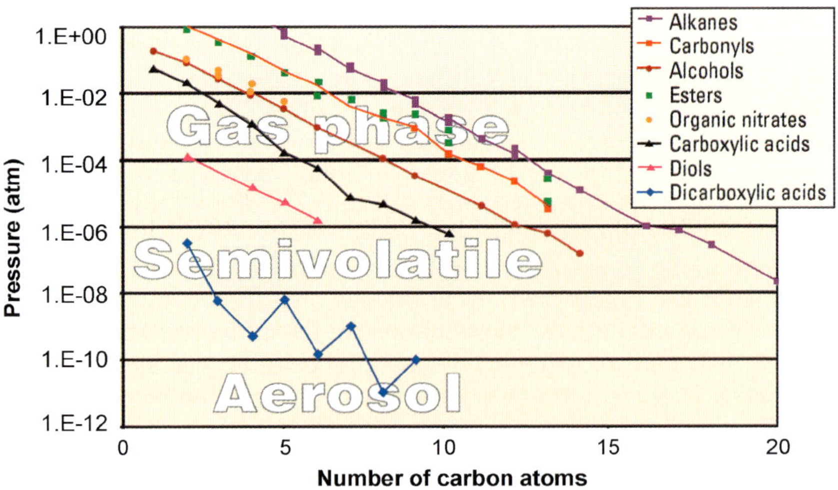

Figure 1 Vapour pressure of organic compounds at 298 K as a function of the number of carbon atoms and functional groups in the molecule [1].

in the atmosphere by photo-oxidation of organic species and the subsequent formation of lower vapour pressure products. Thus, atmospheric aerosol particles comprise a complex mixture of volatile and semivolatile inorganic and organic compounds as well as elemental carbon. Depending on the site and the amount of pollution, organic compounds may represent up to 70% of the total dry fine particle mass in the lower troposphere, and the number of organic compounds in aerosol particles may be several hundreds [9]. Organic compounds in aerosol particles show wide variation in water solubility, reactivity, polarity and volatility. In addition, many compounds are present only at trace levels (ng/m^3). Particulates are usually categorised with respect to their size, referred to as fractions. Table 1 shows typical fractions and their size ranges. Particles below 100 nm, down to the size of individual molecules, are classified as ultrafine particles, and further, particles with diameter below 50 nm are classified as nanoparticles. The nanoparticles and ultrafine particles have received special attention due to their potential effect on human health [6–8].

Although much emphasis has been placed on characterisation of the organic compounds in atmosphere, recent research suggests that a substantial fraction of both gas-phase and aerosol atmospheric organics have not been, or have very rarely been, determined. A major challenge in atmospheric chemistry research will be to elucidate the sources, structure, transformation and formation processes, and fate of the clearly ubiquitous, yet poorly constrained, organic atmospheric constituents.

Determining the chemical composition of air, especially the organic fraction, is challenged by the low concentrations of compounds, the complexity of the composition, and, on the other hand, the wide range of different compounds. Analysis of trace volatile and semivolatile species has been most commonly performed using gas chromatographic techniques. Of the huge number of instruments and methods reported, however, the majority of the methods describe at best the determination of less than some 60 individual compounds. One main reason for this low number is that in the analysis of complex atmospheric samples, one chromatographic step does not often provide sufficient separation efficiency, even when mass spectrometric detection is used. Two approaches can be used to deal with this problem. The traditional approach is to use multistep sample cleanup, such as fractionation before the chromatographic separation. However, sample preparation is not only time consuming, but it can also cause serious errors to the analysis thanks to contamination, sample loss and chemical

Table 1 Classification of particle fractions

Fraction	Size range
PM_{10}	$\leqslant 10\,\mu m$
$PM_{2.5}$	$\leqslant 2.5\,\mu m$
PM_1	$\leqslant 1\,\mu m$
Ultrafine (UFP or UP)	$\leqslant 0.1\,\mu m$
Nanoparticles	$\leqslant 0.05\,\mu m$

alteration of the sample. A more attractive solution is to use minimal sample preparation combined with a very efficient separation method, such as comprehensive two-dimensional chromatographic techniques. Comprehensive two-dimensional gas chromatography (GC×GC) is the most powerful of these techniques, and it is highly useful in the characterisation of the nonpolar fraction of the aerosols.

Over the past few years, GC×GC has been utilised in air and aerosol analysis. Particularly, GC×GC in combination with a fast acquisition mass spectrometer, for example, time-of-flight mass spectrometer (ToF MS) with a unit-mass resolution, provides extremely high analytical resolution with mass spectral information. Thus, the GC×GC–ToF MS is an exceptionally powerful tool in the air and aerosol analysis. It should be noted, however, that, as already mentioned in previous chapters, the system produces large amounts of data, and it is difficult to identify compounds from these datasets even with the structural nature and the mass spectral information. Therefore, automated procedures for data handling have been developed for identification purposes.

1.1 Sources and fate of organic compounds in atmosphere

Two main sources of organic compounds in atmosphere are human activities and biogenic emissions. Anthropogenic emissions arise mainly from motor vehicle exhausts, evaporation of petrol vapours from motor cars, industrial processes, oil refining, petrol storage and distribution, land filled wastes, food manufacture, and agriculture [10]. Natural biogenic processes include the emissions from plants, trees, wild animals, natural forest, and anaerobic processes in bogs and marshes [11]. Biomass burning is also a large source of VOCs worldwide [12] and leads to emissions of numerous VOCs, including many oxygenated species (organic acids, carbonyls, and multifunctional species), nitriles (HCN, CH_3CN), and aromatics (benzene, toluene) [13]. It should be noted that even though more effort has been put into determining pollutants in air and aerosols, the biogenic emissions dominate those from anthropogenic sources by one order of magnitude [14] with estimated global emission rates of 1150 Tg yr^{-1}. Near industrialized areas, however, anthropogenic sources are the most important contributors to ambient aerosol. The biogenic compounds released include isoprene (C_5H_8), monoterpenes ($C_{10}H_{16}$), sesquiterpenes ($C_{15}H_{24}$), and several oxygenated species [15,16]. In addition, many of these biogenic compounds are highly reactive and control photochemistry in many locations. The concentrations of VOCs in the gas phase vary greatly. In urban areas, concentrations of individual species, such as benzene, can be hundreds of $\mu g/m^3$, while in remote areas the concentrations are much lower and show a clear seasonal cycle [17].

The main sources of aerosols are of natural origin, including soil and sea dust, volcanoes, forest and grassland fires, as well as living vegetation. Human activities, such as the burning of fossil fuels in vehicles, power plants, and various industrial processes also generate significant amounts of aerosols. Aerosols can be either primary or secondary, the latter part deriving from the oxidation of VOCs. Urban aerosols are dominated by anthropogenic sources. Aerosol mass concentrations in urban areas range from a few tens of $\mu g/m^3$ to 1 mg/m^3 during air pollution episodes in heavily polluted cities in developing countries. The chemical composition of *fine* and *coarse* particles in the atmosphere differs greatly. Because little mass transfer takes place between fine and coarse particles, they exist together in the atmosphere as two chemically distinct aerosols. As a group, the fine particles are acidic and contain most of the sulphates, ammonium compounds, hydrocarbons, elemental carbon (soot), toxic metals, and water in the atmosphere [18]. The coarse particles are basic and contain mainly silicon, iron, calcium, and aluminium, and their oxides, as well as large sea salt particles and vegetation debris [18].

The fates and behaviour of the organic compounds emitted into the atmosphere are markedly dependent on the physical and chemical properties of the individual organic compound. VOCs are removed by photochemical and deposition processes on timescales varying from minutes to months [19,20]. Removal of VOCs from the atmosphere is mainly initiated by reaction with an OH radical, although reactions with O_3 and NO_3 may also be significant for certain species under specific conditions. Through oxidation, the VOCs are converted into more polar and hydrophilic forms, which make these photo-oxidation products more susceptible to wet removal by rain, formation of SOA, or dry deposition on surfaces.

The SVOCs attached to aerosol particles are removed from the atmosphere largely by hydroxyl radical oxidation followed by removal by settling, deposition, rainout, and washout. Larger particles (PM_{10}) tend to settle to the ground by gravity in a matter of hours, whereas the smallest particles ($< PM_1$) can stay in the atmosphere for weeks and are mostly removed by precipitation. The majority of the particulate mass from soil dust, volcano aerosol emissions, and anthropogenic direct emissions is likely to be large particles that fall out near the source. Lifetimes for SVOCs adsorbed onto aerosol particles are similar to those of aerosol particles themselves, generally about 5 to 10 days.

2. STATE OF THE ART

Different methods are used in sampling VOCs and SVOCs in gaseous and particulate phase. VOCs are typically collected in canisters, on adsorbents, or in cryostats. Whole-air sampling into canisters generally requires relatively simple equipment but is limited to those very volatile compounds that have high concentrations in the atmosphere. The most widely used method is based on sampling by pumping air through an adsorption tube packed with adsorbent(s). The most commonly used adsorbents are Tenax, Carbopack, and Carbosieve.

For the sampling of aerosols, filters are the most commonly used collection substrates, but a variety of films and foils have been used with impactors to collect size-resolved samples. Sampling times vary with ambient loadings, sampling rates, substrate blanks, and analytical sensitivities but typically vary from several hours in urban areas to a day or more under clean background conditions.

Sample pretreatment is dependent on the sampling method. VOCs collected on adsorbents are typically desorbed thermally directly into the GC. The filter and impactor samples typically require extraction into a suitable solvent, such as n-hexane or a mixture of n-hexane:acetone, although thermal desorption can be utilised also for filter samples.

Most atmospheric analyses based on GC×GC are qualitative in nature. The main aim has been to identify chemical patterns of the samples, and only a few quantitative applications have been reported. Table 2 lists current applications together with technical information on the GC×GC systems. Similar GC×GC methods are used for both types of samples, however, more volatile compounds are analysed from gas phase (from ca. C_6) and thus, modulators operating with liquid nitrogen are better suited because of their more efficient trapping of the most volatile compounds (see Chapter 2).

2.1 Air and VOCs

In an early study, Lewis et al. showed the potential of the GC×GC–FID system for characterisation of urban air samples. In this application, more than 500 chemical species of VOCs were separated from the gas phase samples [21]. Interestingly, the baseline noise that was observed in GC–MS was in GC×GC analysis determined to consist of numerous air pollutants. The presence of aliphatic, carbonyl, and aromatic bands in regions of monodimensional (1D) baseline noise was established. Prior to this study, many of these compounds were not known to exist in the atmosphere [21]. The results showed that because of the limited resolving power and sensitivity of conventional 1D GC methods, the contribution of some VOCs to urban air pollution may have been previously underestimated. Figure 2 shows an illustrative example of how traditional techniques fail to detect a large amount of compounds present in an air sample. In 1D GC separation, only ca. 15 peaks could be clearly determined above baseline, whereas in the GC×GC chromatogram around 120 peaks were detected in this part of the chromatogram. Discrimination between aliphatics (band 1), carbonyls (band 2), and aromatics (band 3) on the GC×GC separation indicated that co-elutions are occurring almost continuously on the single column [21,22]. In the volatility range covering C_6–C_{14}, 550 individual peaks were isolated in the same analytical run. This huge number of species was then further classified using retention behaviour, indicating around 100 multisubstituted aromatics and 50 carbonyls, along with many hundreds of aliphatic hydrocarbons. The same group has also studied mono-aromatic complexity in urban air samples [23]. Interestingly, comparison of urban air and gasoline vapour showed very similar mono-aromatic composition. In the air samples, 147 mono-aromatics were detected, and of these 130 compounds were found to be in gasoline as well. The distribution of C_3 and C_4 alkyl-substituted aromatic compounds was almost identical in these two types of samples. The group has also utilised GC×GC for measurements of photo-oxidation products from the reactions of alkyl-benzenes with hydroxyl radicals [24].

Table 2 Summary of the GC×GC applications for atmospheric analyses

Matrix	Analytes	Sample prep.	Column combination (m×mm ID×μm d_f)	Modulator	Detector	Reference
Rural air	reactive hydrocarbons	TD	DB-5 (30×0.25×1)× Carbowax (3×0.1×0.1)	- (Zoex)	FID	[26]
Rural air	isoprene and monoterpenes	TD	HP-5 (30×0.32×0.25)× BP50 (?×?×?)	Jet (LECO)	ToF MS	[25]
Air	VOCs	DTD	DB-5 (30×0.25×1)× Carbowax (1×0.1×0.1)	Four-jets (Zoex)	FID; ToF MS	[33,35]
Urban air	VOCs	TD	BP-1 (50×0.53×5)× BPX50 (2.2×0.15×0.2)	LMCS	FID	[21,39]
Rural aerosols	SVOCs, VOCs	SLE	HP-5 (20×0.25×0.25)× BGB-1701 (0.7×0.1×0.1)	Semi-rot. cryo	FID; ToF MS	[32]
Urban aerosols		DTD	BPX50 (30×0.25×0.25)× BPX1 (1×0.1×0.1)	Jet (LECO)	ToF MS	[38]
Urban aerosols	semi-VOCs	DTD	RTX-1MS (30×0.2×0.5)× SolgelWax (1×0.1×0.1)	Jet (LECO)	ToF MS	[28]
Urban aerosol	partially oxidised comp.	DTD	HP-5 (10×0.18×0.18)× DB-1701 (1.66×0.1×0.1)	Jet (LECO)	ToF MS	[27]
Urban aerosols	SVOCs	?	BPX5 (30×0.25×0.25)× BPX50 (1.5×0.1×0.1)	Four-jets (LECO)	ToF MS	[36]
Urban aerosols	PAHs and oxy-PAHs	SLE	ZB-5 (20×0.25×0.25)× BGB-1701 (1.0×0.1×0.1)	Semi-rot. cryo	FID; qMS	[31]
Roadside nanoparticles (29–58 nm)	SVOCs	DTD	BPX5 (30×0.25×0.25)× BPX50 (1.0×0.10×0.10)	Loop (Zoex)	ToF MS; qMS & NPD	[30]

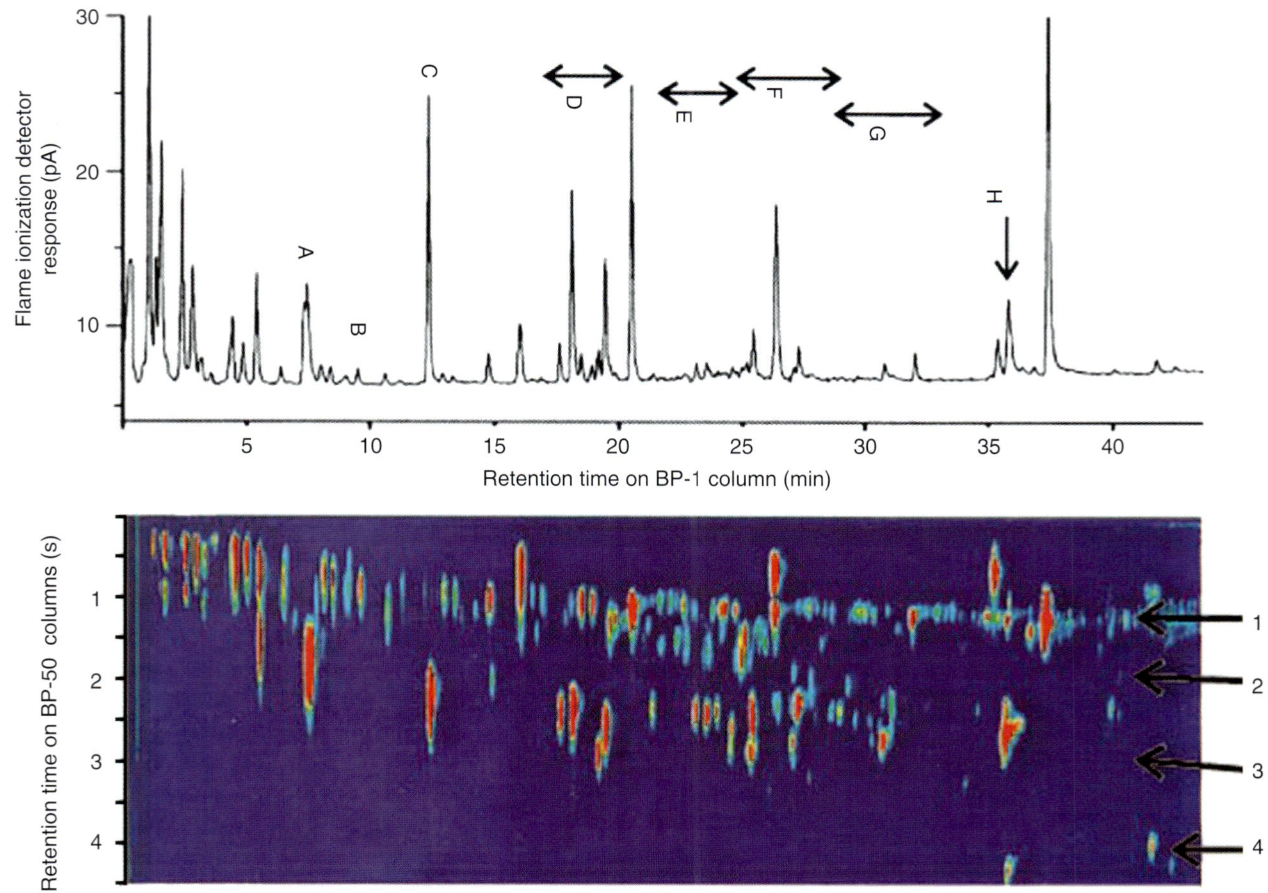

Figure 2 Comparison of sensitivity of 1D GC and GC×GC in the analysis of gaseous air samples. Upper trace, GC–FID. Lower trace, GC×GC–FID. A, C_3-benzenes; B, C_4-benzenes; C, C_5-benzenes. 1, aliphatic band; 2, carbonyl band; 3, aromatic band [21].

A similar direct thermal desorption-based system, DTD–GC×GC, has been used for study of isoprene and mono- and sesquiterpene emissions in rural sites (Benin) [25]. Air samples were collected from the forest canopy and from branch and cuvette enclosure systems by adsorption tubes filled with Carbopack B, Carbopack C, and Carbosieve SIII. The samples were thermally desorbed to the GC×GC–ToF MS system equipped with a liquid nitrogen cooled gas jet midpoint modulator. Cryo-focusing was necessary for the reconcentration of the desorbed compounds. Several monoterpenes, including α-pinene, camphene, β-pinene, p-cymene, and limonene, in addition to a number of unidentified species including monoterpene- and sesquiterpene-type structured compounds, were found in the ambient air samples collected. Definitive identification of the specific compounds detected was limited by the lack of authentic standards and low confidence in library matching due to structural similarities within this class of compounds. Also, the study identified traces of several compounds whose chromatographic retention, estimated from Kovat's retention indices (see also Chapter 3), and mass spectrum library identification were consistent with oxygenated terpenoid compounds such as camphor and menthol. The highest concentrations in ambient air of any species quantified were recorded for limonene (from tens of pptV to over 5000 pptV), with the largest variation observed during daylight hours and into the early evening.

GC×GC–FID with thermal desorption has also been utilised in the analysis of rural samples from Germany [26]. The air samples (3 L) were collected into Tenax TA/Carbograph I tubes and thermally desorbed into the GC×GC system. Nitrogen-cooled modulator was used in the system. In this study, GC×GC–FID measurements were compared with GC–MS measurements, and a total of 162 and 130 measurements were made with the GC–MS and the GC×GC–FID, respectively. Both anthropogenic compounds (n-hexane, n-heptane, n-octane, n-nonane, ethylbenzene, *p-/m*-xylene, and *o*-xylene) and biogenic compounds (α- and α-pinene, 3-carene, camphene, and eucalyptol) were included in the comparison. The study showed that there was generally good agreement between the results and that the slight differences could be attributed to the different sampling periods of the two instruments.

2.2 Particle phase

DTD–GC×GC–ToF MS has also been used for the analysis of organic compounds in ambient aerosol particles [27–29]. In this method, samples were collected on filters, and a piece of the filter was placed into an injector liner, which was put into the cold injector by autosampler instrumentation and thermally desorbed to the GC×GC system. The use of DTD as a sample introduction method simplifies the sample preparation, as no liquid extraction is needed.

Partially oxidized organic compounds associated with up to $PM_{2.5}$ aerosols have been studied in samples collected from London and Leeds (England) [27]. Samples were collected on quartz filter for 24 h. The nonselective DTD combined with GC×GC–ToF MS yielded an extremely complex chromatogram with over

10,000 organic compounds. The method resolved a volatile range equivalent to C_7 to C_{30} and a polarity range from alkanes to monosubstituted acids. Di-acids were seen to have relatively low recovery, and poly-acids were not released. Because of the complexity, ordered structures were not immediately visible, as shown in Figure 3. Oxygenated VOCs (oVOCs) found in a London aerosol were inventoried on the basis of ordered structures visualized by selecting suitable m/z ions. A total of 52 linear, 21 substituted, and 64 cyclic oVOCs were identified, and at least 100 oVOCs with longer chain lengths and increasing substitutions were observed for which insufficient information could be retrieved from the MS library. If compounds have a unique band pattern, they can be identified with some degree of confidence even though they do not have unique spectra. Many of the carbonyl species observed could be linked to gas phase aromatic hydrocarbon oxidation, and there was good agreement in terms of speciation between the urban samples analysed here and those degradation products observed in smog chamber experiments of aromatic oxidation. The presence of partially oxidised species such as linear chain aldehydes and ketones and cyclic products such as furanones suggests that species generated early in the oxidative process may undergo gas to particle partitioning despite their relatively high volatility.

DTD–GC×GC–ToF MS has also been used for the analysis of organic compounds in ambient aerosol particles collected in the city of Augsburg (Germany) [28]. Samples were collected with a sequential sampler on quartz fibre filters for 24 h. The method was compared to similar analysis with GC–ToF MS, and it was observed that GC×GC–ToF MS exhibited a tenfold increase in the number of peaks detected and produced highly structured chromatograms ideal for rapid screening purposes. More importantly, the comprehensive two-dimensional (2D) GC approach reduced the limitations of ToF MS deconvolution observed in 1D; this led to improved library matches and more confident analyte identification. In the GC×GC–ToF MS chromatograms (Figure 4) of the $PM_{2.5}$ samples, about 1500 compounds could be separated (partly by peak deconvolution), 200 compounds were quantified and semiquantified, and n-alkanes, n-alkan-2-ones, n-alkanoic acid methyl esters, acetic acid esters, n-alkanoic acid amides, nitriles, linear alkylbenzenes and 2-alkyl-toluenes, hopanes, PAHs, alkylated PAHs and oxidised PAHs, and several compounds that were not grouped in homologous rows or compound classes were determined.

A similar method has been used to characterise size-resolved particles, including the nanoparticles fraction with a diameter of 29–58 nm in roadside atmosphere [30]. The size-resolved samples were collected with a low-pressure impactor using ungreased aluminium foils as the collection substance. Several detection techniques were used in combination with TD–GC×GC, namely, high-resolution time-of-flight mass spectrometry (HRToF MS) and simultaneous detection with a nitrogen-phosphorous detector (NPD) and a quadrupole mass spectrometer (qMS). The aim of the simultaneous detection with HRToF MS, NPD and qMS detection systems were to obtain exact mass measurements and to elucidate the presence of nitrogen-containing compounds. The exact mass measurement served to increase selectivity and group-type separation of oxy-PAHs. Figure 5 shows the total ion chromatogram (TIC) of S1 (D_p 29–58 nm;

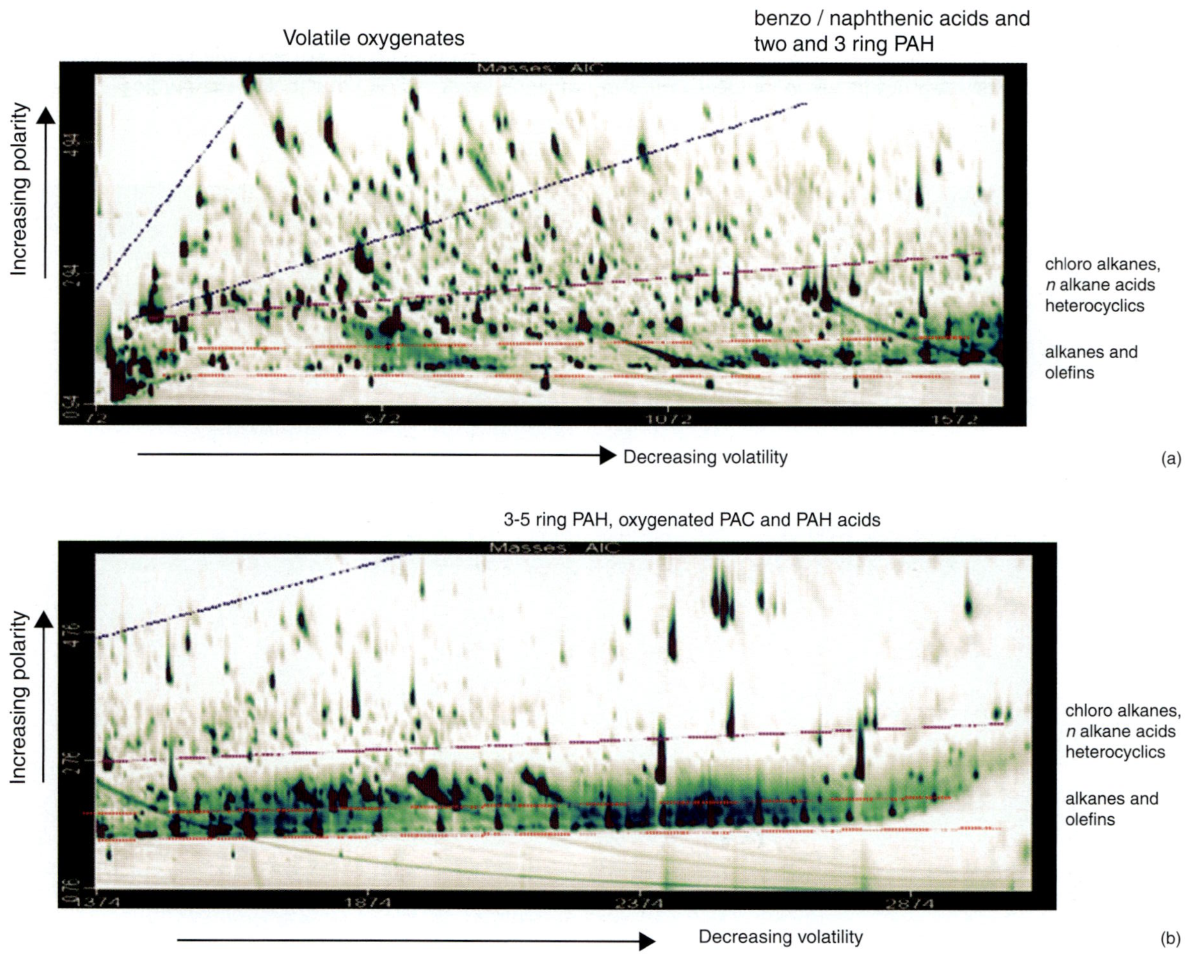

Figure 3 DTD–GC×GC–ToF MS analysis of $PM_{2.5}$ aerosols collected from London (England). (a) First half of separation – volatiles. (b) Two-dimensional total ion chromatogram. Second half of separation – semivolatiles [27].

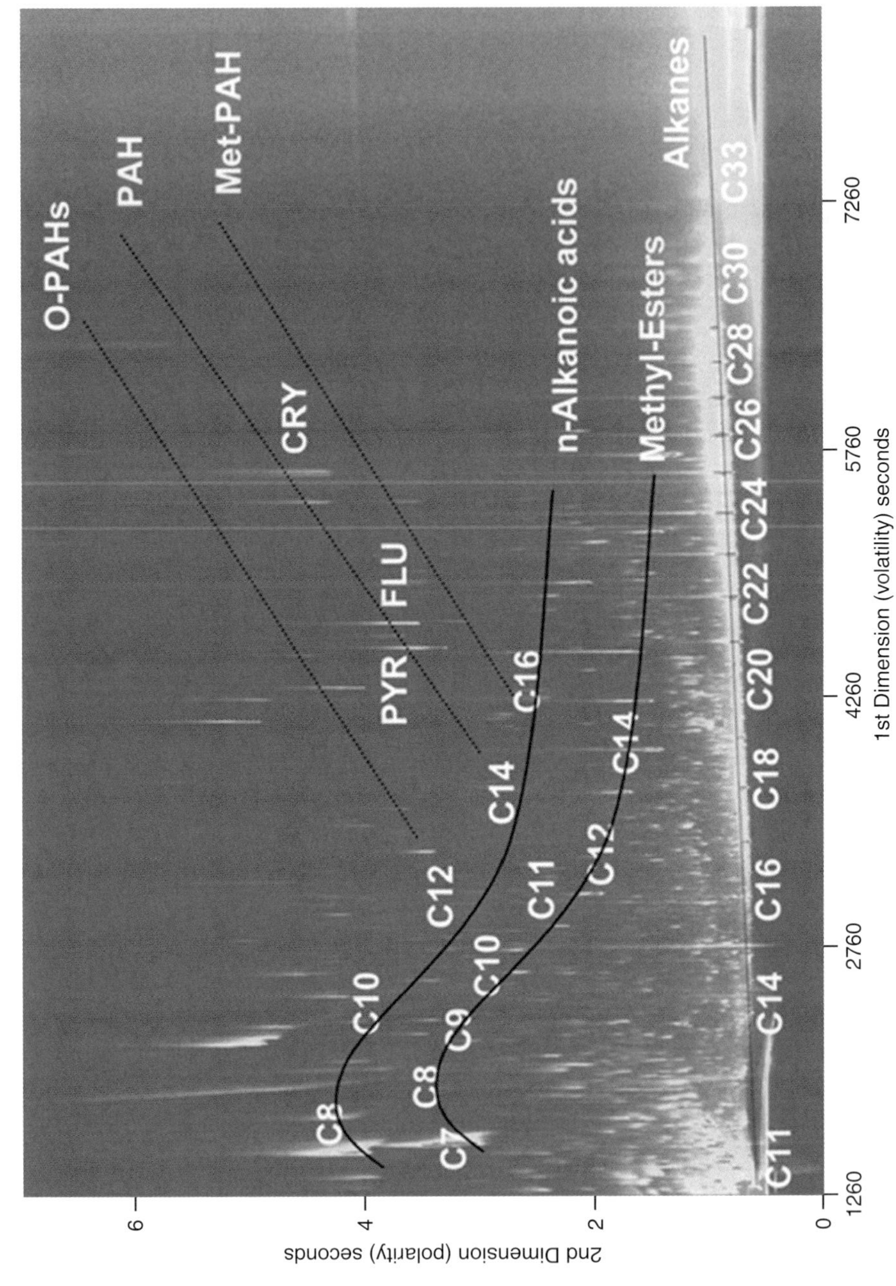

Figure 4 DTD–GC×GC–ToF MS analysis of the $PM_{2.5}$ samples collected in winter from an urban site in Germany. Homologous rows of compounds or compound classes are indicated by black lines [28].

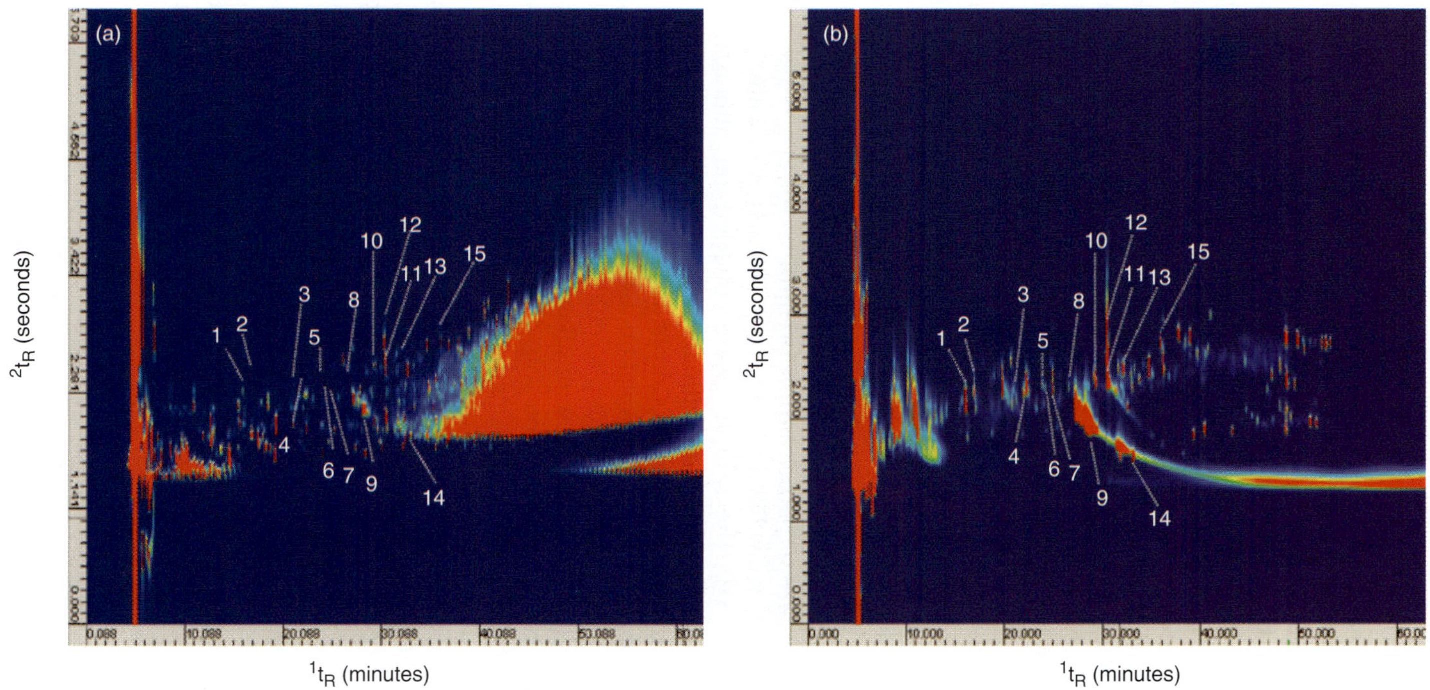

Figure 5 Comparison of the 2D chromatograms of size-resolved roadside particles (D_p 29–58 nm; 2.7 μg-PM) obtained by TD–GC×GC–HRToF MS and GC×GC–NPD/qMS. (a) total ion chromatogram and (b) NPD chromatogram. The marked peaks represent tentatively identified nitrogen-containing compounds [30].

2.7 μg-PM) obtained by the TD–GC×GC–HRToF MS. Using an automated NIST library search, various chemical classes (e.g. alkanes, alkenes, cycloalkanes, long chain carboxylic acids, aldehydes, ketones, substituted aromatics, PAHs and oxy-PAHs, heterocyclic compounds, and heterocyclic aromatic compounds) could be found in the 2D chromatogram.

Most of the candidate compounds presented at trace level, however, were still mismatched because there were large numbers of co-elution and band overlaps in the 2D TIC even with the GC×GC separation. Manual procedure was required and for tentative identification with mass chromatography with a 0.05-Da-wide window, the NIST library search, and a calculation of elemental composition was manually performed for 50 compounds. The simultaneous NPD detection was utilised in the search for nitrogen-containing compounds, and 15 more compounds could be identified. Seven of these were tentatively identified by TD–GC×GC–HRToF MS; the other eight were not identified because the reverse factor and/or the mass errors were outside the acceptance criteria selected by the authors. Quantitative analysis of selected PAHs in several size-resolved particles was also performed by use of the TD–GC×GC–qMS, with limited scan range (*m/z* 177–280) to achieve a data acquisition speed of 27 Hz. The method showed good linearity and high sensitivity (LOQ $<$10 pg) for most of the target PAHs. The concentration of PAHs was found to be considerably higher in nanoparticles than those of larger size of particles ($d_p > 102$ nm).

Urban air samples from Finland have been studied with both GC×GC–FID and GC×GC–qMS [31]. The samples were collected on filters, which were extracted in an ultrasonication bath with an n-hexane:acetone mixture. Column chromatography was used to remove aliphatic alkanes from the extract prior to the GC×GC analysis. The fractionation improved the sensitivity of the method because a larger amount of sample could be injected into the GC without the risk of overloading the column with alkanes, which were present in high quantities in extracts. The modulation was done with a laboratory-constructed two-stage cryogenic modulator. The method allowed the detection of approximately 1500 peaks and identification of target PAHs. The GC×GC–qMS was used with limited scan range (75 to 280 amu, with 18.94 scans/s) for compound identification. Altogether, 23 PAHs were identified from the samples, and of these ten target PAHs were quantified. The PAH concentration range found (0.5–5.5 ng/m^3) was comparable to results obtained by standard methods in other parts of Europe.

GC×GC–FID and GC×GC–ToF MS methods have also been applied to the determination of organic species in rural atmosphere [32]. A highly sensitive method was required because of the short sampling interval needed for study of this phenomenon (4 h) and the very low concentrations of organic species in this rural sampling site. High-volume sampling to quartz filters were used for sample collection, and dynamic ultrasonication-assisted extraction was used for sample pre-treatment. GC×GC–ToF MS with concentrating modulation improved the signal intensity, separation efficiency, and MS spectrum quality relative to normal GC–MS analysis and allowed the identification of several compounds that have not been identified previously in aerosol particles.

An example of the improvement of sensitivity and quality of spectral identification is shown in Figure 6 where, in GC–MS analysis, the small peak of alloaromadendrene was overlapped by the peaks of alkanes or alkenes (not shown in this figure) and undecanal. The GC–MS spectral match was low

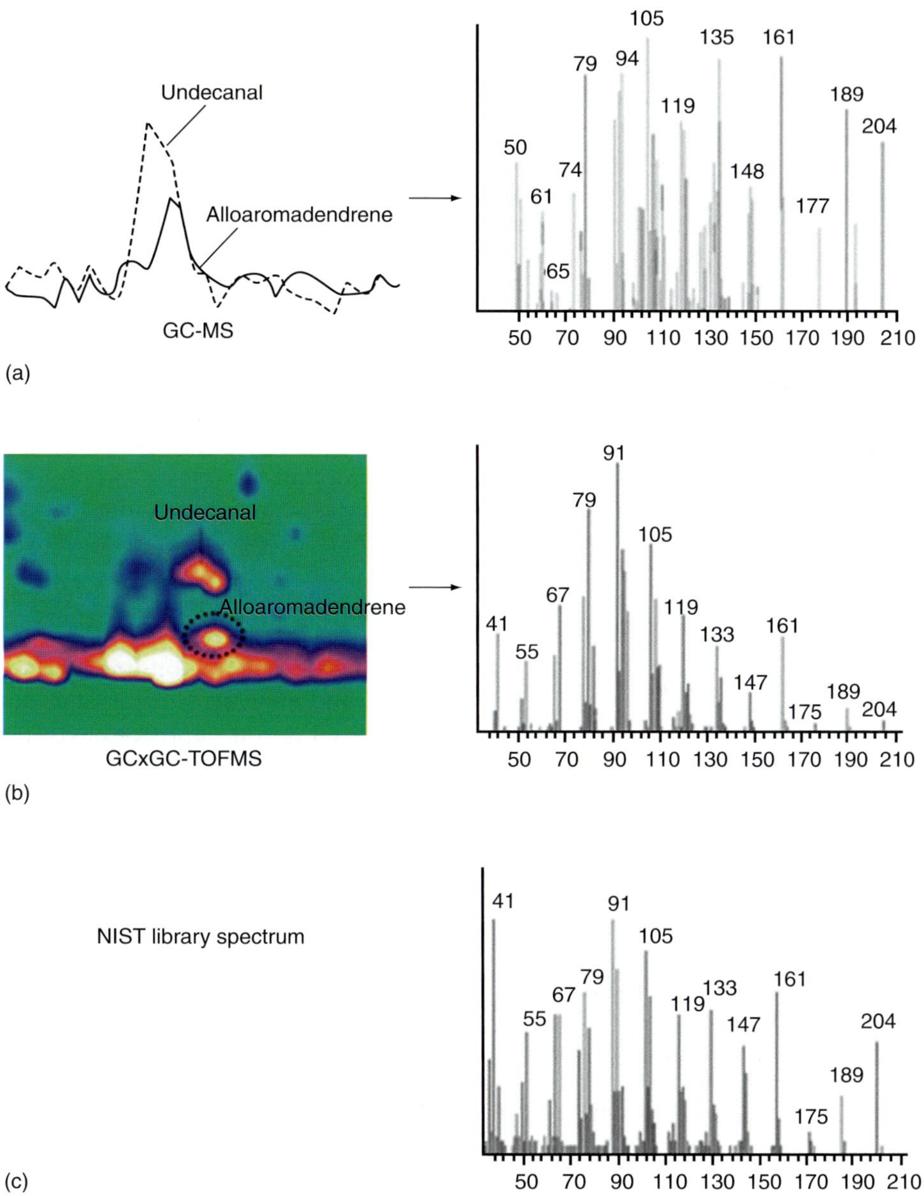

Figure 6 Comparison of GC–MS and GC×GC–ToF MS in the characterisation of rural aerosols. Spectral quality improvement of alloaromadendrene: extracted ion chromatogram and mass spectrum obtained by (a) GC–MS and (b) GC×GC–ToF MS and from (c) NIST library spectrum [32].

because of the many interfering fragments. In the GC×GC–TOF MS analysis, the concentrating modulation (Figure 6a) and the additional separation column in the second dimension improve the separation, and the compound is clearly separated from less polar components but, more importantly, from more polar undecanal. As a result, the spectral match improved substantially, and the intensity pattern was closer to the library spectrum. In addition to alkanes, over 50 compounds could be identified. Some volatile compounds present at trace levels could be identified with GC×GC–ToF MS, while these compounds could not be detected with GC–qMS analyses. For most compounds, the concentrations were at the low ng m^{-3} range ($<10\,ng\,m^{-3}$). An exception was the average concentration of pinonaldehyde, which was remarkably higher, $180\,ng\,m^{-3}$. The results showed that the oxidation products of α-pinene (pinonaldehyde, verbenone, and pinonic acid) comprised an important part of organic aerosols in forest atmosphere.

2.3 In situ analyses

In situ analyses are possible, if sampling is connected in-line with the GC×GC system. At present, two systems have been reported, one for gaseous samples and the other one for particles [33–35].

A schematic of an on-site system for collection of air samples is presented in Figure 7 [33]. In the on-line mode, ambient air is drawn through a link tube and collected directly onto the cold trap of the thermal desorber and analysed immediately after the sampling. The cold trap contains two beds of sorbent (Tenax TA and Carbograph). The sampling flow was ca. $50\,mL\,min^{-1}$, and the sampling time varied from 60 to 80 min.

The air samples were collected from Crete. In addition to the on-line measurements, several cartridge samples were collected and analysed in the laboratory with GC×GC–ToF MS. The on-site GC×GC–FID system was optimised to resolve C_7–C_{14} organic components. About 650 peaks were identified in the 2D contour plot, and of these, 235 of the identifications have been confirmed by an independent identification method, that is, the retention index comparison. Of the 235 confirmed compounds, 150 show up in the C_7-C_{14} range on the chromatogram from the in situ measurement. When comparing in situ and off-line analyses, it was noticed that fewer compounds were visible in the in situ measurements, mainly because the off-line cartridge samples were more concentrated than the on-line samples for the field measurements and the sensitivity of ToF MS was different from that of FID. In addition, sampling artefacts for off-line measurements can have caused some of the observed differences. More than half of the confirmed compounds were hydrocarbons, with alkanes, alkenes, and aromatic hydrocarbons contributing 31%, 10%, and 15%, respectively. Nearly one-third of the compounds were oxygenated species, including alcohols, aldehydes, ketones, esters, and oxygenated aromatic compounds. Other compounds, such as nitriles, halogenated hydrocarbons, and some miscellaneous species made only a small contribution to the total number of confirmed compounds.

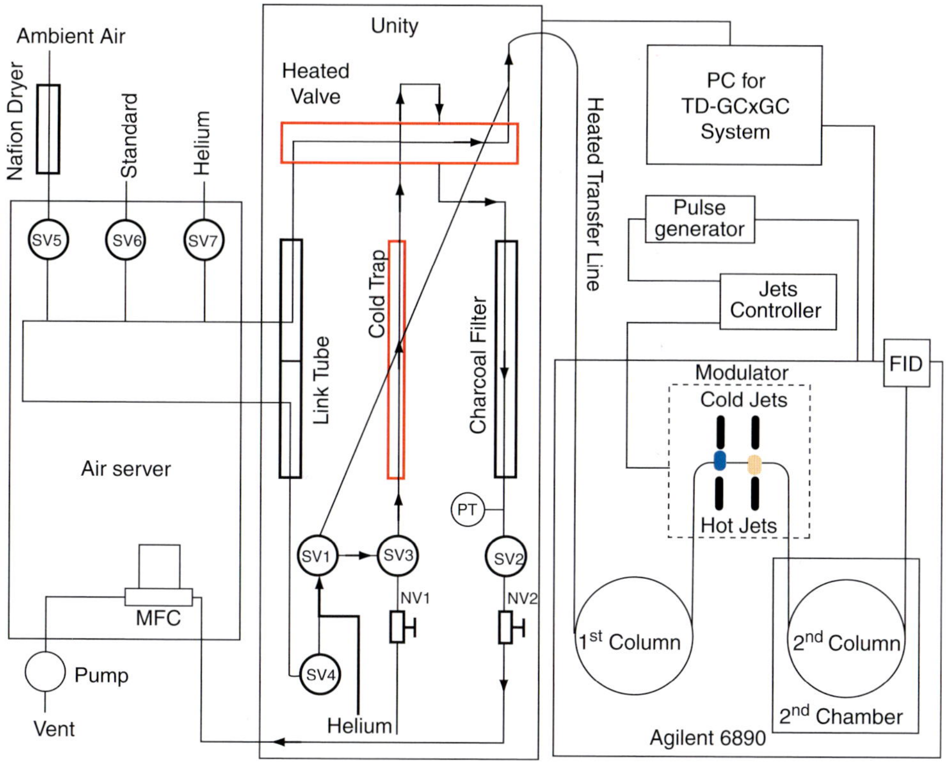

Figure 7 Schematic of the thermal desorber–GC×GC–FID system. The left part shows an air server that contains a sampling manifold and a mass flow controller (MFC). A Nafion dryer is used for removing moisture from ambient air. The middle part shows the thermal desorber in the trap desorption step. The arrows give flow directions of carrier gas (helium). During on-line sampling, the carrier gas flows in the reversed direction. The solid and dotted lines show flow-paths with and without gas flow, respectively. SV, NV, PT, and MFC represent solenoid valve, needle valve, pressure transducer, and mass flow controller, respectively. The right part shows the GC×GC system with its controlling units. In the real design, the hot jet tubes are orthogonal to the cold jet tubes [33].

In a system developed for in situ measurements for particles, an impactor particle collector with thermal desorption was connected to the GC×GC instrument [34]. In the sampling part, particles were deposited into the collection and thermal desorption (CTD) cell by impaction, followed by TD onto the GC×GC system. During the analysis, the collection cell was cooled, and the next sample collected. The impaction collector was preceded by a cyclone to exclude particles above 2.5 μm and by a humidifier to minimize loss by bounce and re-entrainment. The TAG system is shown in Figure 8. During desorption, the valve and transfer lines were heated, while the column was held at 45°C, to achieve cold-trapping of the desorbed compounds. The modulator was a custom-made, air-cooled two-stage thermal modulator that consisted of a short piece of Silcosteel tubing internally coated with a thin layer of polydimethylsiloxane stationary phase

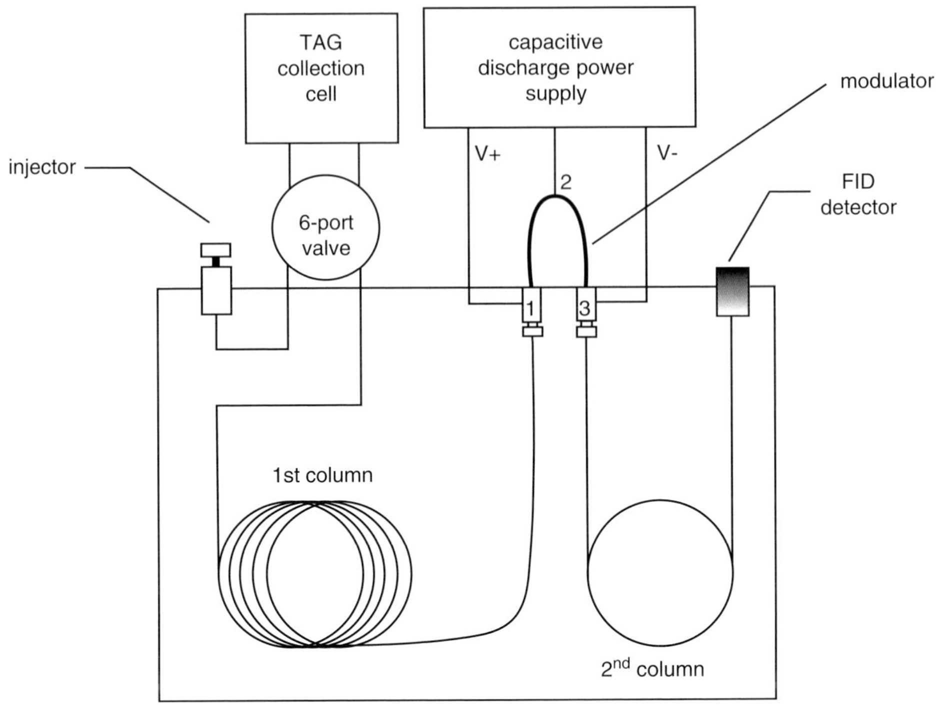

Figure 8 On-site system for collection of particulate samples [34].

(1–3 μm). Dual-stage modulation was achieved through alternatively heating the two segments of the trapping capillary using a custom-built capacitative discharge power supply. The on-site system was applied to the sampling, and analyses of urban air samples were collected in Berkeley, California, with sampling times varying from 90 to 120 min. Compared with GC–MS analyses with the same sampling system, the GC×GC system provided drastically improved compound separation, as was shown by clear separation of the saturated from the unsaturated fatty acid esters, and by the separation of many compounds in the EPA Method 8270 standard, which co-eluted in 1D GC.

2.4 Identification procedures

GC×GC often improves the identification of unknown analytes because structurally related compounds are grouped close to each other (Chapters 2 and 3). However, in air and aerosol analyses, grouping is typically not as clear as in oil and diesel analyses, for example (Chapter 7). Lack of ordered chromatograms for aerosol samples has been explained by Hamilton et al. [23] in terms of the many different groups present in aerosol samples, so that the conventional GC×GC ordering of the chromatogram presents no simple band structures. Thus, identification based on retention times required reference standards and sufficient separation. Most GC×GC analyses in atmospheric determination have been done with mass spectral detection, in order to identify chemical groups and/or individual compounds. However, even with the help of mass spectral data, the identification of unknowns is not easy, and the large amount of information generated by GC×GC–MS can create problems in data handling. Identification can be done manually, but automated procedures are preferable for data analysis of the large dataset created by GC×GC–MS. Two types of automated procedures have been developed. First, automated methodologies for classification of chemically related compounds into groups have been developed [36,38]. Second, procedures for automated identification of individual species have been developed [32]. The implementation of such classification and identification procedures simplifies the interpretation of data, making rapid peak classification, searching, and identification easier.

An example of the group-type identification procedure is the classification of complex GC×GC–ToF MS data based on primary and secondary retention times and MS fragmentation patterns developed by Welthagen et al. [36]. The idea of utilization of fragmentation patterns in mass spectral data to identify compounds and to classify compounds is a well-established technique developed for GC–MS [37]. In GC×GC–MS analyses, classifiers help to identify unknown peaks and aid in statistical analysis because of improved spectral quality due to better separation and the possibility of including first- and second-dimension retention time restrictions on compound class assignments [36]. In the method developed by this group, three types of classifiers were used: domain knowledge classifiers (based on fragmentation pattern and retention time rules), a substructure-type classifier (multivariate classifier based on mathematical transformations), and a category-type classifier (multivariate classifier partly based on domain knowledge transformations) [38]. The compound groups identified can be plotted as bubble plots. In addition to classifying peaks for further statistical analysis, use of a bubble plot can help a GC×GC user to quickly evaluate group-type separations achieved by a particular column combination. The bubble plot can also be used for a rapid visual recognition of pattern changes in monitoring studies. This was convincingly demonstrated in a three-year monitoring programme characterising urban air particulate matter ($PM_{2.5}$) from Augsburg (Germany) [35]. The data consisted of around 15,000 peaks, out of which approximately 700 compounds were identified. Further development of the classification method was utilised for characterising similar urban samples [38]. An example of the GC×GC separation and utilization of the identification procedure is shown in Figure 9.

Kallio et al. developed an automated procedure for the identification of individual compounds in atmospheric samples and used it for the data analysis of rural aerosol samples [32]. In this procedure, retention indexes, quality parameters (minimum required similarity, *S/N* value, allowed *I* difference between experimental and library values), modulation parameters, library files, and retention times of reference compounds were utilized to construct a program for data analysis. As output, the program listed compounds that fulfilled the required criteria. The automated procedure was compared with manual identification, and it was concluded that the automated procedure worked satisfactorily if the concentrations were sufficiently high (above ca. 10 ng/m^3), but for very low concentrations (low ng/m^3) manual search was more accurate.

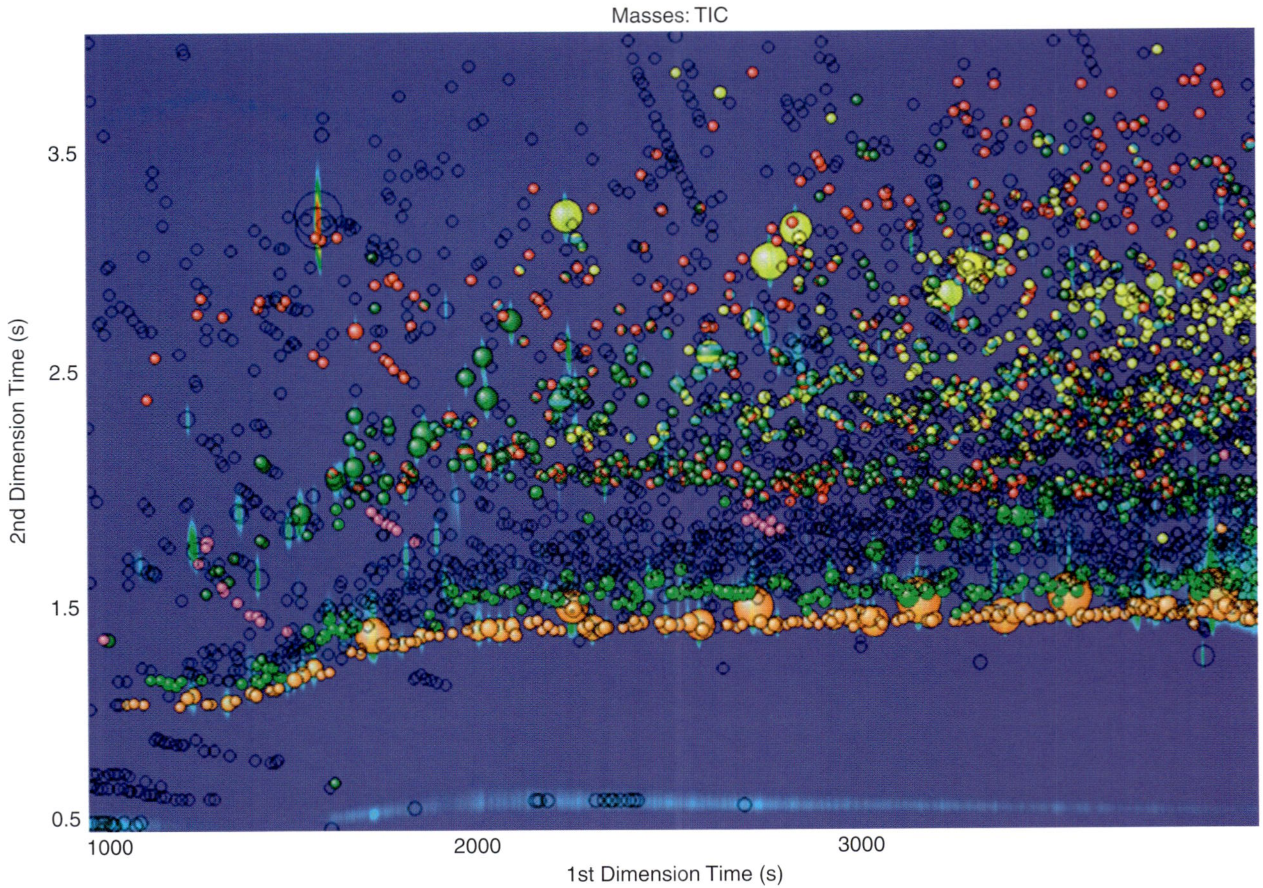

Figure 9 Classification of an aerosol sample ($PM_{2.5}$, Augsburg) (S/N, 100:1) with n-alkanes (orange), alkenes and cycloalkanes (light green), n-alkane acids (purple), partially hydrated naphthalenes and alkenyl-substituted benzenes (light blue), naphthalene and alkyl-substituted naphthalenes (yellow), polar benzenes (red), and alkyl-substituted benzenes (dark green). Bubbles with more than one colour have been identified by the scripts as belonging to more than one class [38].

3. FUTURE TRENDS

Knowledge of organic compounds in air and aerosols, their sources, chemistry, and role in the atmosphere and the earth's climate system is still extremely limited, and the importance of biogenic and anthropogenic precursors for secondary aerosol production is a major current research topic. Because the limited sensitivity and selectivity of conventional analytical methods is the main reason that a large part of the organic compounds still has not been identified, development of novel GC×GC–(ToF MS) methodologies can have a significant impact on the information that can be obtained on volatile and semivolatile organics in the atmosphere.

The GC×GC analyses offer several benefits in the analysis of atmospheric samples. The separation efficiency is superior to conventional analyses with the 1D system. Another important feature is that due to the cold-trap modulation, sensitivity is increased as well, and trace amounts of compounds are also detected [32,39]. The applications have shown that compounds that could not have been detected in 1D analyses can be identified both from gas and particulate phase samples owing to this concentrative modulation and improved separation. Combining GC×GC with DTD for analysis of the particulate phase is advantageous because sample pretreatment is then minimal. Several detection modes have been used, and among these, ToF MS is the most powerful one. A further challenge that the GC×GC–ToF MS system presents in atmospheric analyses is the development of automated identification procedures: manual identification of unknown compounds is extremely time consuming due to the large amount of data created with the system. Identification of unknown compounds in atmospheric samples is challenged by the lack of ordered chromatograms that would aid the identification. From the instrumental point of view, further development of on-site and in situ instruments would be beneficial because many of the compounds present in atmospheric samples are highly reactive, and thus, off-line sampling and analysis can even cause significant errors in the results.

REFERENCES

1 R.E. Hester and R.M. Harrison (Eds.), *Volatile Organic Compounds in the Atmosphere*, Royal Society of Chemistry, England, 1995.
2 A.H. Goldstein and I.E. Galbally, *Environ. Sci. Tech.*, 40 (2007) 1514.
3 C. Pilinis, S.N. Pandis and J.H. Seinfeld, *J. Geophys. Res.*, 100 (1995) 18739.
4 T.S. Twomey, *Atmospheric Environment Part A*, 25 (1991) 2435.
5 M. Kulmala, H. Vehkamäki, T. Petäjä, M. Dal Maso, M. Boy, A. Lauri, V.M. Kerminen, W. Birmili and P.H. McMurry, *J. Aerosol Sci.*, 35 (2004) 143.
6 D.W. Dockery, C.A. Pope, X.P. Xu, J.D. Spengler, J.H. Ware, M.E. Fay, B.G. Ferris and F.E. Speizer, *New England J. Medicine*, 329 (1993) 1753.
7 J.M. Samet, F. Dominici, F.C. Curriero, I. Coursac and S.L. Zeger, 1987–1994, *New England J. Med.*, 343 (2000) 1742.
8 H.E. Wichmann and A. Peters, *Philosophical Transactions of the Royal Society of London Series A*, 358 (2000) 2751.

9 M. Kanakidou, J.H. Seinfeld, S.N. Pandis, I. Barnes, F.J. Dentener, M.C. Facchini, R. van Dingenen, B. Ervens, A. Nenes, C.J. Nielsen, E. Swietlicki, J.P. Putaud, Y. Balkanski, S. Fuzzi, J. Horth, G.K. Moortgat, R. Winterhalter, C.E.I. Myhre, K. Tsigaridis, E. Vignati, E.G. Stephanou and J. Wilson, *Atmos. Chem. Phys.*, 5 (2005) 1053.
10 S.D. Piccot, J.J. Watson and J.W. Jones, *J. Geophys. Res.*, 97 (1992) 9897.
11 T.E. Graedel, T.S. Bates, A.F. Bouwman, D. Cunnold, J. Dignon, I. Fung, D.J. Jacob, B.K. Lamb, J.A. Logan, G. Marland, P. Middleton, J.M. Pacyna, M. Placet and C. Veldt, *J. Biogeochem. Cycles*, 7 (1993) 1.
12 P.J. Crutzen and M.O. Andreae, *Science*, 250 (1990) 1669.
13 M.O. Andreae and P. Merlet, *Global Biogeochem. Cycles*, 15 (2001) 955.
14 A. Guenther, C.N. Hewitt, D. Erickson, R. Fall, C. Geron, T. Graedel, P. Harley, L. Klinger, M. Lerdau, W. McKay, T. Pierce, B. Scholes, R. Steinbrecher, R. Tallamraju, J. Taylor and P. Zimmerman, *J. Geophys. Res.*, 100 (1995) 8873.
15 G. König, M. Brunda, H. Puxbaum, C.N. Hewitt, S.C. Duckham and J. Rudolph, *Atmos. Environ.*, 29 (1995) 861.
16 J. Kesselmeier, L. Schäfer, P. Ciccioli, E. Brancaleoni, A. Cecinato, M. Frattoni, P. Foster, V. Jacob, J. Denis, J.L. Fugit, L. Dutaur and L. Torres, *Atmos. Environ.*, 30 (1997) 1841.
17 H. Hakola, H. Hellen and T. Laurila, *Atmos. Environ*, 40 (2006) 3621.
18 B.J. Finlayson-Pitts and J.N. Pitts, *Atmospheric Chemistry: Fundamentals and Experimental Techniques*, Wiley, New York, 1986.
19 R. Atkinson and J. Arey, *Chem. Rev.*, 103 (2003) 4605.
20 R.G. Derwent, In: R.E. Hester and R.M. Harrison (Eds.), *Volatile Organic Compounds in the Atmosphere*, Royal Society of Chemistry, 1995.
21 A.C. Lewis, N. Carslaw, P.J. Marriott, R.M. Kinghorn, P. Morrison, A.L. Lee, K.D. Bartle and M.J. Pilling, *Nature*, 405 (2000) 778.
22 A.C. Lewis, *Atmos. Environ.*, 34 (2000) 1155.
23 J.F. Hamilton and A.C. Lewis, *Atmos. Environ.*, 37 (2003) 589.
24 J.F. Hamilton, A.C. Lewis, C. Bloss, V. Wagner, A.P. Henderson, B.T. Golding, K. Wirtz, M. Martin-Reviejo and M.J. Pilling, *Atmos. Chem. Phys.*, 3 (2003) 1999.
25 J.E. Saxton, A.C. Lewis, J.H. Kettlewell, M.Z. Ozel, F. Gogus, Y. Boni, S.O.U. Korogone and D. Serca, *Atmos. Chem. Phys.*, 7 (2007) 4095.
26 S. Bartenbach, J. Williams, C. Plass-Duelmer, H. Berresheim and J. Lelieveld, *Atmos. Chem. Phys.*, 7 (2007) 1.
27 J.F. Hamilton, P.J. Webb, A.C. Lewis, J.R. Hopkins, S. Smith and P. Davy, *Atmos. Chem. Phys.*, 4 (2004) 1279.
28 J. Schnelle-Kreis, W. Welthagen, M. Sklorz and R. Zimmermann, *J. Sep. Sci.*, 28 (2005) 1648.
29 O. Pani and T. Górecki, *Anal. Bioanal. Chem.*, 386 (2006) 1013.
30 N. Ochiai, T. Ieda, K. Sasamoto, A. Fushimi, S. Hasegawa, K. Tanabe and S. Kobayashi, *J. Chromatogr. A*, 1150 (2007) 13.
31 M. Kallio, T. Hyötyläinen, M. Lehtonen, M. Jussila, K. Hartonen, M. Shimmo and M.L. Riekkola, *J. Chromatogr. A*, 1019 (2003) 251.
32 M. Kallio, M. Jussila, T. Rissanen, P. Anttila, K. Hartonen, A. Reissel, R. Vreuls, M. Adahchour and T. Hyötyläinen, *J. Chromatogr. A*, 1125 (2006) 234.
33 X. Xu, L.L.P. van Stee, J. Williams, J. Beens, M. Adahchour, R.J.J. Vreuls, U.A.Th. Brinkman and J. Lelieveld, *Atm. Chem. Phys.*, 3 (2003) 665.
34 A.H. Goldstein, D.R. Worton, B.J. Williams, S.V. Hering, N.M. Kreisberg, O. Panić and T. Górecki, *J. Chromatogr. A*, 1186 (2008) 340.
35 X. Xu, J. Williams, C. Plass-Dülmer, H. Berresheim, G. Salisbury, L. Lange and J. Lelieveld, *Atm. Chem. Phys.*, 3 (2003) 1461.
36 W. Welthagen, J. Schnelle-Kreis and R. Zimmermann, *J. Chromatogr. A*, 1019 (2003) 233.
37 F.W. Mclafferty and F. Turecek, *Interpretation of Mass Spectra*, 4th ed, University Science Books, Sausalito, CA, 1993.
38 L. Vogt, T. Gröger and R. Zimmermann, *J. Chromatogr. A*, 1150 (2007) 2.
39 A.L. Lee, K.D. Bartle and A.C. Lewis, *Anal. Chem.*, 73 (2001) 1330.

CHAPTER 9

Volatile Components of Plants, Essential Oils, and Fragrances

Robert A. Shellie

Contents

1. INTRODUCTION

The 20 billion dollar industry [1] that produces the familiar flavours and fragrances that surround us in everyday life relies heavily on separation science. Indeed, the relationship between essential oil analysis and gas chromatography (GC) began within a few years of James and Martin first describing the technique [2]. Gas chromatography, particularly when combined with mass spectrometry (MS) has strongly contributed to the development of the science of essential oils and fragrances in the areas of phytochemistry, chemotaxonomy, olfactory research, biochemistry, plant-insect research, the search for new sources of odoriferous compounds for industry, and quality control. Today thousands of flavour and fragrance compounds derived from plant materials have been characterised, but the need to perform basic work on the chemical composition of essential oils and fragrances still exists and there is growing interest in the use of comprehensive two-dimensional gas chromatography (GC×GC) for these analyses.

Comprehensive Analytical Chemistry, Volume 55
ISSN: 0166-526X, DOI 10.1016/S0166-526X(09)05509-3

Essential oils are analysed for a variety of reasons [3] such as (i) to determine the qualitative and/or quantitative composition of the product, (ii) to control the quality and authenticity of the product, or perhaps (iii) to detect the presence of adulteration or contamination. Perfumes generally comprise both volatile and nonvolatile fractions, and it is commonplace that they are characterised through their volatile fraction using gas chromatography [4]. The level of sophistication of the chromatographic technique employed depends on the purpose of the analysis, which ranges from routine control to dedicated analysis [4], which may be, for example, an exhaustive compositional study for perfume formulation or competitor analysis, or for the analysis of particular trace components such as suspected allergenic compounds. Although routine analyses can be readily carried out by employing established methods supported with automated instrumentation, dedicated analyses often require complex and time-consuming methods and skilled operators [4]. Essential oils and fragrances often exhibit overwhelming complexity in terms of number of components present in them. This complexity has driven those studying these sample types to embrace technological advances in gas chromatography, to the extent that most gas chromatographic techniques have been applied in some way to the study of essential oils and fragrances. Having first appeared in the literature in 1991 [5], GC×GC appealed almost immediately to researchers in the petrochemical industry and to those interested in analysing environmental pollutants. The potential to employ GC×GC to plant essential oil analysis was not shown until 2000 [6] when the analyses of spearmint and peppermint essential oils were reported, illustrating a two- to threefold increase in the separation power of GC×GC over monodimensional (1D) GC, followed by reports describing the separation of vetiver [7] and tea tree and lavender essential oils [8]. These first reports led the way to steady development of GC×GC for essential oils and fragrance analysis, and the approach can now essentially be routinely employed. GC×GC usage is increasing within the field, and it is likely to be an important addition to the suite of routine approaches already employed regularly by essential oil and fragrance chemists. The perfume industry has recognised the importance of GC×GC to the field, one expert stating that "the most promising development in GC research in the last decade has without a doubt been the introduction of two-dimensional comprehensive GC" [9]. It is within this framework that selected applications of GC×GC for essential oils and fragrances analysis will be discussed in this chapter.

2. APPLICATIONS AND CHROMATOGRAPHIC CONDITIONS

Table 1 provides a list of GC×GC references, employing either flame-ionization (FID) or mass spectrometry (MS) detection, that are concerned with the analysis of volatile plant extracts. Details of the column sets and modulation interface used are also listed.

The majority of these applications have utilised a low-polarity stationary phase column to affect displacement of the sample components along the *x*-axis

Table 1 GC×GC and GC×GC–MS analysis of plant extracts, essential oils, and chiral compounds

Application	Column combination (m×mm ID×μm d_f)	Modulator	Reference
Separation of peppermint and spearmint essential oil components and comparison with GC–MS	DB-1 (1×0.100×3.50)× OV-1701 (2×0.100×0.500)	Thermal modulation unit (Zoex Corporation)	[6]
Illustration of GC×GC to separate vetiver essential oil and comparison with monodimensional GC	BPX-5 (25×0.250×0.250)× BPX-50 (0.8×0.100×0.100) BPX-5 (25×0.250×0.250)× BP-20 (1×0.100×0.100)	Cryogenic (Chromatography Concepts)	[7]
Characterisation and comparison of tea tree and lavender essential oil components	BPX-5 (30×0.250×0.250)× BP-20 (0.8×0.100×0.100)	Cryogenic (Chromatography Concepts)	[8]
GC×GC–ToF MS lavender essential oil	BPX-5 (30×0.250×0.250)× BP-20 (2×0.100×0.100)	Cryogenic (Chromatography Concepts)	[10]
Enantioselective analysis of tea tree essential oil	EtTBS-β-CD (25×0.250×0.250)× BP-20 (0.8×0.100×0.100)	Cryogenic (Chromatography Concepts)	[11]
Tea tree oil — demonstration of retention time repeatability	BPX-5 (30×0.250×0.250)× BP-20 (1×0.100×0.100)	Cryogenic (Chromatography Concepts)	[12]
Characterisation of lavender oil and comparison with GC–MS	BPX-5 (30×0.250×0.250)× BP-20 (1×0.100×0.100)	Cryogenic (Chromatography Concepts)	[13]
GC×GC with fast ^{2}enantioseparation of bergamot essential oil	DB-5 (10×0.100×0.100)× EtTBS-β-CD (1×0.250×0.250)	Cryogenic (Chromatography Concepts)	[14]
SPME-GC×GC of ginger volatiles	BPX-5 (30×0.250×0.250)× BP-20 (1×0.100×0.100)	Cryogenic (Chromatography Concepts)	[15]
Characterisation of geranium essential oil with GC×GC–qMS	HP-5 (30×0.250×0.250)× BP-20 (0.5×0.320×0.250)	Cryogenic (Chromatography Concepts)	[16]

Table 1 *(Continued)*

Application	Column combination (m×mm ID×μm d_f)	Modulator	Reference
Characterisation of volatile oil from traditional Chinese medicine	Various stationary phases (50–60 m)×Various stationary phases (2.5–3×0.100)	4-jet (Leco)	[17]
Characterisation of hop essential oil	DB-5 (10×0.180×0.180)× DB-17 (1.9×0.180×0.180)	4-jet (Leco)	[18]
Characterisation of *Origanum onites* essential oil with GC×GC–ToF MS	DB-5 (10×0.180×0.180)× DB-17 (1.6×0.180×0.180)	4-jet (Leco)	[19]
Characterisation of pistachio essential oil with GC×GC–ToF MS	DB-5 (10×0.180×0.180)× DB-17 (1.6×0.180×0.180)	4-jet (Leco)	[20]
Analysis of sandalwood oil	BPX-5 (30×0.250×0.250)× BP-20 (0.5×0.100×0.100)	Cryogenic (Chromatography Concepts)	[21]
Characterisation of citrus oils	Supelcowax-10 (30×0.250×0.250)× SPB-5 (1×0.100×0.100)	Cryogenic (Chromatography Concepts)	[22]
Analysis of *Ziziphora taurica* with GC×GC–ToF MS	DB-5 (30×0.320×0.250)× DB-17 (1.9×0.100×0.100)	4-jet (Leco)	[23]
Identification of character-impact odorants in coriander	BPX-5 (30×0.250×0.250)× BP-20 (1×0.100×0.100)	Cryogenic (Chromatography Concepts)	[24]
Quantitative determination of tobacco essential oil components	DB-petro (50×0.200×0.500)× DB-17ht (2.5×0.100×0.100) DB-WAX (60×0.250×0.250)× DB-1701 (3×0.100×0.400)	Cold-jet (Zoex)	[25]

GC×GC–ToF MS analysis of tobacco essential oil	DB-petro (50×0.200×0.500)× DB-17ht (2.5×0.100×0.100) DB-WAX (60×0.250×0.250)× DB-1701 (3×0.100×0.400)	Cold-jet (Zoex)	[26]
Characterisation of *Achillea monocephala* essential oil with GC×GC–ToF MS	DB-5 (10×0.180×0.180)× DB-17 (1.6×0.180×0.180)	4-jet (Leco)	[27]
Characterisation of rose oil with GC×GC–ToF MS	DB-5 (30×0.320×0.250)× DB-17 (1.9×0.100×0.100)	4-jet (Leco)	[28]
Analysis of Lemon thyme oil using a single-stage liquid cooled modulator	RTX-1 (30×0.320×0.250)× RTX-Wax (2×0.100×0.100)	Custom	[29]
Fast GC×GC separations (chiral and achiral)	β-cyclodextrin derivative (10×0.100×0.100)× BP-20 (0.3×0.050×0.050) BP-20 (5×0.100×0.100)× BGB-1701 (0.3×0.050×0.050)	Cryogenic (Chromatography Concepts)	[30]
Separation of peppermint essential oil GC×GC–qMS and GC×GC–FID	OV-1 (30×0.250×0.250)× OV-1701 (1×0.100×0.100)	CO_2 jet (Thermo Electron Corporation)	[31]
Characterisation of pepper volatiles	BPX-5 (30×0.250×0.250)× BP-20 (1.5×0.100×0.100)	Cryogenic (Chromatography Concepts)	[32]
Separation of peppermint essential oil	OV-1 or CW20M (30×0.250×0.250)×A range of custom made mixed stationary phase columns	CO_2 jet (Thermo Electron Corporation)	[33]
Characterization of tobacco leaf extract	DB-Petro (50×0.200×0.500)× DB-17 (2.5×0.100×0.100)	4-jet (Leco)	[34]

Table 1 *(Continued)*

Application	Column combination (m×mm ID×μm d_f)	Modulator	Reference
GC×GC–TOF MS for characterisation of odor active compounds in coriander and spicy fraction of hop oil	BPX-5 (30×0.250×0.250)× BP-20 (0.8×0.100×0.100)	Cryogenic (Chromatography Concepts)	[35]
Analysis of odor active compounds in hop essential oil	BPX-5 (25×0.250×0.250)× BP-20 (1.1×0.100×0.100)	Cryogenic (Chromatography Concepts)	[36]
Basil essential oil characterisation	DB5-MS (30×0.250×0.250)× Supelcowax-10 (1.25×0.100×0.100)	4-jet (Leco)	[37]
Comparison of Asian and American ginseng extracts by GC×GC and GC×GC–qMS	BPX-5 (30×0.250×0.250)× BP-20 (1.2×0.100×0.100)	Cryogenic (Chromatography Concepts)	[38]
SPME–GC×GC for chemical profiling of volatile herbal mixtures	HP-5 (30×0.320×0.250)× BP-20 (1.0×0.100×0.100)	Cryogenic (Chromatography Concepts)	[39]
Enantioselective separation of ephedrine-type alakaloids in herbal materials	Cyclodex-β (30×0.250×0.250)× BP-20 (1×0.100×0.100)	Cryogenic (Chromatography Concepts)	[40]

of the two-dimensional separation plane and employed a polyethylene glycol (wax) second-dimension stationary phase column (chiral analysis, which is also possible, is described separately in this chapter). Low-polar and polar wax-type stationary phases are also common for monodimensional analysis of essential oils and fragrances, so it is not surprising that they are most regularly employed for two-dimensional analysis. Phenomenal separation gains over monodimensional analysis arise from selecting the most suitable columns, as illustrated in the expanded modulation pulses depicted in Figure 1. Despite being one of the earliest published GC×GC separations of an essential oil, this chromatogram of vetiver essential oil by Marriott et al. [7] probably illustrates the benefit of the comprehensive two-dimensional analysis for this class of samples as well as any application. The second-dimension peak capacity of around ten is typical for applications of GC×GC, leading to a tenfold enhancement in peak capacity over monodimensional analysis. This substantial peak-capacity enhancement greatly increases the chance of isolating single-component peaks in the separation space and has been enjoyed by many researchers engaged in the analysis of essential oils and fragrances.

The sample composition of plant extracts is not one of multiple homologous series, so GC×GC separations of essential oils do not exhibit the same degree of chromatographic structure widely illustrated in the GC×GC analysis of

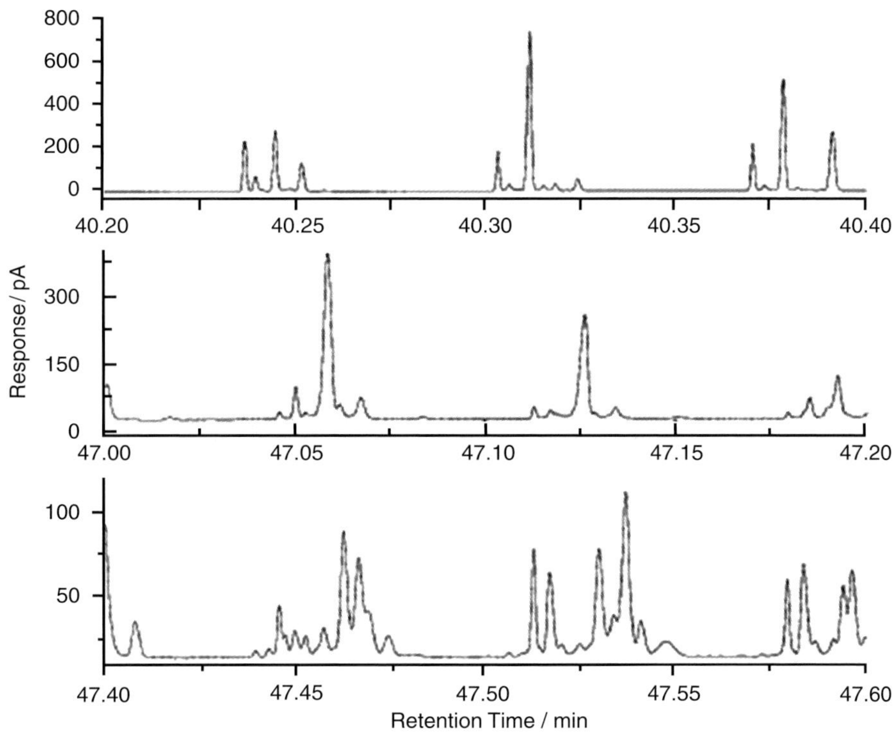

Figure 1 Expanded 0.2-min sections of the GC×GC separation of vetiver essential oil, each showing three successive modulation events [7].

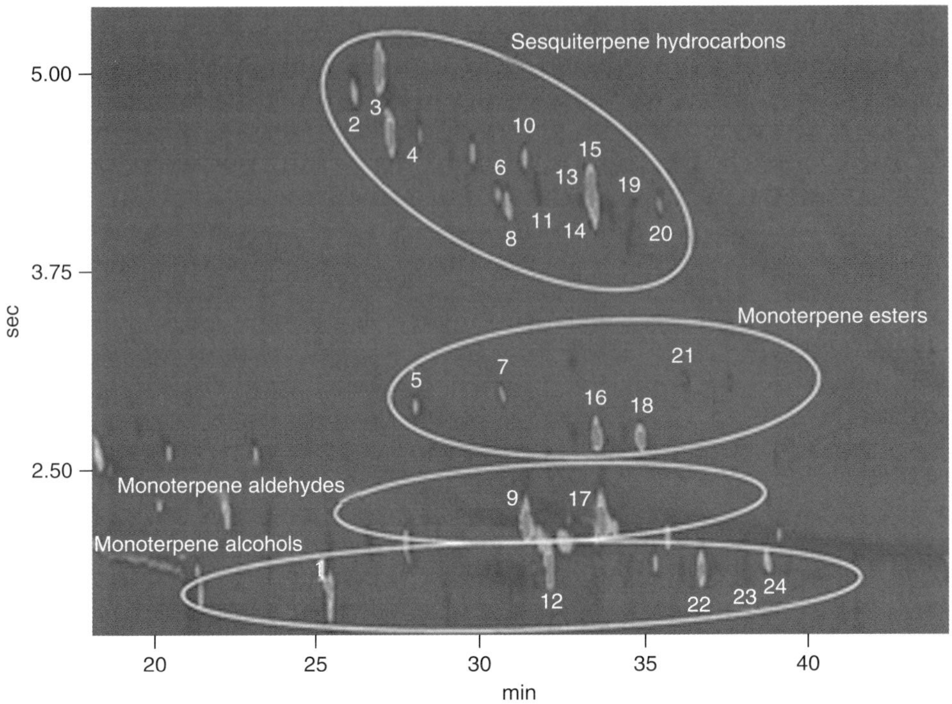

Figure 2 A 26-min expansion of a 2D chromatogram of lemon essential oil. 1, linalool, 2, *cis*-α-bergamotene; 3, (E)-caryophyllene; 4, *trans*-α-bergamotene; 5, linalyl acetate; 6, (Z)-β-santalene; 7, citronellyl acetate; 8, α-humulene; 9, neral; 10, (Z)-β-farnesene; 11, germacrene D; 12, α-terpineol; 13, valencene; 14, bicyclogermacrene; 15, β-bisabolene; 16, neryl acetate; 17, geranial; 18, geranyl aceate; 19, *cis*-α-bisabolene; 20, *trans*-α-bisabolene; 21, citronellyl formate; 22, nerol; 23, p-cymen-8-ol; 24, geraniol [22].

petroleum-based samples (see Chapter 7). However, similar compounds do fall within regions or bands in the two-dimensional separation space, and this assists characterisation. A fine illustration of chromatographic structure is illustrated in Figure 2, where an expanded region of the separation of lemon essential oil is shown [22]. In this case the "regular" apolar-polar column arrangement used in most studies has been reversed, and a polar polyethylene glycol stationary phase was used in the first-dimension followed by a low-polarity second-dimension column. Clearly, there are benefits in reversing the column polarity arrangement, and it has become apparent over recent years that the most appropriate polarity regime is somewhat sample dependent. (For further discussion on this theme, see Chapters 3, 6, and 7.) GC×GC users must be skilled in column installation and should take necessary precautions to ensure that all fittings are made correctly to avoid extended peak tailing that is sometimes observed in GC×GC chromatograms of essential oils and fragrances. Beens and co-workers first described these extended peak tails and suggested that this observation could be used as a diagnostic tool for injector performance [41]. The familiar shape of these peak tails later led to approaches for the determination of retention indices (Chapter 3), but in the case of sample characterisation they are clearly undesirable. Initially, the observed peak tailing in essential oil analysis was attributed to being an injector effect [13] following Beens's observations. The peak tailing certainly does not seem to be a column effect; however, it is most likely that the peak tailing in many applications is caused by unswept volumes around column connections or by active sites within these connections. These peak tails are not readily observed in monodimensional GC (such behaviour results in increased background/ baseline response in monodimensional analysis), even if the same column set is employed (but with the modulator disengaged), but they are exacerbated by the peak focusing and enhanced resolution.

3. MASS SPECTOMETRIC DETECTION

One of the major aims of plant extract analysis by GC×GC is to determine sample composition. Characterisation of volatile plant extracts relies heavily on retention index-based identification of the separated components, and this is usually supported by mass spectrometry. The impact of GC–MS on the current knowledge of essential oil composition should not be underestimated. Over the years, many hundreds of manuscripts presenting the composition of essential oils, determined by GC–MS, have been published in the general literature. GC–MS is accepted as a true two-dimensional analysis system; hyphenation of GC×GC with mass spectrometry therefore represents triple-dimensional analysis. While identifying and reporting the $n^{\mathrm{th}+1}$ component in a complex sample may not always be the primary focus of a particular analysis, the basic requirements of peak capacity and resolution are still critical. The GC×GC resolution advantage increases the chance of isolating more individual compounds, leading to improved analysis. Owing to the fast second-dimension elution in GC×GC, time-of-flight mass spectrometry (ToF MS) is the most viable technology (see Chapter 2). The hyphenated approach was illustrated in 2001 for the characterisation of lavender essential oil [10] and has been further optimised to fully exploit the three analytical dimensions. Very high-resolution separations with accurate full mass range spectra are now readily achievable using GC×GC–ToF MS. Wu et al. detected 394 components in *Pogostemon cablin Benth* volatile oil, each with a library spectrum similarity of over 800, which is quite remarkable given that monodimensional separation resulted in 27 positive identifications with similarity exceeding 800 [17]. Although there are numerous reasons for greater numbers of identified peaks, improved resolution is a major factor. Figure 3 shows a small portion of a GC×GC–ToF MS total ion current chromatogram and compares the same region from monodimensional GC separation. At least six components are located under what appears to be no more than two poorly resolved peaks in the single-column experiment. Ozel and co-workers have used GC×GC–ToF MS to characterise a wide range of plant-derived samples [19,20,23,27,28].

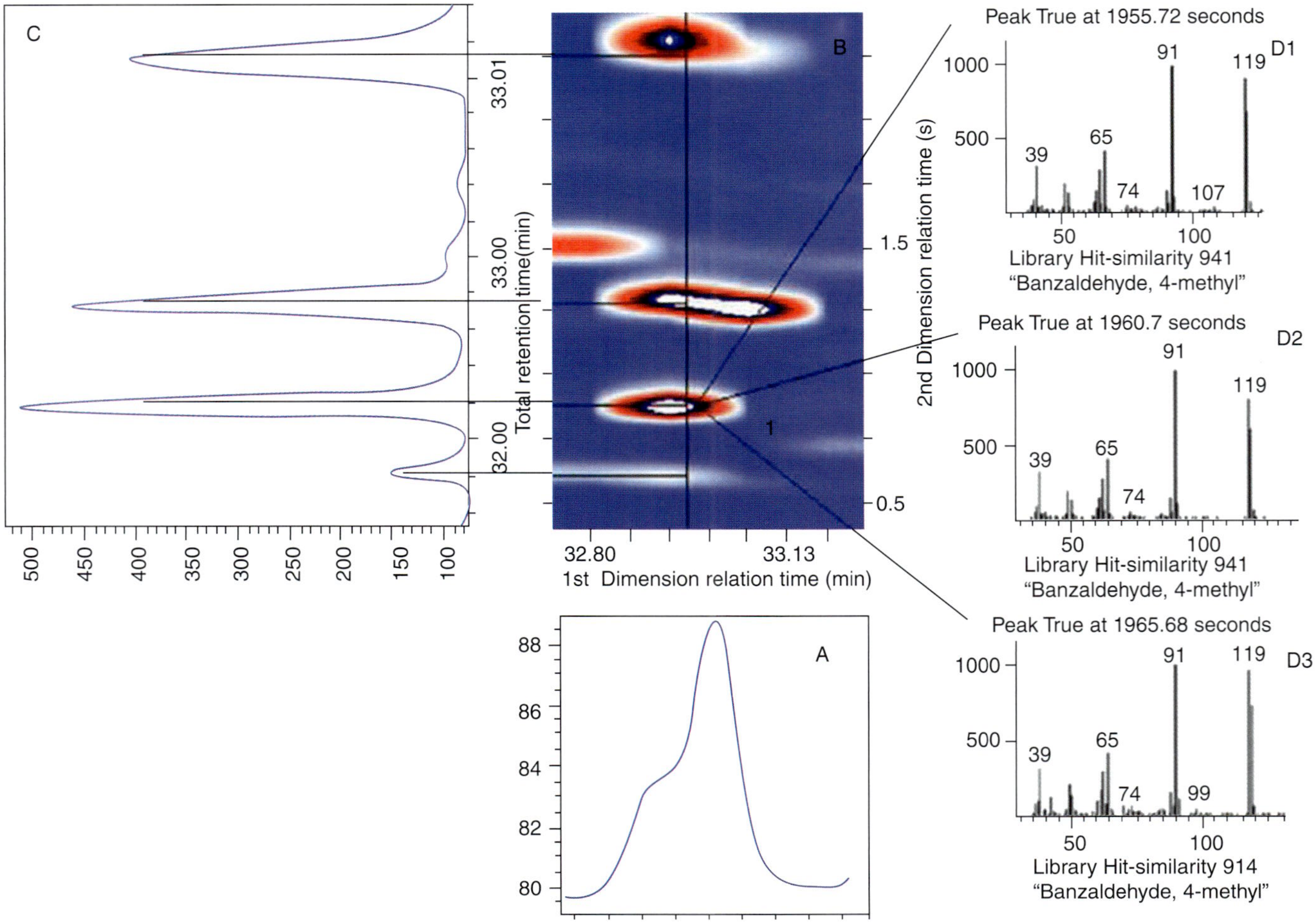

Figure 3 Detailed comparison of 1D GC with GC×GC acquired using the same injection amount, split ratio, and the same carrier gas velocity. (A) Detail of 1D GC chromatogram; (B) detail of GC×GC contour plot; (C) detail of GC×GC chromatogram. The vertical line at 32.96 min indicates the second-dimension chromatogram that is shown in (C). In (B), peak 1 was modulated three times by GC×GC; therefore, identified three times by ToF MS. The corresponding deconvoluted mass spectra for these three modulation slices are shown in D1, D2 and D3 [17].

The infiltration of GC×GC–ToF MS into widespread routine application may be hindered by the fact that at the time of writing essentially only one instrument company offers a fast GC×GC–ToF MS instrument. The expense of a top-end instrument may also be prohibitive for some researchers, and investigations into the suitability of quadrupole MS (qMS) for GC×GC–qMS have been made. For some time, owing to its slow data acquisition restriction, the more common cousin of the ToF MS, the bench-top quadrupole mass spectrometer, was largely overlooked regarding its suitability. The low-molecular-weight range of essential oil and fragrance compounds is advantageous for implementation of qMS technology because narrow mass range scanning increases the data acquisition rate. A nominal mass spectral acquisition rate of 20 Hz was achieved by using a reduced mass scan range of 188 u using a 4000 u/s qMS [16]. This data acquisition rate can now be achieved over a wider mass range using rapid scanning instrumentation. As explained in Chapter 2, modern rapid-scanning qMS instruments have 10,000 u/s full scan capability. Fast-scanning qMS cannot match ToF MS performance. Nonetheless, very good qualitative data are amenable by GC×GC–qMS, which produces fast, accurate spectra in which scan-to-scan mass bias or skewing is essentially nonexistent [42]. While this relatively slow data acquisition speed may not be suitable for quantitative analysis, a considerably improved quality of uncorrected spectra, arising from the enhanced separation over single-column GC–qMS analysis, is apparent. In conjunction with retention data, the qMS data permit identification of separated components. The analysis of Egyptian geranium (*Pelargonium graveolens*) essential oil was reported using GC×GC–qMS [16]. Sixty-five components were identified within the two-dimensional separation space. Very high-quality mass spectra were obtained, which facilitated accurate library matching without the need for background correction. In fact, spectral matches as high as 99% were found for a number of components.

A simple procedure that permits a direct method translation of retention times in two dimensions for GC×GC–FID and GC×GC–ToF MS analyses has been described [43]. By use of a Tee-union with independent pressure control, it is possible to adjust the outlet pressure of the second column in the ToF MS instance to provide a suitable pressure drop across the column set, which allows correction of total flow in the two-column GC×GC–ToF MS experiment and certainty of retention time. The adjustment is achieved by obtaining the correct holdup time for the unretained peak. Shellie et al. showed that the average absolute retention time differences were 42 ms in the second-dimension and 3.7 s in the first-dimension for a sample of 18 components of varying chemical class [43]. The two sandalwood essential oil GC×GC chromatograms shown in Figure 4 illustrate sufficiently well-correlated retentions that should permit validation of peak identities in GC×GC–FID analyses using an independent GC×GC–ToF MS when a matched column combination is employed in each experiment. By extension, it should also be possible to accommodate variations in column dimensions to obtain the same chromatogram if an additional Tee-union and gas supply is used.

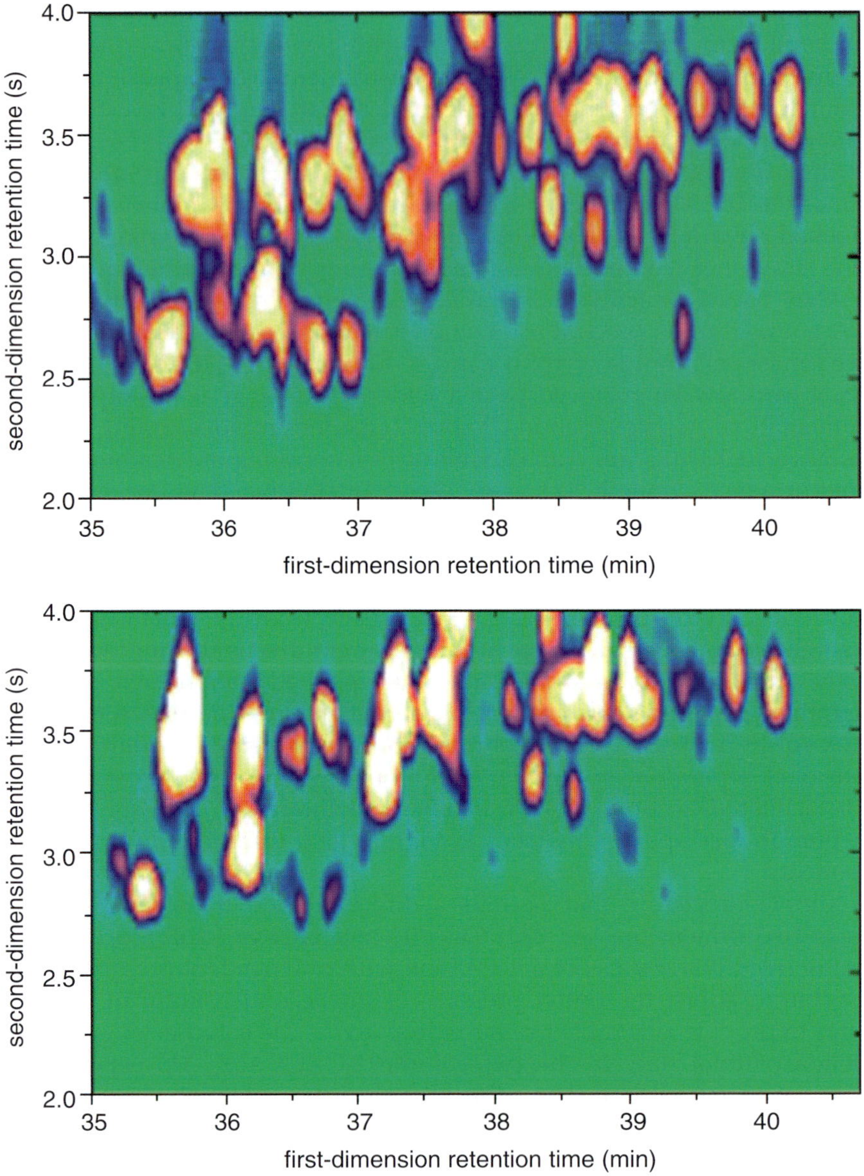

Figure 4 Side-by-side comparison of an expanded region of the GC×GC–FID (upper) and GC×GC–ToF MS (lower) chromatograms of sandalwood essential oil [43].

4. USE OF RETENTION INDICES IN GC×GC ANALYSIS OF ESSENTIAL OILS AND FRAGRANCE COMPOUNDS

Where exact peak position matching is not made, then retention indices are an excellent alternative for porting retention data. Retention indices are extremely important in identification strategies of essential oil and fragrance components because mass spectral data alone are generally insufficient to provide positive identification. Neither GC nor MS evidence on its own can be a satisfactory basis for the conclusive component identification of individual components [44]. The similarity of mass spectra for a series of terpenes of the same nominal empirical formula makes identification based on MS data highly problematic. Identification of sesquiterpene hydrocarbons, of which approximately 400 are known, is almost impossible using only MS data since many of these compounds yield identical, or almost identical, mass spectra [45]. By using the combined information content of the mass spectrum and retention index, the probability of correct assignment increases significantly because there are two independent parameters on which to base the identity of the compound. The identification process using automated library searching itself is also simplified somewhat because a smaller set of library spectra (only of components that are known to produce the given retention index of the unknown) are chosen for comparison.

Column selection in GC×GC for essential oils or fragrance analysis has been driven largely by ensuring compatibility with retention index databases. Although specialist laboratories often develop their own compilation of reference data under their own standard conditions, many rely on published compilations, which list the relative retention indices of compounds [46–52]. If appropriate first-dimension stationary phases are employed for GC×GC, historic retention index databases can be utilised directly for identification along the *x*-axis of the separation plane (the first-dimension separation) in GC×GC analysis [16]. This in part describes why the choices for column selection listed in Table 1 are very consistent. The Adams retention data [47] were acquired on a 5% phenyl polysilphenylene-siloxane stationary phase column (30 m×250 μm), which was temperature programmed from 60°C to 246°C at 3°C/min. By using a 5% phenyl polysilphenylene-siloxane stationary phase column (30 m×250 μm), which was temperature programmed from 60°C to 246°C at 3°C/min according to Adams's method, Shellie and Marriott reported the apparently facile characterization of geranium (*P. graveolens*) essential oil using GC×GC–qMS [16]. Similarly, Eyres et al. employed GC×GC–ToF MS to tentatively identify peaks eluted in the odor-active regions of a selection of hop essential oils [36]. Here retention indices served a dual purpose: (1) they were utilised to locate the odor active regions in the GC×GC chromatogram by matching linear retention indices with GC–olfactometry chromatograms and (ii) they were used for identification of selected compounds.

A properly tuned GC×GC separation offers two independent retention data for each separated compound, and it has been widely proposed that reliable identification should be possible without recourse to MS. In a comprehensive investigation of the reliability of peak identifications using retention index data, Bicchi and co-workers [53] revealed that the percentage of correct identifications made using solely retention indexes increases from approximately 65% to approximately 80% when two different polarity columns are used. Such findings are highly supportive of the aforementioned concept of MS-free identification. Chapter 3 discusses how a GC×GC retention space can be defined by two different types of retention index. Two approaches are considered here for their suitability for flavour and fragrance analysis.

The approach described by Arey and co-workers [54], which uses n-alkanes as reference solutes in both dimensions, is acceptable for nonpolar stationary phases, since the difference between the temperature-programmed retention index and the isothermal retention index for terpenes does not generally exceed 1% [49]. However, there is a marked temperature dependence of retention indices on polar polyethylene glycol stationary phase columns [55], so this approach may not be suitable for the "reversed polarity" column GC×GC arrangement that is sometimes employed for essential oil analysis.

Bieri and Marriott [56] proposed an arrangement that simultaneously produces three retention indices that employed a dual second-column approach (GC×2GC). GC×2GC separations produce a pair of GC×GC chromatograms because the modulated first-dimension chromatogram is split into two second-dimension columns, which have different selectivity to the first column and also exhibit different selectivity to one another [57].

This approach makes it possible to have alkane-based retention indices in the second-dimension if low-polarity second-dimension columns are employed; therefore, most reference data for essential oil and fragrance compounds are compatible with GC×GC. Theoretical details for this method of calculating retention indices can be found in Chapter 3. The research group from the Dalian Institute of Chemical Physics has demonstrated the qualitative determination of 60 compounds in tobacco leaf extract based on two-dimensional retention indices alone [34], confirming these identities with mass spectrometry.

5. COSMETICS, FRAGRANCES, AND ALLERGENS

When the seventh amendment of the European Cosmetic Directive 76/768/EEC [58] was adopted, there was increased activity for the development of quantitative analytical methods that could determine low levels of certain fragrance ingredients. These raw materials were identified by the Scientific Committee on Cosmetic Products and Non-Food Products intended for consumers as likely to cause contact allergy when above a certain trigger level. The directive requires that the presence of any of 26 raw materials higher than 10 mg/kg in any cosmetic product intended to remain on the skin or above 100 mg/kg in any cosmetic product to be rinsed off the skin must be declared on the package label. This problem continues to challenge researchers because the methods must be able to detect low levels of the raw materials in the presence of often highly complex matrices such as fragrances and their ingredients. A list of the compounds amenable to analysis by GC is given in Table 2.

Identification of these suspected allergens according to the International Fragrance Association (IFRA) official method [60] relies on a combination of retention time and ratio of selected ions acquired in selected ion monitoring mode. A single *m/z* channel is used for quantification, and the ratio of this ion with two additional characteristic is used for identification confirmation

Table 2 List of the 24 allergens and characteristic ion required for detection/quantification using GC x GC–MS and GC–MS

Name	CAS Reg. no.	Ions (GCxGC–MS) [59]	Ions (GC–MS) [60]
Amyl cinnamic alcohol	[101-85-9]	133	91, 115, 133, 204
Amyl cinnamic aldehyde	[122-49-7]	202	145, 115, 129, 202
Anisyl alcohol	[105-13-5]	138	138, 109, 121, 137
Benzyl alcohol	[100-51-6]	108	108, 79, 107
Benzyl benzoate	[120-51-4]	105	105, 91, 212
Benzyl cinnamate	[103-41-3]	131	131, 91, 192
Benzyl salicylate	[118-51-1]	91	91, 65, 228
Cinnamic alcohol	[104-54-1]	92	134, 92, 105, 115
Cinnamic aldehyde	[104-55-2]	131	131, 77, 103, 132
Citral[a]	[5392-40-5]	69	69, 84, 94, 109
Citronellol	[106-22-9]	69	81, 95, 123
Coumarine	[91-64-5]	146	118, 89, 90, 146
Estragole[b]	[140-67-0]	148	
Eugenol	[97-53-0]	164	164, 103, 149
Farnesol[c]	[106-28-5]	Isomer 1: 81	69, 81, 93
		Isomer 2: 93	93, 69, 81
Geraniol	[106-24-1]	69	93, 69, 123
Hexyl cinnamic aldehyde	[101-86-0]	216	145, 117, 129
Hydroxycitronellal	[107-75-75]	59	59, 71, 81, 96
Isouegenol	[97-54-1]	164	164, 103, 149
Butylphenyl methylpropional	[80-54-6]	189	189, 117, 131, 147
Limonene	[5989-27-5]	68	68, 67, 93, 121
Linalool	[78-70-6]	93	71, 80, 93, 121
Hydroxyisohexyl-3-cyclohexene carboxaldehyde	[31906-04-4]	Isomer 1: 136; Isomer 2: 136	136, 93, 149, 192
Methyl-2-nonynoate[b]	[111-80-8]	79	
Methyl-2-ocynoate	[111-12-6]	95	123, 67, 79, 95
Methyl eugenol[b]	[93-15-2]	178	
Phenylacetaldehyde[b]	[122-78-1]	91	
α-Isomethylionone	[127-51-5]	135	150, 107, 135

[a]Mixture of neral and geranial.
[b]Other compounds, not in the list of 24 allergens.
[c]Only isomers 1 and 2 (*ZE* and *ZZ*), isomers 3 and 4 were not considered due to their low abundance.

purposes. The analytical working group of IFRA developed a GC–MS method for allergen analysis; calibration curves were linear in the range of 2–100 mg/L, with coefficients of determination in excess of 0.99 [61]. Typical analysis times are on the order of 30 min, although fast GC–MS with full-scan mass spectrometry has also been employed to determine the presence of allergens [62]. Although these methods work well for fragrance products that can be separated without overlap of interfering compounds containing the same characteristic ions as the target compound, on many occasions sample complexity limits the applicability of the monodimensional separation approach. Thus, alternative approaches such as GC–ToF MS using computerized algorithms to deconvolute overlapping spectra [62], sequential analysis using dual-column, and twin injection GC–MS [63] have been applied to more challenging analyses.

GC×GC methods for the determination of allergens have been reported, but peak capacity can fall short of providing complete separation and cannot guarantee unambiguous determination of all target compounds [64], even with the enhanced peak capacity of GC×GC. Thus, GC×GC–MS techniques appear to be the most appropriate. A significant amount of work has been performed analysing these suspected allergens using GC×GC–MS [31,33,59,65–66]. Debonneville and Chaintreau determined that qMS coupled with GC×GC shows good performance when applied to the evaluation of allergens in fragrance concentrates [59]. GC×GC–qMS used in selected ion monitoring detection mode was shown to provide superior selectivity over both GC–qMS and GC×GC–FID, and linear calibrations were achieved despite the data acquisition rate providing fewer than 10 data points per second-dimension peak. Figure 5 illustrates the enhanced selectivity of GC×GC–qMS vs. GC×GC–FID for hexyl cinnamaldehyde in a fragrance concentrate. The individual second-dimension peaks of hexyl cinnamaldehyde (marked with arrows and a question mark) within a real fragrance in the upper GC×GC–FID chromatogram demonstrate that unambiguous detection without MS is not possible. The lower trace, acquired using GC×GC–qMS in selected ion monitoring mode, provides a reliable response for the target compound.

GC×GC–qMS analysis revealed nearly 870 resolved and partially overlapped peaks in a perfume sample using 20 Hz full-scan data acquisition (Figure 6) [65]. Twelve suspected allergens were detected among these 870 compounds, two of which could not be detected in a single-column separation. Adahchour et al. showed that a modern qMS can produce very impressive results in scan mode, offering pg detection limits at 33 Hz data acquisition rate and linear calibrations for the suspected allergens [66].

6. ANALYSIS OF CHIRAL COMPOUNDS

The development of stable enantioselective stationary phases for gas chromatography (based mostly on modified cyclodextrins) has permitted the detailed study of the enantiomeric compositions of terpenoids and a host of other

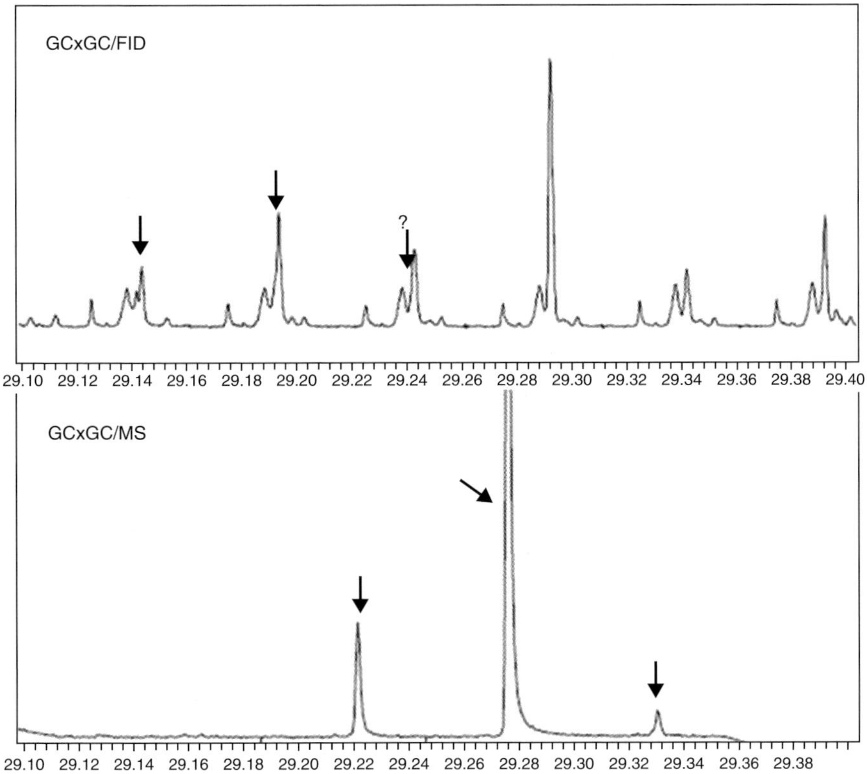

Figure 5 Peaks of hexyl cinnamaldehyde (marked with arrows and a question mark) within a real fragrance using FID and MS in the SIM mode [59].

compounds. The separation of enantiomers by GC using an enantioselective stationary phase was discovered in 1966 [67], and today, more than 2400 publications document some 24,000 enantiomeric separations involving 8000 chiral compounds [68].

The quantification of chiral substances in flavours, aromas, and food is one of the three areas from which more than 90% of the published multidimensional gas chromatography (MDGC) material is drawn [69]. MDGC with an enantioselective separation step (enantio-MDGC) was first described in 1984 [70] and, since the introduction of capillary GC columns containing cyclodextrin derivatives [71,72], has become an almost routine analysis in many specialised laboratories. The enantiomeric distribution of optically active compounds can be very useful for identifying adulterated foods and beverages, for controlling and monitoring fermentation processes and products, and for evaluating age and storage effects [3]. The use of enantio-MDGC for the analysis of lavender and peppermint essential oils has been reported [73], highlighting the ease of direct enantiomer measurement. Mosandl and co-workers have performed many applications employing enantio-MDGC-MS, as well as utilising the enhanced peak purity

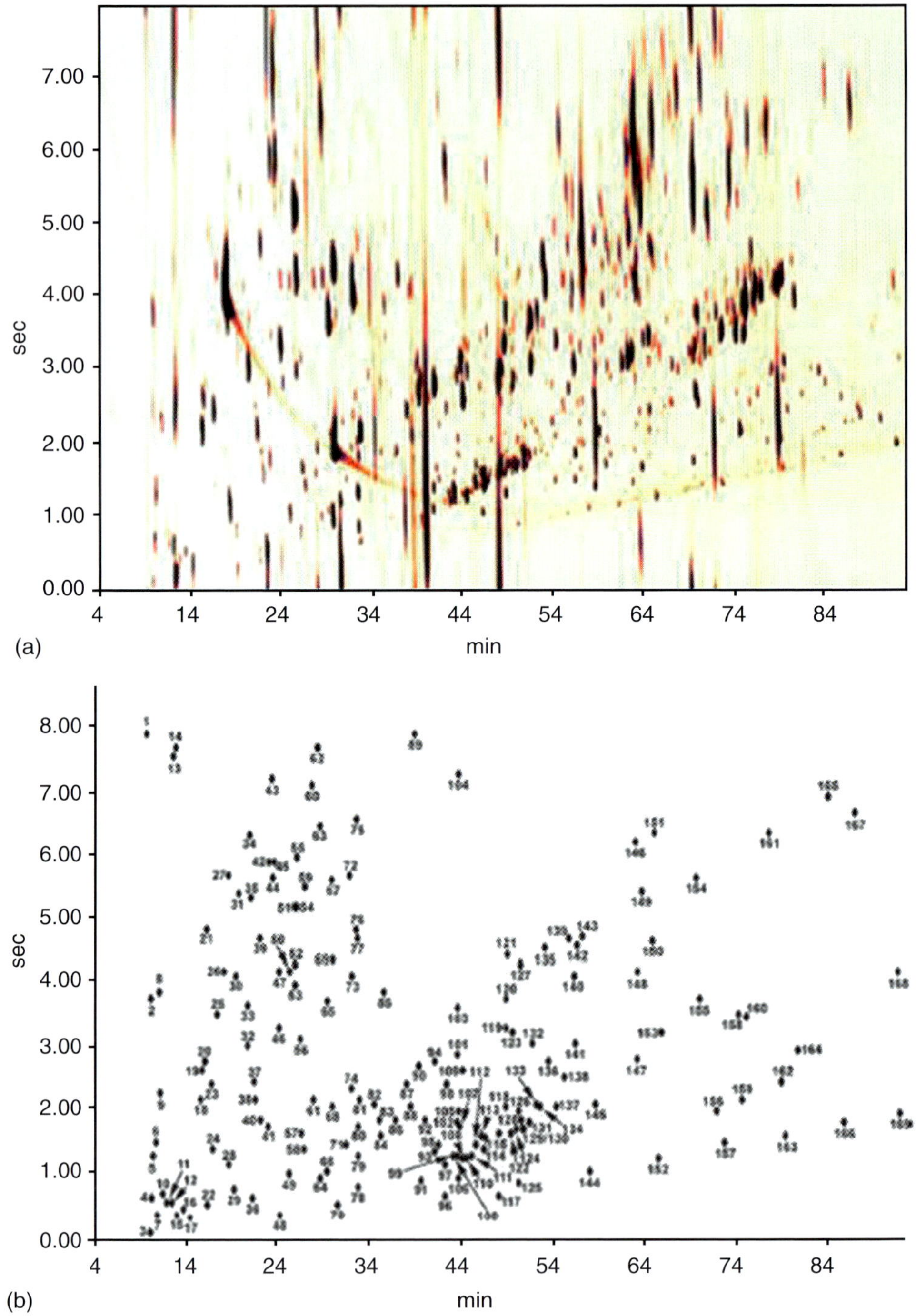

Figure 6 GC×GC–qMS perfume sample 2D contour plot result (upper) and peak apex plot (lower) containing the identified peaks [66].

required for isotope ratio mass spectrometry (IRMS), by using MDGC–IRMS [74]. Mondello and Dugo have reported a number of studies utilising MDGC and enantio-MDGC for the study of citrus essential oils [75–78]. The only suitable experimental arrangement for enantio-MDGC involves an achiral primary column and an enantioselective stationary phase second-dimension analytical column. Unresolved target components are heart-cut from the primary column and delivered to the analytical column, which provides resolution and quantitative measurement of the individual enantiomers. A very small number of publications have now reported the potential of GC×GC to perform this specialised analysis task.

There are two distinctly different approaches to performing the enantioselective GC×GC experiment, which may be distinguished as follows. The first approach uses the enantioselective column as the first-dimension column, which shall be denoted "enantio-GC×GC>". The second uses a short enantioselective column in the second-dimension position; the latter approach shall be called "GC×enantio-GC>". Both techniques should provide a valid result, and although interpretation of the GC×enantio-GC chromatogram is easier, the experimental setup and operation of the enantio-GC×GC experiment is technically less demanding.

Since there is only one enantioselective step in the enantio-GC×GC configuration, the resolution of chiral isomers is essentially a monodimensional chromatographic process, and improved enantiomer separation is not possible over a conventional monodimensional analysis. However, the second-dimension column significantly reduces the chance that other sample components (or matrix components) will co-elute or overlap with the target chiral components, which would otherwise compromise quantitation of the target components. This implies that optimisation of the first-dimension separation is the most crucial to producing a successful result. Harju and Haglund implemented a dual GC-oven enantio-GC×GC system [79], which allowed careful optimisation of the first-dimension enantioselective separation in the first GC oven, using a slow-temperature program ramp. The second GC oven was temperature programmed with a +40°C temperature offset above the first oven, which was shown to improve the overall two-dimensional resolution. For a long time the thinking was that there should ideally be three to four modulation pulses/first-dimension per peak in the first-dimension to ensure that the separated peaks remain resolved throughout the entire process [80]. Recently, Marriott and co-workers introduced the term *modulation ratio* (M_R), which describes the sampling rate of the first-dimension separation [81]. The modulation ratio is defined as the ratio of four times the first column peak standard deviation divided by the modulation period. It has been shown that for the analysis of trace components where precise quantitative measurements are being made, the comprehensive two-dimensional separation should be conducted with $M_R \geqslant 3$ [81]. For the analysis of abundant solutes, or for semiquantitative analysis, an $M_R \sim 1.5$ is sufficient [81]. In the case of enantio-GC×GC, the chiral components will have identical interaction with the achiral second-column stationary phase, and thus are not resolvable in this column. Therefore, a high modulation ratio is needed to maintain resolution of optical isomers separated in the first-dimension. By operating the second-dimension column at a higher temperature, it is possible to ensure that a sufficiently high sampling rate of the first-column effluent into the second column is achieved because this allows a faster second-dimension separation. Enantio-GC×GC is useful as an alternative to enantio-MDGC which may require multiple heart-cuts and individual temperature programmed analysis of these heart-cuts. The main advantage of GC×GC here is the possibility of reducing the total analysis time.

The GC×GC resolution advantage is known to improve the efficiency of enantioselective essential oil analysis (in contrast to one-dimensional analysis). In a single temperature-programmed analysis, the individual antipodes of optically active components can be separated and are effectively free from matrix interferences. The enantiomeric compositions of a number of monoterpene hydrocarbons and oxygenated monoterpenes in Australian tea tree (*Melaleuca alternifolia*), including sabinene, α-pinene, β-phellandrene, limonene, *trans*-sabinene hydrate, *cis*-sabinene hydrate, linalool, terpinen-4-ol, and α-terpineol shown in Figure 7,

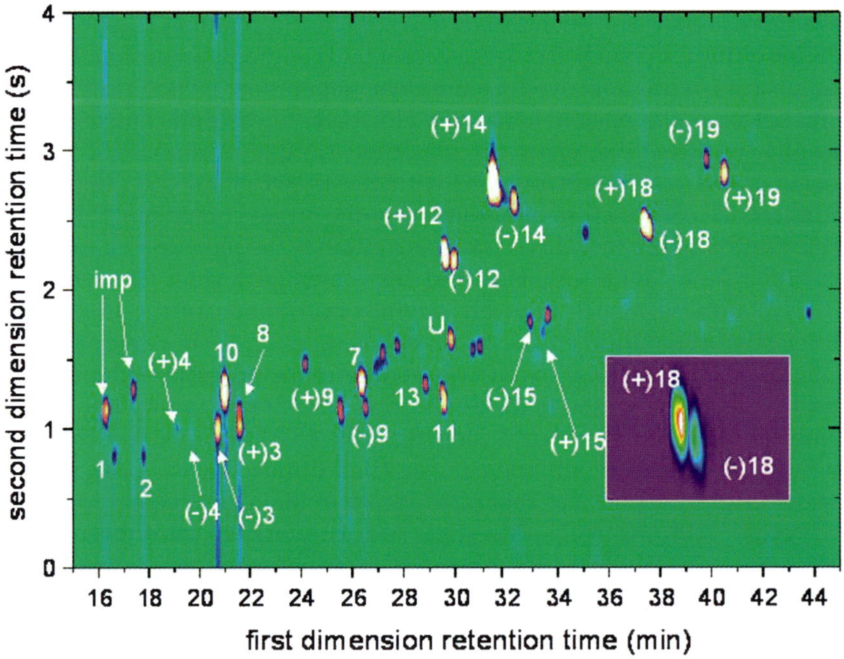

Figure 7 2D contour plot for the enantio-GC×GC analysis of flush growth *Melaleuca alternifolia*. Individual isomers are differentiated by a (+) or (−) sign. Component identity 1, α-thujene; 2, α-pinene; 3, sabinene; 4, β-pinene; 5, myrcene; 6, α-phellandrene; 7, α-terpinene; 8, p-cymeme; 9, limonene; 10, 1,8-cineole; 11, γ-terpinene; 12, *trans*-sabinene hydrate; 13, terpinolene; 14, *cis*-sabinene hydrate; 15, linalool; 16, *cis*-p-menth-2-en-1-ol; 17, *trans*-p-menth-2-en-1-ol; 18, terpinen-4-ol; 19, α-terpineol. Imp is due to a solvent impurity or system peak, and U is also an unknown compound that would interfere with the quantification of *trans*-sabinene hydrate in 1D analysis. The expanded section shows (+)- and (−)-terpinen-4-ol (18) [11].

were determined by enantio-GC×GC [11]. Wang and co-workers also implemented an enantio-GC×GC approach for the separation of optically active ephedrine-type alkaloids in the herb Ma Huang [40], which is commonly used in traditional Chinese medicine. The methodology described was used for chemical fingerprinting as well as for quantitative determination of the active components, (−)-ephedrine, (+)-pseudoephedrine, (−)-N-methylephedrine, and (−)-norephedrine. Good linearity ($R^2 \geqslant 0.999$) over one to two orders of magnitude in concentration and limits of detection in the order of 0.1–1.3 mg/L were reported.

A second GC×GC approach (GC×enantio-GC) employs a short, fast enantioselective stationary phase column as the second-dimension column. Experiments have been performed to determine the suitability of a narrow-bore column with a thin film of the enantioselective stationary phase to achieve sufficiently fast GC separations for use in GC×enantio-GC, but the performance of such columns is unsuitable. However, faster separations (for example, <8 s for (±)-limonene) can be achieved using the principle of low-pressure GC operation in the second-dimension [14]. In this experimental setup, a wider bore second-dimension column was used with mass spectrometry detection. A narrow-bore first-dimension column provided the necessary restriction for low-pressure second-dimension column operation. GC×enantio-GC analysis of bergamot essential oil was performed using a 12-min analysis in which the individual isomers of (±)-sabinene, (±)-limonene, (±)-linalool, (±)-α-terpineol, and (±)-linalyl acetate were successfully separated (Figure 8). Faster enantioseparations are required to

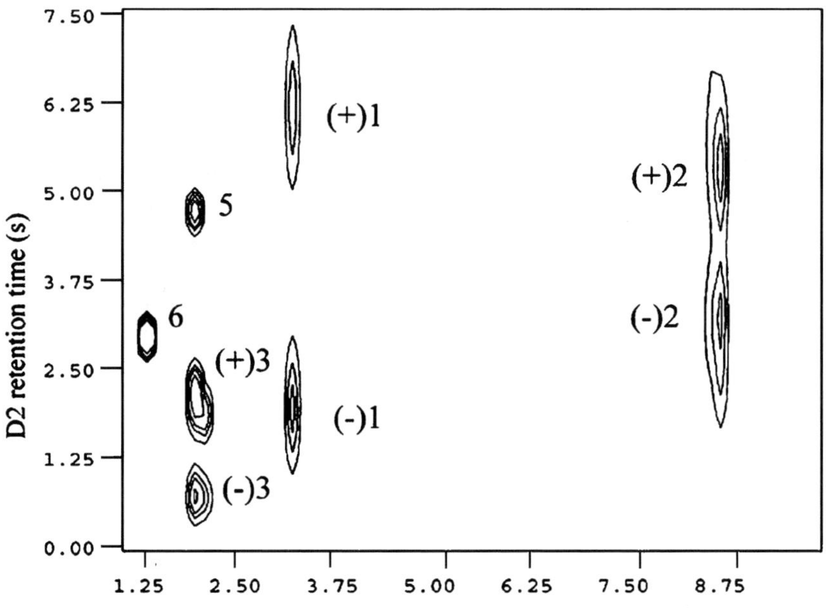

Figure 8 2D contour plot for the GC×enantio-GC–MS analysis of a standard mixture. Identity of components: linalool (1), linalyl acetate (2), limonene (3), 1,8-cineole (5), and α-pinene (6) [14].

improve the efficiency of the GC×enantio-GC experiment because high selectivity factors are required to permit the separation to be effectively sped up; increasing temperature is deleterious to enantioselectivity so it is currently very difficult to achieve fast, efficient chiral separations.

Junge et al. claimed to have developed fast enantioselective analysis using GC×GC with a narrow-bore enantioselective first-dimension column [30], using a temperature-programmed analysis of 2°C/min with a 4-s modulation period. In enantio-GC×GC, the overall speed of analysis is limited by the first-dimension separation compared to fast achiral GC×GC separations [30], which are possible with temperature programming up to 35°C/min and total analysis time of <10 min. Conventional enantio-MDGC holds a major advantage over enantio-GC×GC for fast analysis in that the enantioselective step is in the second-dimension. Thus, fast enantio-MDGC [82] was demonstrated by Mondello and co-workers, isolating 18 enantiomers within 10 min, as shown in Figure 9. At this time, GC×GC cannot match conventional MDGC performance in terms of speed or resolution for enantioselective applications.

7. FUTURE TRENDS

This decade has seen a number of significant developments that have made GC×GC appealing for essential oils and fragrance analysis. Most of the advances have been generic advances, such as the development of new modulation interfaces and software advances, discussed elsewhere in this book, but two of the most useful specifically for this class of sample are the coupling with MS as well as the development of retention index approaches, this field being heavily dependent on both MS and retention index systems. Further development of retention index systems will be highly beneficial in the future for characterisation of essential oils and fragrances using GC×GC. This is perhaps the area that still needs the greatest attention. It is possible that GC×GC will never match MDGC's capabilities for enantioselective analysis, but the development of new stationary phases with high selectivity factors may permit further development of largely underdeveloped GC×enantio-GC approaches. A striking difference between the practice of GC–MS and GC×GC/GC×GC–MS is that GC–MS uses largely standardised methods. One can perform a GC–MS separation of an essential oil, for instance, and quite readily compare the results with literature chromatograms. On the other hand, almost every GC×GC approach to date in the literature has been slightly different. This can make comparison with published results somewhat difficult. It is perhaps not unexpected that there is so much variation in methods: GC×GC has only been applied to this type of sample for a little over eight years and is really still undergoing a development stage. Nonetheless, the differences in each approach make it difficult to compare results from laboratory to laboratory and can give the perception that GC×GC is not straightforward to implement. From an experienced GC×GC user's perspective, this is certainly

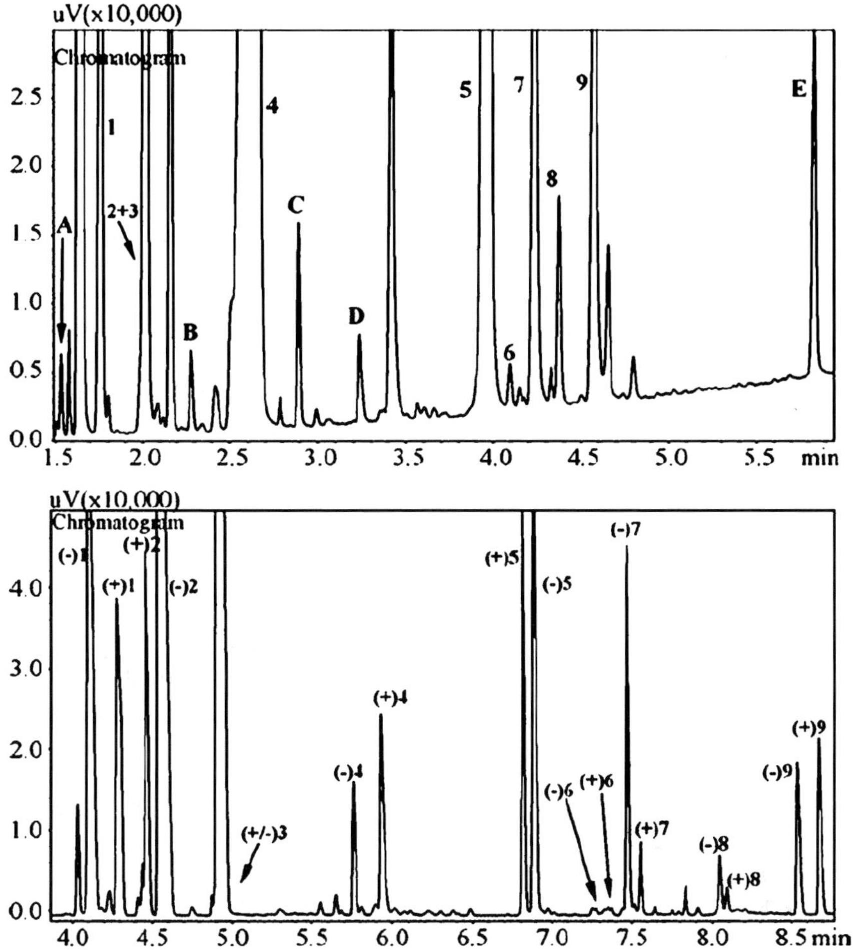

Figure 9 Upper chromatogram: a 4.5-min chromatographic expansion from the fast MDGC rosemary oil application with the transfer system in the standby position; peak identification: tricyclene (peak A); α-phellandrene (peak B); unknown (peak C); α-terpinolene (peak D); bornyl acetate (peak E). Lower chromatogram: a 5-min second-dimension chromatographic expansion from the fast MDGC rosemary oil application. Other compound identities are 1, camphene; 2, β-pinene; 3, sabinene; 4, limonene; 5, camphor; 6, isoborneol; 7, borneol; 8, terpinen-4-ol; 9, α-terpineol [82].

not the case. In fact, GC×GC has reached such a level of maturity that there is no reason GC×GC cannot be readily and routinely applied, especially now that dedicated software has become available for questioning GC×GC results. In the next five years some consolidation must take place if more and more users, rather than developers, are to embrace this marvelous technology.

REFERENCES

1 http://www.leffingwell.com/top_10.htm
2 A. Liberti and G. Conti, Possibilita di applicazione della chromatografia in fase gassosa allo studio della essenze, Presented at 1° Convegno internazionale di studi e ricerche sulle essenze, Reggio Calabria, Italy, 1956.
3 G. Dugo, P. Dugo and L. Mondello, In: L. Mondello, A.C. Lewis and K.D. Bartle (Eds.), *Multidimensional Chromatography*, John Wiley & Sons, Chichester, England, 2001, pp. 217–249.
4 C. Bicchi, P. Rubiolo and C. Cordero, *Anal. Bioanal. Chem.*, 384 (2006) 53.
5 Z. Liu and J.B. Phillips, *J. Chromatogr. Sci.*, 29 (1991) 227.
6 J.M.D. Dimandja, S.B. Stanfill, J. Grainger and D.G. Patterson, Jr., *J. High Resolut. Chromatogr.*, 23 (2000) 208.
7 P. Marriott, R. Shellie, J. Fergeus, R. Ong and P. Morrison, *Flavour Fragr. J.*, 15 (2000) 225.
8 R. Shellie, P. Marriott and C. Cornwell, *J. High Resolut. Chromatogr.*, 23 (2000) 554.
9 A. van Asten, *Trends Anal. Chem.*, 21 (2002) 698.
10 R. Shellie, P. Marriott and P. Morrison, *Anal. Chem.*, 73 (2001) 1336.
11 R. Shellie, P. Marriott and C. Cornwell, *J. Sep. Sci.*, 24 (2001) 823.
12 R.A. Shellie, L.L. Xie and P.J. Marriott, *J. Chromatogr. A*, 968 (2002) 161.
13 R. Shellie, L. Mondello, P. Marriott and G. Dugo, *J. Chromatogr. A*, 970 (2002) 225.
14 R. Shellie and P.J. Marriott, *Anal. Chem.*, 74 (2002) 5426.
15 Y. Shao, P. Marriott, R. Shellie and H. Hügel, *Flavour Fragr. J.*, 18 (2003) 5.
16 R.A. Shellie and P.J. Marriott, *Analyst*, 128 (2003) 879.
17 J. Wu, X. Lu, W. Tang, H. Kong, S. Zhou and G. Xu, *J. Chromatogr. A*, 1034 (2004) 199.
18 M.T. Roberts, J.P. Dufour and A.C. Lewis, *J. Sep. Sci.*, 27 (2004) 473.
19 M.Z. Özel and H. Kaymaz, *Anal. Bioanal. Chem.*, 379 (2004) 1127.
20 M.Z. Özel, F. Göğüş, J.F. Hamilton and A.C. Lewis, *Chromatographia*, 60 (2004) 79.
21 R. Shellie, P. Marriott and P. Morrison, *J. Chromatogr. Sci.*, 42 (2004) 417.
22 L. Mondello, A. Casilli, P.Q. Tranchida, P. Dugo and G. Dugo, *Flavour Fragr. J.*, 20 (2005) 136.
23 M.Z. Özel, F. Göğüş and A.C. Lewis, *Anal. Bioanal. Chem.*, 382 (2005) 115.
24 G. Eyres, J.P. Dufour, G. Hallifax, S. Sotheeswaran and P.J. Marriott, *J. Sep. Sci.*, 28 (2005) 1061.
25 S. Zhu, X. Lu, L. Dong, J. Xing, X. Su, H. Kong, G. Xu and C. Wu, *J. Chromatogr. A*, 1086 (2005) 107.
26 S. Zhu, X. Lu, J. Xing, S. Zhang, H. Kong, G. Xu and C. Wu, *Anal. Chim. Acta*, 545 (2005) 224.
27 F. Göğüş, M.Z. Özel and A.C. Lewis, *Flavour Fragr. J.*, 21 (2006) 122.
28 M.Z. Özel, F. Göğüş and A.C. Lewis, *Anal. Chim. Acta.*, 566 (2006) 172.
29 M. Libardoni, E. Hasselbrink, J.H. Waite and R. Sacks, *J. Sep. Sci.*, 29 (2006) 1001.
30 M. Junge, S. Bieri, H. Huegel and P.J. Marriott, *Anal. Chem.*, 79 (2007) 4448.
31 C. Cordero, C. Bicchi, D. Joulain and P. Rubiolo, *J. Chromatogr. A*, 1150 (2007) 37.
32 Z.L. Cardeal, M.D.R. Gomes da Silva and P.J. Marriott, *Rapid Commun. Mass Spectrom.*, 20 (2006) 2823.
33 C. Cordero, P. Rubiolo, B. Sgorbini, M. Galli and C. Bicchi, *J. Chromatogr. A*, 1132 (2006) 268.
34 T. Pang, S. Zhu, X. Liu and G. Xu, *J. Sep. Sci.*, 30 (2007) 868.
35 G. Eyres, P.J. Marriott and J.P. Dufour, *J. Chromatogr. A*, 1150 (2007) 70.
36 G.T. Eyres, P.J. Marriott and J.P. Dufour, *J. Agric. Food Chem.*, 55 (2007) 6252.
37 E. Klimánková, K. Holadová, J. Hajšlová, T. Čajka, J. Poustka and M. Koudela, *Food Chem*, 107 (2008) 464.
38 R.A. Shellie, P.J. Marriott and C.W. Huie, *J. Sep. Sci.*, 26 (2003) 1185.
39 X. Di, R.A. Shellie, P.J. Marriott and C.W. Huie, *J. Sep. Sci.*, 27 (2004) 451.
40 M. Wang, P.J. Marriott, W.H. Chan, A.W.M. Lee and C.W. Huie, *J. Chromatogr. A*, 1112 (2006) 361.
41 J. Beens, H. Boelens, R. Tijssen and J. Blomberg, *J. High Resolut. Chromatogr.*, 21 (1998) 63.
42 L. Mondello, D. Sciarrone, A. Casilli, P.Q. Tranchida, P. Dugo and G. Dugo, *J. Sep. Sci.*, 30 (2007) 1905.
43 R. Shellie, P. Marriott, P. Morrison and L. Mondello, *J. Sep. Sci.*, 27 (2004) 504.
44 International Organisation of the Flavor Industry (IOFI), *Z. Lebensm. Unters. Forsch.*, 192 (1991) 530.
45 W.A. König, N. Bülow and Y. Saritas, *Flavour Fragr. J.*, 14 (1999) 367.

46 V.I. Babushok, P.J. Linstrom, J.J. Reed, I.G. Zenkevich, R.L. Brown, W.G. Mallard and S.E. Stein, *J. Chromatogr. A*, 1157 (2007) 414.
47 R.P. Adams, *Identification of Essential Oil Components by Gas Chromatography/Quadrupole Mass Spectrometry*, 3rd ed, Allured Publ., Carol Stream, IL, 2001.
48 http://www.flavornet.org
49 W.G. Jennings and T. Shibamoto, *Qualitative Analysis of Flavor and Fragrance Volatiles by Glass Capillary Gas Chromatography*, Academic Press, New York, 1980.
50 N. Kondjoyan and J.L. Berdague, *A Compilation of Relative Retention Indices for the Analysis of Aromatic Compounds*, Laboratorie Flavour, Theix, France, 1996.
51 http://wwwleffingwell.com/baciseso.htm
52 http://www.crec.ifas.ufl.edu/crec_websites/Rouseff/index.htm#
53 C. Bicchi, A. Binello, A. D'Amato and P. Rubiolo, *J. Chromatogr. Sci.*, 37 (1999) 288.
54 J.S. Arey, R.K. Nelson, L. Xu and C.M. Reddy, *Anal. Chem.*, 77 (2005) 7172.
55 N.W. Davies, *J. Chromatogr.*, 503 (1990) 1.
56 S. Bieri and P.J. Marriott, *Anal. Chem.*, 78 (2006) 8089.
57 J.V. Seeley, F.J. Kramp, K.S. Sharpe and S.K. Seeley, *J. Sep. Sci.*, 25 (2002) 53.
58 EC (2003) Directive 2003/15/EC (relating to cosmetic products). European Community, Brussels.
59 C. Debonneville and A. Chaintreau, *J. Chromatogr. A*, 1027 (2004) 109.
60 *Analytical Procedure for the GC/MS Quantitation of Potential Fragrance Allergens in Fragrance Compounds; version 1. Analytical Procedures*, International Fragrance Associations, Brussels, Belgium, 2003.
61 A. Chaintreau, D. Joulain, C. Marin, C.O. Schmidt and M. Vey, *J. Agric. Food Chem.*, 51 (2003) 6398.
62 Rapid Determination and Quantification of Sensitizers and Skin Irritants in Fragrances by GC-TOFMS. LECO Separation Science Application Note, Form 203-821-191, 3/03-REV2, 1.
63 H. Leijs, J. Broekhans, L. van Pelt and C. Mussinan, *J. Agric. Food. Chem.*, 53 (2005) 5487.
64 R. Shellie, P.J. Marriott and A. Chaintreau, *Flavour Fragr. J.*, 19 (2004) 91.
65 L. Mondello, A. Casilli, P.Q. Tranchida, G. Dugo and P. Dugo, *J. Chromatogr. A*, 1067 (2005) 235.
66 M. Adahchour, M. Brandt, H.U. Baier, R.J.J. Vreuls, A.M. Batenburg and U.A.Th. Brinkman, *J. Chromatogr. A*, 1067 (2005) 245.
67 R. Charles-Sigler and E. Gil-Av, *Tetrahedron Lett*, 7 (1966) 4231.
68 V. Schurig, *Trends Anal. Chem.*, 21 (2002) 647.
69 W. Bertsch, *J. High Resolut. Chromatogr.*, 22 (1999) 647.
70 G. Schomburg, H. Husmann, E. Hübinger and W.A. König, *J. High Resolut. Chromatogr.*, 7 (1984) 404.
71 V. Schurig and H.P. Nowotny, *J. Chromatogr.*, 441 (1988) 155.
72 W. König, S. Lutz, P. Mischnick-Lübbecke and B. Brassat, *J. Chromatogr.*, 447 (1988) 193.
73 C. Bicchi and A. Pisciotta, *J. Chromatogr.*, 508 (1990) 341.
74 S. Reichert, D. Fischer, S. Asche and A. Mosandl, *Flavour Fragr. J.*, 15 (2000) 303.
75 L. Mondello, K.D. Bartle, G. Dugo and P. Dugo, *J. High Resolut. Chromatogr.*, 17 (1994) 312.
76 L. Mondello, M. Catalfamo, A.R. Proteggente, I. Bonaccorsi and G. Dugo, *J. Agric. Food Chem.*, 46 (1998) 54.
77 L. Mondello, A. Verzera, P. Previti, F. Crispo and G. Dugo, *J. Agric. Food Chem.*, 46 (1998) 4275.
78 L. Mondello, M. Catalfamo, G. Dugo and P. Dugo, *J. Chromatogr. Sci.*, 36 (1998) 201.
79 M. Harju and P. Haglund, *J. Microcolumn Sep.*, 13 (2001) 300.
80 R.E. Murphy, M.R. Schure and J.P. Foley, *Anal. Chem.*, 70 (1998) 1585.
81 W. Khummueng, J. Harynuk and P.J. Marriott, *Anal. Chem.*, 78 (2006) 4578.
82 L. Mondello, A. Casilli, P.Q. Tranchida, M. Furukawa, K. Komori, K. Miseki, P. Dugo and G. Dugo, *J. Chromatogr. A*, 1105 (2006) 11.

Table 1 GC×GC food applications, detector(s), column combination(s), and modulator

Application	Column combination(s) (m×mm ID×μm d_f)	Modulator	Detector(s)	Reference
Lipids				
Vegetable and fish oil FAMEs	HP-1 (9.0×0.2×0.33)×CP-Wax-52 (0.3×0.1×0.2)	Thermal sweeper	FID	[7]
Vegetable, marine and land animal FAMEs	a) BPX-5 (30×0.25×0.25)×SupelcoWax-10 (1×0.1×0.1) b) SupelcoWax-10 (30×0.25×0.25)×SPB-5 (1×0.1×0.1)	LMCS[a]	FID	[8]
Blue-mussel marine oil FAMEs	a) BPX-5 (30×0.25×0.25)×BP-20 (0.8×0.1×0.1) b) BP-1 (25×0.22×0.2)×BPX-70 (1×0.1×0.2)	LMCS	FID	[9]
Cod liver oil FAMEs	Equity-5 (30×0.25×0.25)×SupelcoWax-10 (1×0.1×0.1)	LMCS	FID	[10]
cis, *trans*-FAME isomers	CP-WAX (30×0.25×0.25)×VF-23 ms (3×0.1×0.1)		FID	[11]
Cod liver oil FAMEs (GC×GC gas flow optimization)	a) Equity-5MS (30×0.25×0.25)×SupelcoWax-10 (1×0.1×0.1) b) Equity-5MS (30×0.25×0.25)×SupelcoWax-10 (1×0.1×0.1) + retention gap (0.3×0.1)	LMCS	FID	[14]
Lanolin	a) XTI-5 (10×0.25×0.25)×BPX-50 (1×0.1×0.1) b) XTI-5 (10×0.25×0.25)×BGB-Wax (0.5×0.05×0.1) c) BP-1 (15×0.25×0.25)×HT-8 (1×0.1×0.1) d) ZB-5 (30×0.25×0.25)×BPX-50 (1×0.1×0.1) e) DB-WAX (25×0.32×0.25)×BPX-35 (1×0.1×0.1)	Ambient air twin-jet modulator	ToF MS	[18]
Milk fat FAMEs	a) CP-7420 (100×0.25×0.25)×HP-5MS (1.5×0.1×0.1) b) CP-7420 (100×0.25×0.25)×HP-1 (1×0.1×0.1) c) HP-1 (30×0.25×0.25)×cyano test column (1×0.1×0.1) d) HP-1 (30×0.25×0.25)×DB-Wax (0.35×0.05×0.1)	Semirotating cryogenic modulator	FID	[20]

Table 1 *(Continued)*

Application	Column combination(s) (m×mm ID×μm d_f)	Modulator	Detector(s)	Reference
FAMEs (structured retention study)	a) BPX-5 (30×0.25×0.25)×BP-20 (0.8×0.1×0.1) b) BPX-5 (30×0.25×0.25)×BPX-80 (0.8×0.1×0.1) c) BPX-60 (30×0.25×0.25)×BP-1 (1×0.1×0.1) d) BPX-70 (30×0.25×0.25)×BP-1 (0.6×0.1×0.1) e) BPX-80 (30×0.25×0.25)×BP1 (0.45×0.1×0.1) f) BPX-90 (30×0.25×0.25)×BP-1 (0.25×0.1×0.1)		FID	[21]
Volatiles				
Garlic powder volatiles	a) HP-1 (30×0.32×0.25)×DB-5 (1.5×0.1×0.1) b) HP-1 (30×0.32×0.25)×BPX-50 (1.5×0.1×0.1)	In-lab constructed moving cryogenic modulator	ToF MS, FID	[17]
Olive oil flavours	BP-1 (27×0.32×0.32)×VF-23ms (1×0.1×0.1)	In-lab constructed CO_2 twin-jet modulator	qMS	[22]
Butter flavours	BP-21 (30×0.25×0.25)×BPX-35 (1×0.1×0.1)	In-lab constructed CO_2 twin-jet modulator	ToF MS, FID	[23]
Dairy spread and sour cream flavours	CP-Sil 5 CB low bleed/MS (15×0.25×0.25)×BPX-50 (0.8×0.1×0.1)	LMCS	ToF MS	[24]
Olive oil flavours	a) DB-1 (30×0.25×0.25)×BP-20 (1×0.1×0.1) b) BP-21 (30×0.25×0.25)×BPX-35 (1×0.1×0.1)	In-lab constructed CO_2 twin-jet modulator	FID	[25]

Vanilla extract	BP-21 (30×0.25×0.25)×BPX-35 (1×0.1×0.1)	In-lab constructed CO_2 twin-jet modulator	ToF MS, FID	[25]
Roasted coffee bean volatiles	a) Solgel Wax (30×0.25×0.25)×BPX-5 (1×0.1×0.1) b) BPX-5 (30×0.25×0.25)×BP-20 (0.8×0.1×0.1) c) SupelcoWax-10 (30×0.25×0.25)×SPB-5 (1×0.1×0.1)	LMCS	ToF MS, qMS	[26]
Roasted coffee bean volatiles	SupelcoWax-10 (30×0.25×0.25)×SPB-5 (1×0.1×0.1)	LMCS	FID	[27]
Roasted coffee bean volatiles	BPX-5 (30×0.25×0.25)×BP-20 (1×0.1×0.1)	LMCS	NPD, ToF MS, FID	[28]
Strawberry enantiomer volatiles	EtTBS-β-CD (20×0.25×0.25) - Cyclosil B (26×0.25×0.25)×BPX-50 (1×0.1×0.1)	LMCS	FID	[32]
Pepper volatiles	a) BPX-5 (30×0.25×0.25)×BP-20 (1.5×0.1×0.1) b) BPX-5 (30×0.25×0.25)×BP-20 (1×0.1×0.1)	LMCS	ToF MS, qMS, FID	[33]
Oven roast beef sulphur volatiles	DB-1 (20×0.18×0.18)×DB-225 (1×0.1×0.1)	Cryogenic two-stage quad-jet modulator	ToF MS	[34]
Honey volatiles	a) DB-5ms (30×0.25×0.25)×BPX-50 (1.25×0.1×0.1) b) DB-5ms (30×0.25×0.25)×SupelcoWax-10 (1.25×0.1×0.1) c) HP-INNOWax (30×0.25×0.25)×BPX-50 (1.25×0.1×0.1)	Cryogenic two-stage quad-jet modulator	ToF MS	[35]
Wine volatiles	EtTBS-β-CD (24×0.25×0.25) - Cyclosil B (30×0.25×0.25)×BPX-50 (1×0.1×0.1)	LMCS	FID	[36]
Wine methoxypyrazines	BPX-5 (30×0.25×0.25)×BP-20 (1×0.1×0.1)	LMCS	NPD, ToF MS	[37]
Cheddar cheese volatiles	DB-5 (10×0.18×0.18)×DB-17 (1.6×0.18×0.18)	Cryogenic two-stage quad-jet modulator	ToF-MS	[38]

Table 1 (*Continued*)

Application	Column combination(s) (m×mm ID×μm d_f)	Modulator	Detector(s)	Reference
Ginger volatiles	BPX-5 (30×0.25×0.25)×BP-20 (0.8×0.1×0.1)	LMCS	qMS	[39]
Lemon juice and lemon-flavoured drinks	SPB-1 (15×0.25×1.0)×SupelcoWax-10 (0.7×0.1×0.1)	KT2003 loop modulator	FID	[40]
Grape monoterpenes	Equity-5 (60×0.25×1.0)×SupelcoWax-10 (2.5×0.1×0.1)	Cryogenic two-stage jet modulator	ToF MS	[41]
Roasted barley volatiles	DB-Wax (30×0.32×0.5)×BPX-50 (0.75×0.1×0.1)	In-lab constructed CO_2 twin-jet modulator	ToF MS	[42]
Other				
Wine, beer, and honey amino acids	a) BPX-5 (30×0.25×0.25)×BPX-50 (2×0.1×0.1) b) SolgelWax (30×0.25×0.25)×BP-1 (1.5×0.1×0.1)	LMCS	ToF MS	[43]
Beer amino acid enantiomers	a) Chirasil-L-Val (25×0.25×0.16)×BPX-50 (1×0.1×0.1) b) Chirasil-L-Val (25×0.25×0.16)×BPX-50 (3×0.1×0.1) c) Chirasil-L-Val (25×0.25×0.16)×BP-1 (1×0.1×0.1) d) Chirasil-L-Val (25×0.25×0.16)×BP-1 (3×0.1×0.1)		FID	[44]

[a]LMCS = longitudinally modulated cryogenic system.

the lipidic intake and metabolism (e.g., excessive consumption of saturate FAs, cholesterol, etc.). Furthermore, fats and oils are a fundamental source of vitamins, and contribute greatly to the physical structure of foods as well as to the solubilisation of several aroma constituents. In consideration of these briefly described features, the availability of technologies for the thorough unveiling of lipid profiles is fundamental in industrial, academic, and clinical fields.

Conventional GC has been, by far, the most common approach used for the unveiling of FA profiles in food matrices, which are considered as simple to moderately complex mixtures (see Figure 2). In order to analyse these lipidic constituents, it is well known that their polarity must be reduced and their volatility increased. Normally, they are transformed in FAMEs and separated on polyethylene glycol columns ($30\,m \times 0.25\,mm$ ID$\times 0.25\,\mu m$ d_f is a typical dimension). The latter have a proven effectiveness for the separation of the most important saturated and unsaturated FAs, with apparently few co-elution problems. Longer capillaries with high-polarity phases (e.g., bis-cyanopropyl polysiloxane) are required for the complete separation of conjugated and/or geometrical FA isomers [6].

In general, GC×GC lipid applications have revealed the higher-than-known complexity of several of these matrices. The enhanced sensitivity has enabled the detection of many trace-quantity components (e.g., certain odd carbon number FAMEs), while the formation of FAME group-type patterns has been exploited for identification purposes. The first research on GC×GC fatty acid analysis was reported by de Geus et al. in 2001 [7]. In this study, which highlighted the sensitivity and group-type pattern advantages provided by GC×GC, the authors used two ovens, a thermal sweeper as modulation system and an "orthogonal" column combination. An application study was carried out on herring oil-derived FAMEs and, although "wraparound" appeared to be rather evident (the $C_{16:0}$ peak touches the first-dimension retention-time axis), the resulting contour plot was well-structured (Figure 3).

In fact, FAMEs with the same carbon number were located in specific clusters, while those with the same number of double bonds were positioned along distinct curves. It must be added that on the basis of the aims of any GC×GC analysis, wraparound is quite acceptable as long as the two-dimensional (2D) structure is maintained, if the highest number of compounds are separated or if target analytes remain entirely resolved (Chapter 2). In terms of sensitivity, several odd-number FAs, hardly detected in conventional 1D GC analysis, were quite visible on the 2D chromatogram. Several C_{19} FAMEs, in the 0–3 double-bond range, were tentatively identified on the basis of their specific bidimensional locations in the contour plot. Obviously, no standard compounds were available for identity-confirmation of such "unusual" FAMEs. Olive oil was also subjected to analysis, highlighting the well-known greater complexity of marine oils. Relative quantitative data was derived for vegetable and fish oils by using a commercial software.

Vegetable, marine, and land animal FAMEs have also been analysed by Mondello et al., using GC×GC [8]. The experiments, as confirmed by the authors, were considered a development of previous published research [7,9]. The

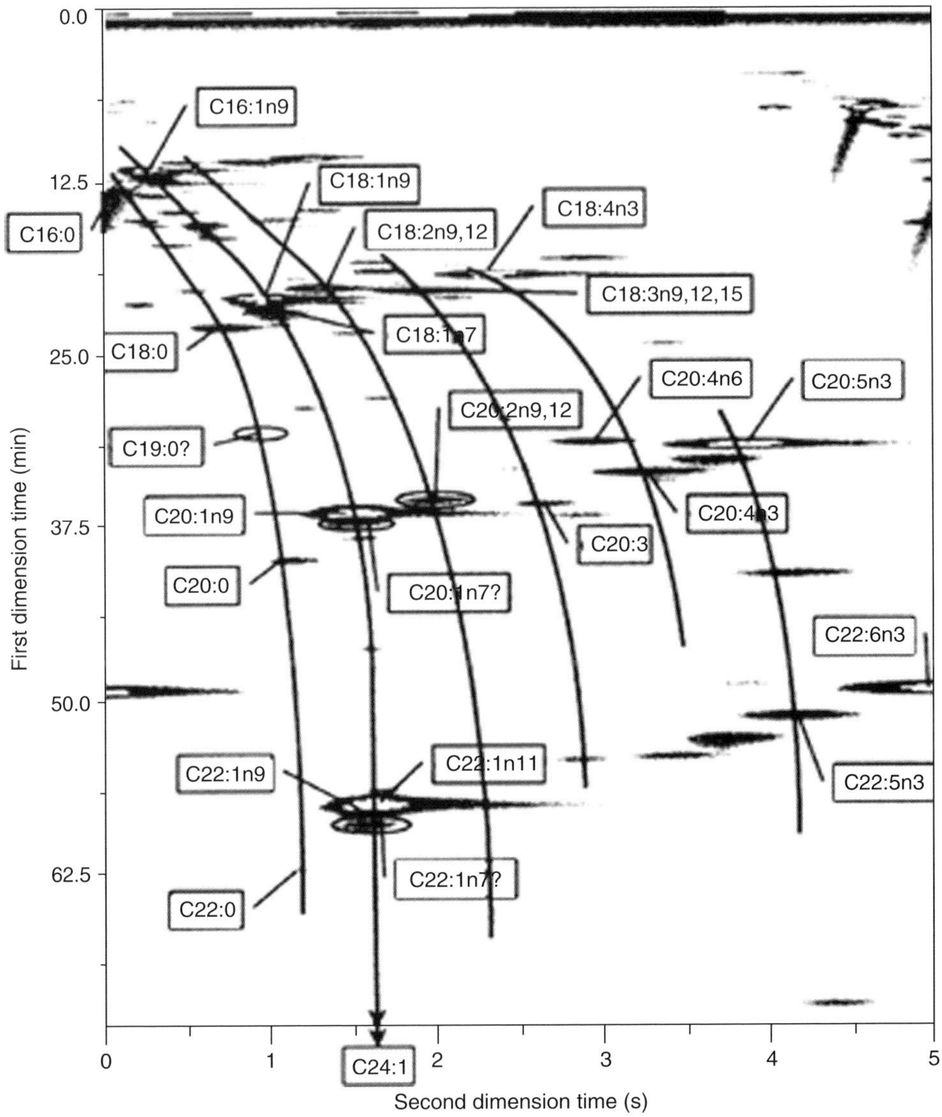

Figure 3 GC×GC–FID chromatogram of a herring oil sample; peaks were identified by using available standards and elution bands of homologous groups [7].

samples were subjected to analysis on both apolar-polar and "reverse" (polar-apolar) column sets, with the former the preferred choice owing to the generation of more orderly, structured, bidimensional contour plots. Among various applications, a nicely structured chromatogram, unveiling the complex profile of menhaden oil, was attained: exponential functions were derived for FAME groups characterized by the same number of double bonds and, in all cases, with a good fit (Figure 4).

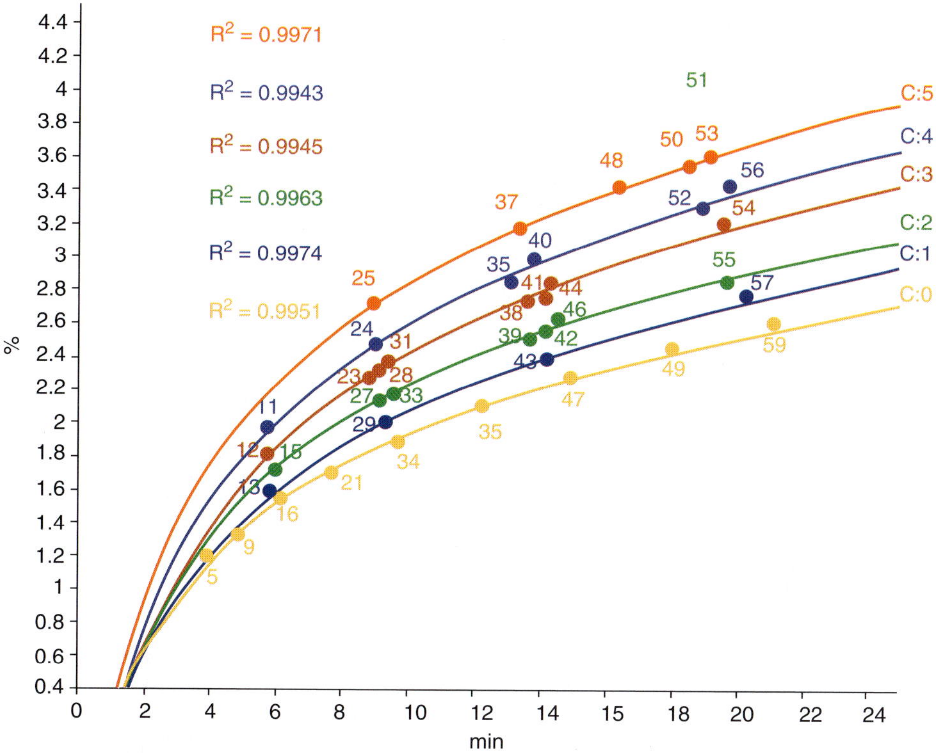

Figure 4 Six exponential curves corresponding to menhaden oil FAMEs families grouped on an equal double-bond number basis (0 to 5 range) [8]. Peak identification: 5) $C_{14:0}$; 9) $C_{15:0}$; 11) $C_{16:4\omega1}$; 12) $C_{16:3\omega4}$; 13) $C_{16:1\omega7}$; 15) $C_{16:2\omega4}$; 16) $C_{16:0}$; 21) $C_{17:0}$; 23) $C_{18:3\omega6}$; 24) $C_{18:4\omega3}$; 25) $C_{18:5\omega3}$; 27) $C_{18:2\omega6}$; 28) $C_{18:3\omega4}$; 29) $C_{18:1\omega9}$; 31) $C_{18:3\omega3}$; 33) $C_{18:2\omega4}$; 34) $C_{18:0}$; 35) $C_{19:0}$; 36) $C_{20:4\omega6}$; 37) $C_{20:5\omega3}$; 38) $C_{20:3\omega6}$; 39) $C_{20:2\omega6}$; 40) $C_{20:4\omega3}$; 41) $C_{20:3\omega4}$; 42) $C_{20:2\omega4}$; 43) $C_{20:1\omega9}$; 44) $C_{20:3\omega3}$; 46) $C_{20:2\omega3}$; 47) $C_{20:0}$; 48) $C_{21:5\omega3}$; 49) $C_{21:0}$; 50) $C_{22:5\omega6}$; 51) $C_{22:6\omega3}$; 52) $C_{22:4\omega6}$; 53) $C_{22:5\omega3}$; 54) $C_{22:3\omega4}$; 55) $C_{22:2\omega6}$; 56) $C_{22:4\omega3}$; 57) $C_{22:1\omega9}$; 59) $C_{22:0}$.

The same research group has carried out other GC×GC applications on one of the most complex lipid foods, cod liver oil [10]. The highly ordered 2D chromatogram attained was exploited for peak assignment in separate mono-dimensional applications. The authors observed that FAMEs with an equal double-bond number were aligned along slanting horizontal bands, while the same happened, along slanting vertical bands, for esters with the same ω number. As can be observed in Figure 5, it was possible to identify FAMEs, at specific intersection points by simply drawing diagonals through compounds with the same number of double bonds and through those with the same ω value. These elution patterns, observed also for other FAMEs groups (C_{16}, C_{18}, etc.), were confirmed by injecting pure standard compounds.

A very nice GC×GC–FID separation of *cis/trans* FAME isomers has been achieved by de Koning et al. [11]. The first-dimension polyethylene glycol column achieved separation on the basis of FAME chain length and double-bond

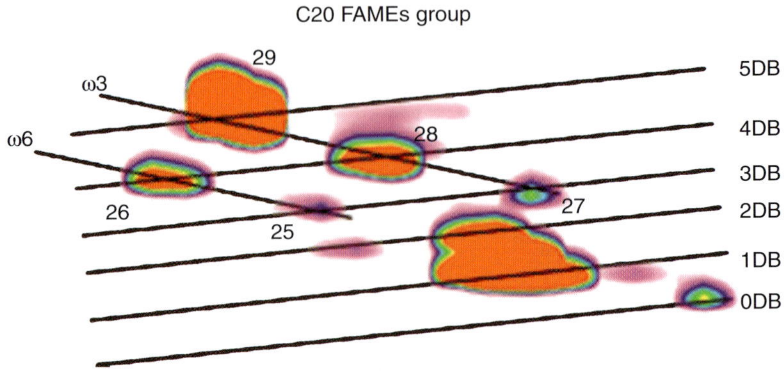

Figure 5 Expansion derived from a cod liver oil FAMEs contour plot, relative to the C_{20} group. The five numbered peaks can be identified considering the intersection points [10].

number/position. A series of overlapping $C_{18:1}$ *cis/trans* isomers (a test sample was injected) was fully resolved on a 3-m segment of cyano-based stationary phase micro-bore capillary under isothermal conditions (Figure 6).

The enhanced separation power of GC×GC does not appear to have been exploited, or even to be really necessary, in the aforementioned lipid applications. This opinion can be altered if the GC×GC result for cod liver oil is observed (Figure 7) [2]. The number of separated FAMEs is much higher if compared to the monodimensional result (Figure 2), with literally hundreds of FAMEs distributed on the space plane. Obviously, the sample is much more complex than it appeared to be. Again, if the FAME groups from C_{10}–C_{24} are observed, the chromatogram is nicely structured. Looking closer at the contour plot, the C_{10}–C_{13}, C_{19} (~25 peaks) and C_{23} groups (~10 peaks), practically undetected in the monodimensional application, are readily visible. Moreover, a series of aligned n-hydrocarbons (eluted earlier on the secondary polar column), altogether invisible in the monodimensional application, benefit from the increased sensitivity and appear below the widened FAMEs band.

The analyte distribution presented in the chromatogram illustrated in Figure 7 is commonly observed in the field of GC×GC. In fact, several separations reported in the literature are characterized by curved or horizontal widened bands of closely eluting (or overlapping) analytes, stretching across different lengths of the space plane. The amount of unexploited chromatogram space above and below these analyte bands is usually considerable. This negative aspect is dependent on two main features: the low second-dimension column efficiency and the lack of orthogonality (see Chapter 3). Neglecting the latter issue, it can be affirmed that many GC×GC applications are carried out at gas velocities that are ideal (or slightly less than ideal) in the first dimension and excessively high in the second dimension [12,13]. Such operational conditions are generated by using a long conventional + the short fast column combination, by far the primary choice made by researchers in this field. An alternative GC×GC system (defined as "split-flow" GC×GC), operated at improved gas

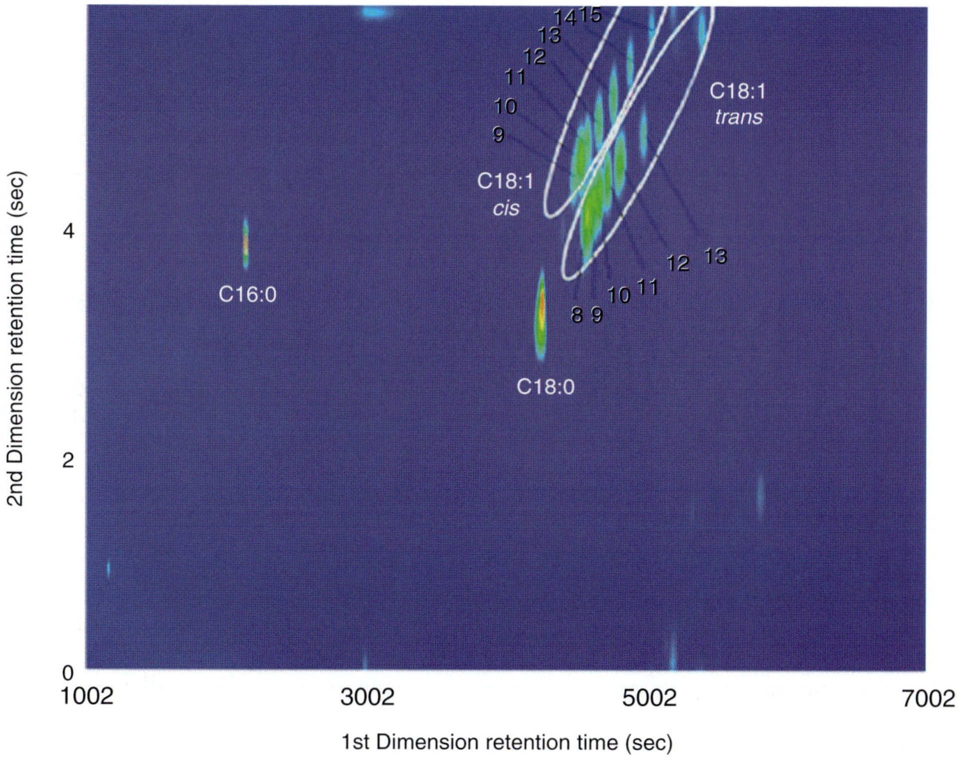

Figure 6 Contour plot expansion of a test sample of a partly hydrogenated vegetable oil consisting mainly of C_{16} and C_{18} FA chains. The numbers indicate double-bond positions [11].

linear velocities, has been developed and tested on cod liver oil FAMEs [14]. The setup consisted of a 30 m×0.25 mm I.D. primary column connected by means of a Y-press fit to an FID-linked 1 m×0.1 mm I.D. analytical column (which passed through the cryogenic modulator) and to a 30 cm×0.1 mm I.D. retention gap, which was connected to a manually operated split valve. This modification enabled gas-flow optimization in the second dimension simply by adjusting the split valve. Initially, a traditional GC×GC experiment was carried out on a cod liver oil sample by closing the split valve. The authors calculated gas linear velocities of 35.3 cm/s and 333.2 cm/s in the first and second dimension, respectively. A chromatogram expansion, relative to the C_{16} group, is illustrated in Figure 8a, where peaks are numbered to enable a direct comparison with the optimized split-flow result. The width of the C_{16} group along the second-dimension axis, which was derived by simply considering the retention-time difference between peaks 1 and 4, equaled 0.688 s. Moreover, it can be noted that peak 3 partially overlaps with peak 2. A single "raw" 2D chromatogram, showing the second-dimension separation of peaks 2 and 3, is reported in Figure 8c; a resolution value of 0.8 was calculated for these compounds. At this point, an optimized split-flow experiment was carried out: 35% of the

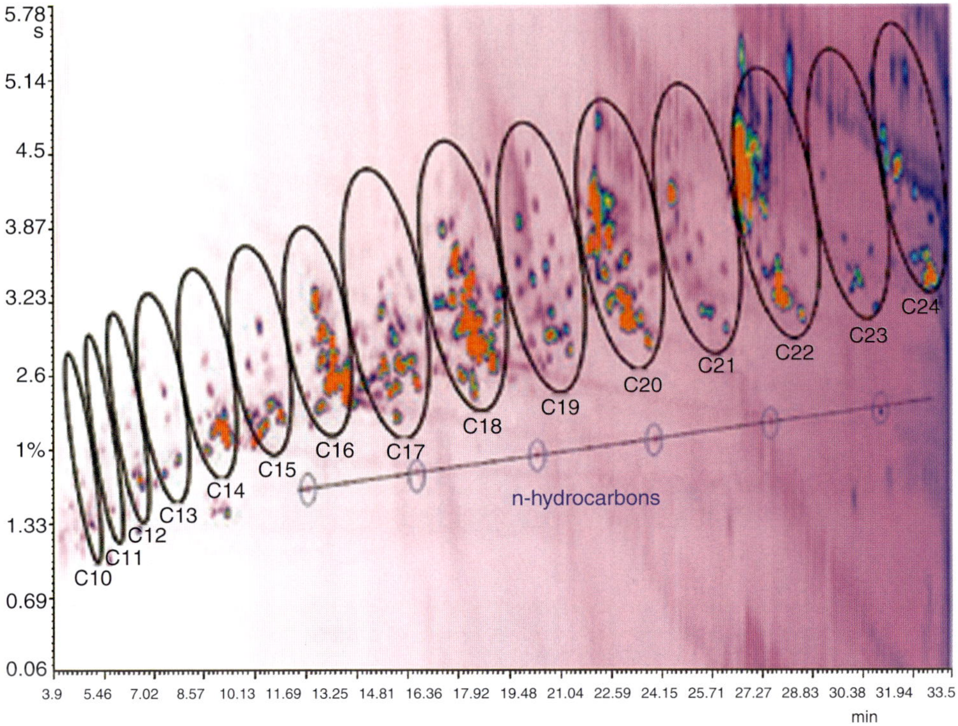

Figure 7 GC×GC chromatogram for cod liver oil FAMEs [2].

gas flow was diverted to waste at the column junction point, with the loss of sample counterbalanced by injecting a higher sample amount. Gas linear velocities of 35.4 and 213.5 cm/s were calculated in the first and second dimension, respectively. A chromatogram expansion, relative to this experiment, is illustrated in Figure 8b: the width of the C_{16} group equaled 1.104 s (+60%). Furthermore, peak 2 is nicely separated from peak 3 (Figure 8d), with a resolution value of 1.2 (+50%).

It is interesting to note that J. Phillips compared GC×GC to GC–MS [15]: both bidimensional methods enable the delivery of analyte bands from a primary to a secondary analytical system, as already explained in Chapter 1. In the case of GC–MS, the mass spectrometer subjects the analyte bands to fragmentation and separates the generated fragments along an m/z axis. In GC×GC, the second dimension consists of a high-speed chromatograph, in which the analyte bands are separated along a secondary retention-time axis. The combination of a mass spectrometer and a GC×GC system introduces a third-separation dimension, producing the most powerful analytical tool available today for food volatile analysis. In particular, the suitability of time-of-flight mass spectrometers (ToF MS) for the detection of the very narrow GC×GC peaks was demonstrated in a petrochemical experiment carried out in 2000 [16], while the first GC×GC–ToF MS food flavour application appeared two years later [17]. In recent years,

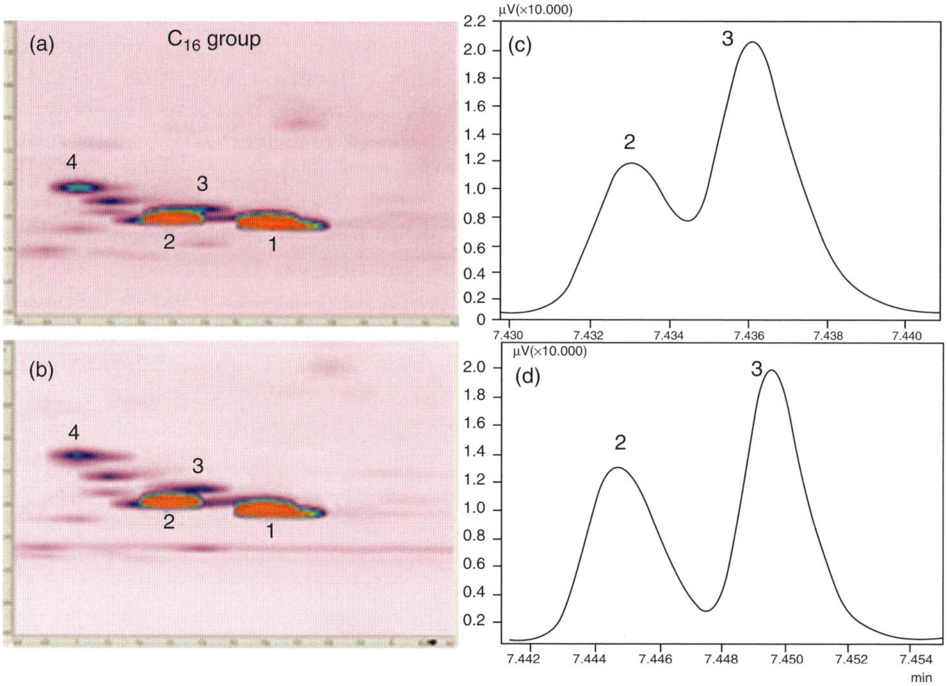

Figure 8 (a) Contour plot expansion derived from a traditional GC×GC cod liver oil application relative to the C_{16} group; (b) expansion derived from a 35:65 (FID) split-flow application on the same sample; (c) traditional GC×GC untransformed chromatogram expansion with peaks 2 and 3; and (d) 35:65 split-flow GC×GC untransformed chromatogram expansion with peaks 2 and 3. Peak identification: (1) $C_{16:0}$; (2) $C_{16:1\omega7}$; (3) $C_{16:1\omega5}$; (4) $C_{16:4\omega3}$.

the use of this three-dimensional method in the food analysis field has gradually grown, although it is still far from being well established.

The employment of GC×GC–ToF MS for the analysis of lipids in lanolin (wool wax) has been reported [18]. Lanolin, among many uses, is employed as a glazing agent (E913) and, hence, can be found in food products. This specific application may not be strictly within the marked boundaries of the present chapter but is worth reporting because it is one of the very few GC×GC–MS lipid applications. Wool wax consists of a complex mixture of esters, di-esters, and hydroxy esters of high-molecular-weight (MW) alcohols (aliphatic and steroidal) and high MW acids; free fatty alcohols and acids are also present [19]. In the wool wax GC×GC–ToF MS experiment, various derivatization processes and column sets were tested. The best results were attained with a dual-step derivatization procedure (methylation for the acidic functions; silylation for the alcoholic groups) and an apolar-polar column combination. The two selected columns presented a high thermal stability and, hence, were suitable for the analysis of less volatile derivatives. The mass spectrometer was operated at a 50-Hz spectral acquisition frequency. In general, the attainment of 50 data points/s is assumed to be the minimum number required for the proper reconstruction of GC×GC peaks (Chapter 2). The application of lower-than-necessary acquisition rates will be the cause of errors during analyte quantitation and of inconsistent retention times. A total ion current GC×GC–ToF MS space plane expansion plus three extracted ion chromatogram expansions are shown in Figure 9. As can be seen, the chromatograms are characterized by horizontal analyte bands, with several overlapping first-dimension peaks nicely separated in the second dimension. Moreover, analytes (diols, FAMEs, hydroxyl acids, fatty alcohols) belonging to the same chemical family tend to form ordered patterns, as can be observed in what the authors defined as fragmentograms. The deconvolution software of the Pegasus II ToF system (LECO) used in the study produced a peak table with 9999 hits (the maximum number). Even considering that modulation produces four to five fractions for each first-dimension peak, the number of different compounds was still considerable, and still up into the thousands mark (the authors did not report the exact number). However, the number of lipid components with an acceptable similarity [this factor expresses the similarity (0–999 range) between the experimental and library spectra, considering all masses], namely, higher than 800, was only 30. The explanation provided by the authors was that the MS library used was rather incomplete with regard to high MW and odd-chain-length components. Furthermore, the MS detector provided poor-quality mass spectra for the heavier ions (scan range: 70 to 800 m/z), with a considerable reduction in ion intensity. For example, cholesterol, a major lanolin constituent, was identified with a good similarity (896) but presented a m/z 458 ion (molecular mass) four times lower than expected.

2.2 Miscellaneous food flavour applications

GC×GC in combination with mass spectrometry has been used for the analysis of flavours contained in lipid foods [22–24]. In particular, Adahchour et al. [22] tested, evaluated, and used a rapid-scanning quadrupole MS (qMS) instrument (Shimadzu MS-QP2010) in the GC×GC analysis of olive oil flavour compounds. The detector, which neared GC×GC requirements, was characterized by a scan speed of up to 10,000 amu/s and could even reach the ultimate goal of 50 spectra/s, at a very restricted mass range (95 amu). Quadrupole mass spectrometers are less expensive than ToF MS systems, and, hence, quite a lot of research work has been devoted to the evaluation of the feasibility of such systems for GC×GC analysis. In the olive oil experiment, the qMS instrument was operated at a scan frequency of 33 Hz and a 50–245 m/z mass range: the latter might be satisfactory for many GC×GC applications, while the former parameter is less than sufficient for the narrower GC×GC peaks (50–100 ms). The authors reported, as a typical example, the determination of 3-octen-2-one in an olive oil extract derived by using high-vacuum degassing (Figure 10). As can be observed in the full-scan GC×GC–qMS chromatogram shown in Figure 10b, the extract is rather complex and contains both high- and low-concentration compounds. Determining many of the low-concentration compounds by using GC–qMS (Figure 10a) would have certainly been a cumbersome task. For

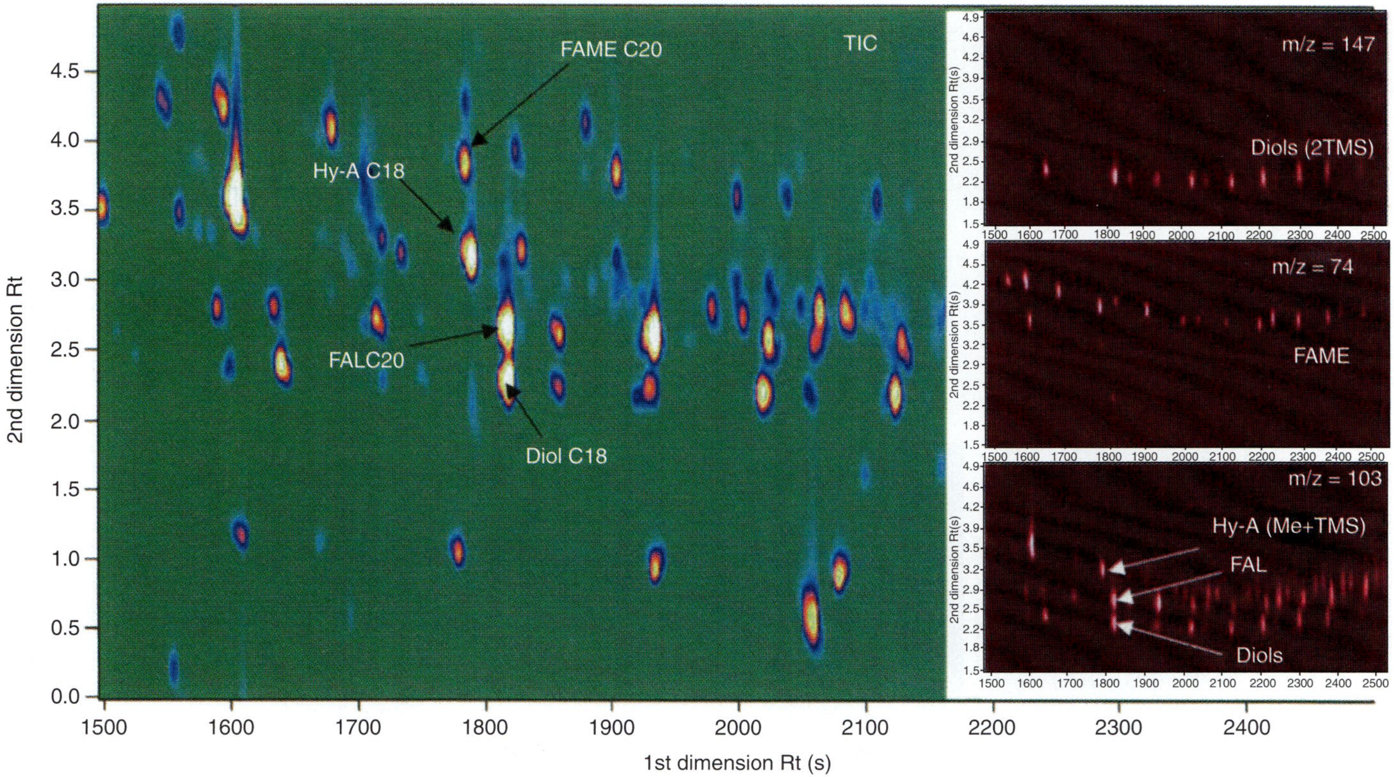

Figure 9 TIC GC×GC–ToF MS contour plot expansion of methylated + silylated lanolin, and inserts of *m/z* 74, 103 and 147 fragmentograms. Abbreviations: Hy-A, Hydroxyl acids; FAL, Fatty alcohols; ME, Methyl derivative; TMS, Trimethylsilyl derivative.

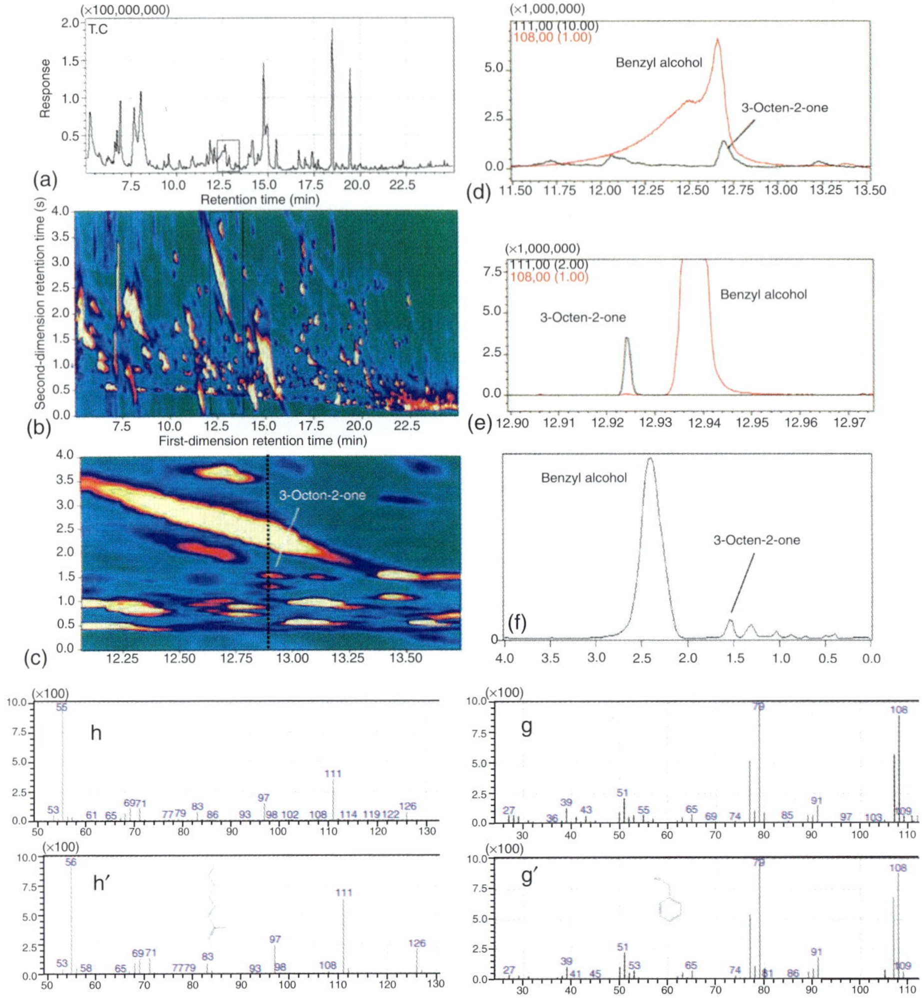

Figure 10 (a) GC–qMS and (b) GC×GC–qMS TIC chromatograms (m/z 50–245) of an olive oil extract; (c) expansion of the area marked in (b); (d and e) m/z 108 and 111 traces of the marked area in (a) without and with modulation, respectively; (f) single two-dimension chromatogram relative to the dotted line in (c); (g) mass spectra of the dual-compound peak in (a); (h) mass spectra of 3-octen-2-one in (b); (g′) and (h′) are the corresponding library spectra [22].

example, 3-octen-2-one overlaps entirely with benzyl alcohol and was not identified by using GC–qMS; the dual-component peak was identified as the alcohol (see spectra in Figure 10g and 10g′). The usefulness of using a high-resolution power GC method prior to the qMS detector is fully demonstrated in Figure 10c and 10f, which show the total isolation of the target flavour from the alcohol, as well as from other minor compounds. Moreover, the mass spectral quality was excellent, as 3-octen-2-one was identified with a similarity value of 98% (Figure 10h and 10h′). Finally, the authors reported the attainment of 12 data points for the flavour compound, sufficient for correct peak reconstruction. However, quantification was carried out by using a single mass (*m/z* 111).

In the food industry, the determination of key odorants in highly complex extracts is a common analytical challenge. Both the suitability of GC×GC–ToF MS and the inadequacy of GC–ToF MS in target flavour analysis were shown by Adahchour et al. [24]. Two odour-impact compounds, methional and sotolon, were identified and quantified in a very complex dairy spread extract by using the three-dimensional approach. In contrast, GC–ToF MS failed in the isolation of these two minor compounds from high-concentration interferences.

During trial-and-error GC×GC method optimization, the primary-apolar and secondary-polar combination is usually the first tested and the most commonly used: isovolatiles are separated in the second dimension on a polarity basis (Chapter 2). It must be emphasized, however, that the "orthogonal" option is not always the most suitable choice and that other combinations can provide a better performance. In the field of food flavour analysis, the main objective usually consists in the separation of the highest number of components per unit of time and/or the isolation of target analytes. This aspect was highlighted by Adahchour et al. [25] in GC×GC–FID and GC×GC–ToF MS experiments on olive oil and vanilla extracts. Considering the latter application, the authors reported that the reversed approach was the most successful in the separation of the highest number of compounds. Moreover, the formation of chemical-class patterns, commonly not attained in flavour applications, was not observed with any of the column sets used. However, even the enhanced separation power of GC×GC did not succeed in the satisfactory resolution of target flavours: a high amount of propylene glycol, an essential solvent in the preparation of vanilla extract, flooded the chromatogram, covering several target analytes (top part of Figure 11).

The potential of a highly selective third ToF MS dimension was exploited for the localization of four target flavours, as can be observed in the lower reconstructed extracted ion chromatogram illustrated in Figure 11. Limits of detection for the four trace-amount analytes, calculated in GC×GC–FID experiments at a 3:1 signal-to-noise ratio, were between 5 and 10 pg, demonstrating the high sensitivity of the method.

Coffee, after tea, is the most popular beverage in the world, with enormous economic importance. Roasting green coffee beans is essential to producing the typical aroma of coffee. Throughout the years, extensive research has been devoted to elucidating the volatile fraction of green and, particularly, of roasted beans. The raw bean volatile composition is less complex than that of the roasted bean, which is characterized by hundreds of components in a wide concentration range. The use of GC–MS has been widely reported in this specific field (a typical GC–MS chromatogram, relative to Arabica bean volatiles, is illustrated in Figure 1) [1].

The enhanced resolving power of GC×GC has been demonstrated to be particularly suited to this type of food matrix [26,27]. In particular, Ryan et al.

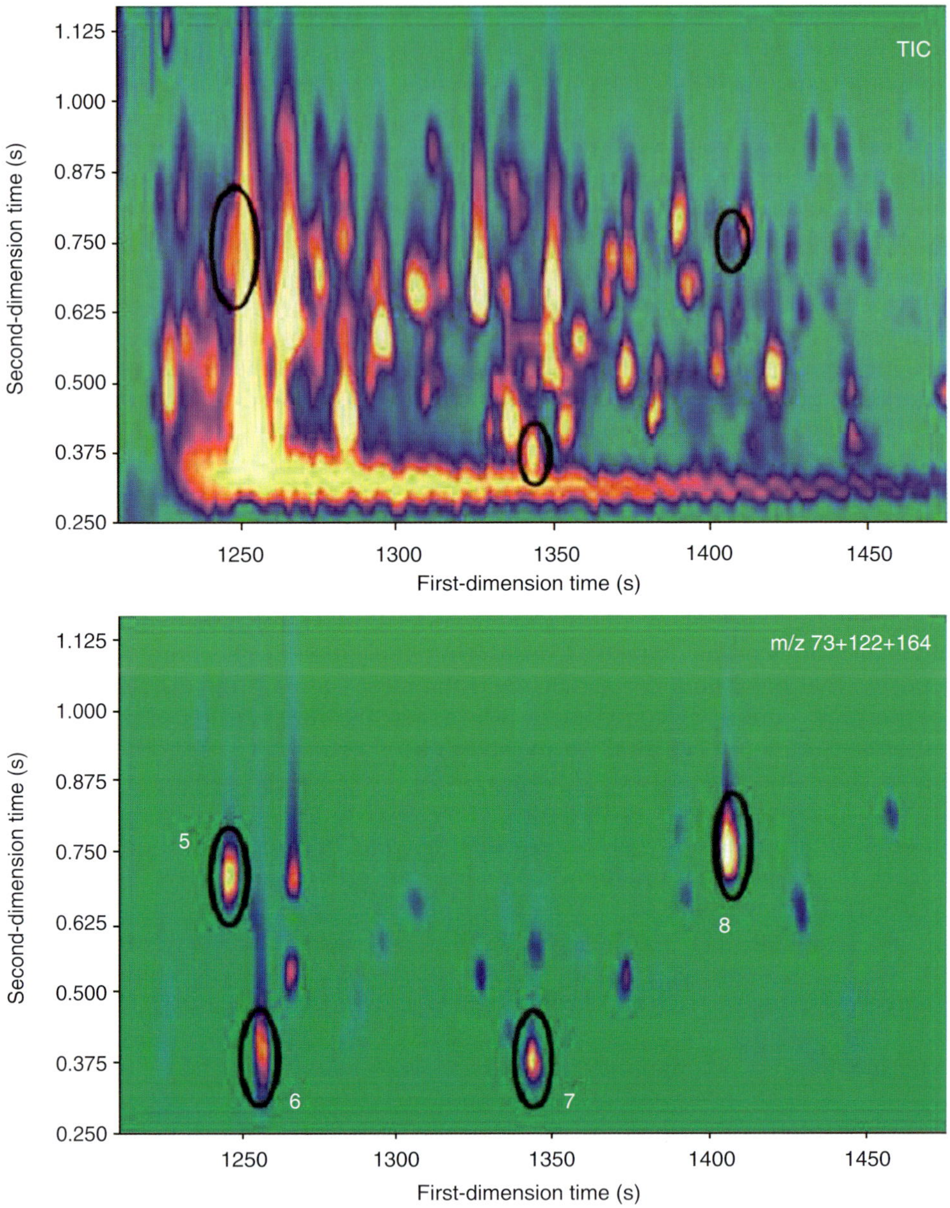

Figure 11 Expansion of the vanilla extract GC×GC–ToF MS chromatogram of the critical propylene glycol region: (top) total ion chromatogram and (bottom) combined reconstructed ion chromatograms of *m/z* 73, 122, and 164 [25]. Peak assignment: (5) benzoic acid ethyl ester; (6) 2-methylbutanoic acid; (7) pentanoic acid; (8) 2-phenylethyl acetate.

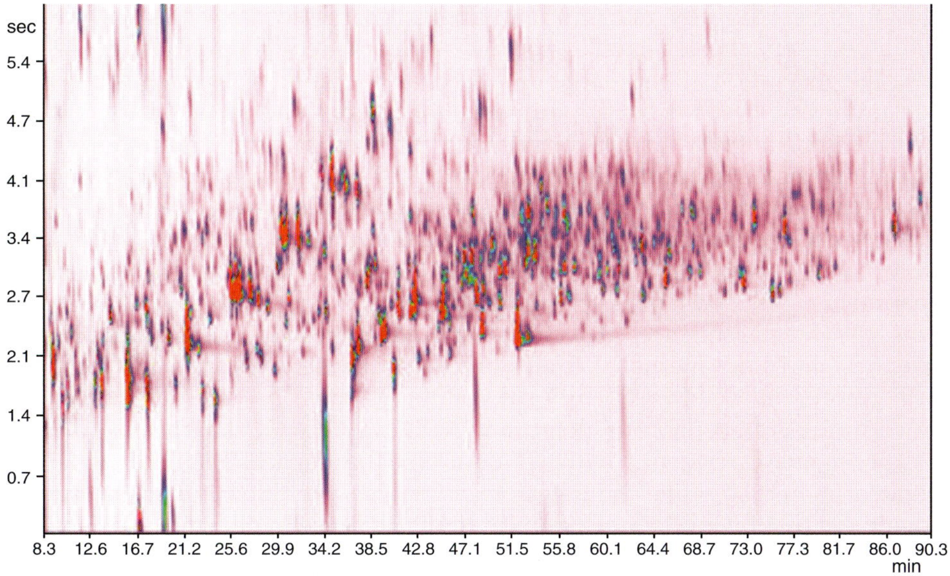

Figure 12 SPME–GC×GC–qMS result for roasted Arabica coffee bean volatiles [28].

used rapid-scanning qMS (MS-QP2010) and ToF MS (Pegasus III) systems in GC×GC experiments on coffee bean volatiles extracted by using SPME [26]. A "reversed" polar-apolar column combination was used because it provided a more satisfactory separation than the more conventional apolar-polar setup. The very high complexity of Arabica aroma was revealed by the literally thousands of peaks scattered across the space plane, as can be seen in the qMS result illustrated in Figure 12 (this chromatogram was not published in Ref. [26] but in a later review paper [28]).

The formation of group-type patterns was also observed. The boiling point-based secondary separations enabled the isolation of pyrazines, well-known coffee aroma constituents, from other sample compounds (the degree of overlap in the first dimension was considerable). Moreover, the pyrazines were situated in distinct horizontal bands on the basis of their degree of carbon substitution (e.g., dimethylpyrazines and ethylpyrazine).

Although high-quality mass spectra were generated, group-specific pyrazines were characterized by very similar MS fragmentation patterns. Positive peak identification was achieved with the support of additional information, namely, monodimensional linear retention indices (LRI) and contour-plot pyrazine positions. Such a considerable amount of valuable analytical information could not have been derived from a GC–qMS application. The qMS instrument was operated at a "normal" m/z 40–400 amu mass range, with a scanning rate of 20 spectra/s. The latter parameter guaranteed reliable peak assignment but was not sufficient, in many cases, for correct peak reconstruction. The ToF MS instrument, on the other hand, generated 100 spectra/s over a m/z 41–415 range. TIC chromatograms were automatically processed with the Chrom-TOF software

(LECO). In the study, the maximum number of processed peaks was restricted to 1000, a necessary choice related to the excessive time required for data processing (8 h). The generation of very large data files, as the authors stressed, is a general problem in GC×GC–ToF MS experiments (Chapters 4 and 5): powerful PCs and fully integrated software are mandatory for instrumental control, to locate and identify peaks in automatically generated 2D space planes and, in general, to process the enormous amount of data generated in these experiments.

As mentioned earlier, the GC×GC modulation process generates very narrow solute bands, altogether comparable with those generated in high-speed GC experiments, with some peaks reaching the detector with a base width as low as 50 ms. As a consequence, detectors with rapid responses, high acquisition capacities, and low internal volumes are necessary, as explained in Chapter 2. Furthermore, the detector operational conditions must be carefully tuned. The use of nitrogen-phosphorous detection (NPD) in GC×GC has been evaluated in experiments carried out with coffee aroma headspace [29]. The NPD was operated at 100 Hz, while the detector gas flows were finely tuned in order to provide the best overall performance in terms of peak shape, width, and height. GC×GC–NPD was directly compared with GC×GC–ToF MS in a HS-SPME roasted coffee aroma experiment. The same benefits that derive from the use of such a detector in monodimensional GC were observed in GC×GC, notably, high selectivity for N-containing compounds (in this case, pyridine, pyrazines, etc.) and enhanced sensitivity. In general, the use of a second GC dimension does not always succeed in fully resolving first-dimension peak co-elution; thus, the selectivity and high sensibility of selective detection are two valuable aspects that can be widely exploited in GC×GC.

Use of an enantioselective stationary phase in either of the dimensions of a GC×GC system is much rarer than in classical multidimensional gas chromatography (MDGC), with the first experiments appearing in 2001–2002 (essential oil applications) [30,31]. In a later experiment, SPME-extracted strawberry volatiles were subjected to GC×GC–FID analysis on two serially connected chiral capillaries in the first dimension and on a 50% phenyl polysilphenylene-siloxane micro-bore column segment in the second dimension [32]. The (-)2,5-dimethyl-4-hydroxy-3[2H]-furanone levorotatory enantiomer, well separated from all other sample constituents, was detected in all samples. The obvious advantage of this type of approach over classical MDGC is that it enables (ideally) the full resolution of enantiomers and nonchiral analytes.

The headspace composition of 13 different pepper varieties, extracted by using HS-SPME, has been elucidated by using both GC×GC–qMS and GC×GC–ToF-MS [33]. A total of 309 compounds were identified with the additional support of monodimensional LRI. In several cases, it was found that the influence of the secondary polyethylene glycol column was rather considerable. Hence, the maximum LRI difference allowed by the authors (with respect to literature values) was rather large, namely, 30 units. HS-SPME–GC×GC–ToF MS analysis has also been used in an interesting experiment involving roast beef aroma [34]. A piece of sirloin beef was cooked in a dedicated

30 R. Shellie, P. Marriott and C. Cornwell, *J. Sep. Sci.*, 24 (2001) 11.
31 R. Shellie and P. Marriott, *Anal. Chem.*, 74 (2002) 5426.
32 A. Williams, D. Ryan, A.O. Guasca, P. Marriott and E. Pang, *J. Chromatogr. B*, 817 (2005) 97.
33 Z.L. Cardeal, M.D.R. Gomes da Silva and P.J. Marriott, *Rapid Commun. Mass Spectrom.*, 20 (2006) 2823.
34 S. Rochat, J.Y. de Saint Laumer and A. Chaintreau, *J. Chromatogr. A*, 1147 (2007) 85.
35 T. Čajka, J. Hajšlová, J. Cochran, K. Holadová and E. Klimánková, *J. Sep. Sci.*, 30 (2007) 534.
36 Y. Shao and P. Marriott, *Anal. Bioanal. Chem.*, 375 (2003) 635.
37 D. Ryan, P Watkins, J. Smith, M. Allen and P. Marriott, *J. Sep. Sci.*, 28 (2005) 1075.
38 F. Gogus, M.Z. Ozel and A.C. Lewis, *J. Sep. Sci.*, 29 (2006) 1217.
39 Y. Shao, P. Marriott, R. Shellie and H. Hügel, *Flav. Fragr. J.*, 18 (2003) 5.
40 H. Komura, *J. Sep. Sci*, 29 (2006) 2350.
41 S.M. Rocha, E. Coelho, J. Zrostlíková, I. Delgadillo and M.A. Coimbra, *J. Chromatogr. A*, 1161 (2007) 292.
42 F. Bianchi, M. Careri, C. Conti, M. Musci and R. Vreuls, *J. Sep. Sci.*, 30 (2007) 527.
43 R. Mayadunne, T.T. Nguyen and P.J. Marriott, *Anal. Bioanal. Chem.*, 382 (2005) 836.
44 M. Junge, H Huegel and P.J. Marriott, *Chirality*, 19 (2007) 228.
45 K. Grob, *On-Line Coupled LC-GC*, Hüthig, Heidelberg, 1991.
46 W.W.C. Quigley, C.G. Fraga and R.E. Synovec, *J. Microcol. Sep.*, 12 (2000) 160.
47 H.G. Janssen, W. Boers, H. Steenbergen, R. Horsten and E. Flöter, *J. Chromatogr. A*, 1000 (2003) 385.
48 S. de Konig, H.G. Janssen, M. van Deursen and U.A.Th. Brinkman, *J. Sep. Sci.*, 27 (2004) 397.

CHAPTER 11

Environmental Analysis

Juan Jose Ramos, Miren Pena-Abaurrea and **Lourdes Ramos**

Contents

1. INTRODUCTION

Environmental samples, understood here as any liquid or solid environmental matrix, are extremely complex mixtures in which a large variety of compounds with related and unrelated structures are simultaneously present at widely variable levels of concentration. A large majority of these compounds can mutually interfere during their instrumental analysis even if a highly selective separation-plus-detection technique is selected for final determination. Organic micropollutants are entrapped in these complex mixtures, but typically at very low concentrations (typically in the μg/g–pg/g range) and frequently absorbed or strongly bound to other matrix components. In addition, it is now generally accepted that, because of the several anthropogenic impacts simultaneously registered by ecosystems, organic micro-contaminants are usually present as complex mixtures rather than as individual entities in these samples. This is especially true for some well-known families of industrial pollutants that have been used — and enter the ecosystem webs — as mixtures of isomers in which

Comprehensive Analytical Chemistry, Volume 55
ISSN: 0166-526X, DOI 10.1016/S0166-526X(09)05511-1

individual components exhibit different levels of toxicity. These considerations justify the main features of the analytical procedures in use for analysis of trace organic pollutants in environmental liquid and (semi-)solid matrices. First, the extremely low concentrations at which the target compounds should accurately be detected makes mandatory an exhaustive extraction of the analytes from the matrix to ensure their quantitative recovery and proper detection. The essentially nonselective character of this first step makes mandatory the development of a subsequent laborious — and frequently manual — multistep protocol for purification and fractionation of the target analytes, first from unrelated co-extracted material to avoid matrix effect and then from other structure-related compounds present in the sample that can interfere with their final determination. Despite these tedious and sophisticated sample preparation procedures, and the highly selective and sensitive analytical techniques typically chosen for final instrumental determination of pollutants, the complexity of many of the environmental extracts and the simultaneous presence of known (and unknown) compounds can yield inaccurate and inconsistent results and compromise the validity of the final result.

The extended risk of overlap and the environmental field's interest in determining the actual concentration of a particular analyte, rather than that of a family of pollutants, explain the need for enhanced separation power. The resolution provided by gas chromatography (GC) and its feasibility for direct coupling with a number of selective and sensible detectors explain the general preference for this technique over liquid chromatography and capillary electrophoresis for the analysis of medium- to low-volatility organic pollutants. The significant simplification achieved in the monodimensional (1D) GC chromatograms by adding an extra dimension either by using mass spectrometric detectors (MS) or other chromatographic-based multidimensional approaches has been demonstrated in the literature and illustrated by self-explicative chromatograms (Chapter 1). But the overwhelming separation power of comprehensive two-dimensional gas chromatography (GC×GC) as compared to any other monodimensional and multidimensional GC-based techniques and the sensitivity enhancement achieved through the modulation process were considered especially interesting features that could help solve some of the most pressing challenges in this research area.

This chapter highlights the main advantages and remaining limitations regarding the use of GC×GC for the analysis of organic (semi-)volatile micropollutants in environmental matrices through selected applications that represent the current state of the-art in three main research fields: the analysis of organohalogenated pollutants, the determination of other nonhalogenated toxic compounds, and the enantiomeric analysis of chiral contaminants.

2. ORGANOHALOGENATED POLLUTANTS

2.1 Aromatic organohalogenated pollutants

Probably the field in which GC×GC was more rapidly incorporated after its development in the petrochemical field was the analysis of organohalogenated compounds [1]. The aim in these studies has usually been rather different from that of petrochemical studies. While group-type separation and fingerprinting are usually the goals of petrochemical studies, most environmental studies that have been reported up to now focused on target-compound analysis. The reason is clear: even in the case of numerous families of pollutants containing isomers that are closely related structurally, such as polychlorinated dibenzo-*p*-dioxins and furans (PCDD/Fs; 135 isomers) or polychlorinated biphenyls (PCBs; 209 possible isomers), variable toxic effects have been reported for the several congeners present in the mixtures [2]. Congener-specific determination then becomes mandatory to determine if a particular matrix accomplishes with the maximum residue levels (MRLs) set in current legislation. The extremely low concentration levels at which these particular congeners should accurately be determined in real-life samples and the inherent complexity of the extracts have typically required parallel analysis of the extracts in at least two GC column with different selectivity and, in some cases, the use of high-resolution mass spectrometry (HRMS) for final determination of the target compounds [3]. The enhanced separation power provided by GC×GC was immediately recognised as a possible analytical alternative to these tedious and expensive instrumental analysis protocols. Early studies involving GC×GC–HRMS reported promising results and an impressive low limit of detection (LOD) for 2,3,7,8-TCDD of about 200 at [4] using a conventional nonpolar×polar column set. The chromatographic times, initially similar to those required in conventional 1D GC–HRMS, were reduced to less than 20 min in subsequent studies [5]. Despite its evident advantages, progress involving this coupling has been somehow limited [6], probably because of the high price of the detector and its relatively slow acquisition speed. In fact, most of the research dealing with the analysis of 2,3,7,8-PCDD/Fs and toxic PCBs (non-*ortho*-CBs, 77, 81, 126, and 169; and mono-*ortho*-CBs, 105, 114, 118, 123, 156, 157, 167, and 189 [7]) has been carried out with alternative (less expensive and easy to handle) detectors, such as the flame-ionization detector (FID) [8,9] and the electron-capture detector (ECD) [10] in the early stage of development of the technique, and with a clear preference for the micro electron-capture detector (μECD) and time-of-flight mass spectrometry (ToF MS) in the last decade (Table 1).

The feasibility of FID for fast determination of toxic non- and mono-*ortho*-CBs in the technical mixture Clophen A50 was demonstrated by Haglund et al. [8] using a smectic liquid-crystal phase as first dimension (^{1}D) and a nonpolar column, BPX-5, as second dimension (^{2}D). The special selectivity of liquid-crystal phases for planar compounds resulted in a relatively fast elution of non-planar CBs from the column as compared to toxic congeners. To reduce the retention of the later analytes in this phase, a short (10 m) and narrow (0.15 mm) ^{1}D column, with a thin phase of 0.1 μm, was used. The column was also operated at a higher than usual carrier gas velocity. Altogether leaded to an elution of analytes from the ^{1}D at relatively low temperature which, consequently, experienced a strong retention in the ^{2}D, housed in the main oven. This effect was compensated by using as short a ^{2}D column as possible (i.e., 0.4 m, corresponding to an effective length of 0.25 m). Under these conditions, temperature ramps as fast as 18°C/min

Table 1 Selected GC×GC applications involving the analysis of persistent organohalogenated pollutants. For simplicity, only optimised experimental setups or those providing the most conclusive results have been mentioned

Analytes	Sample	Column combination (m×mm ID×μm d_f)	Modulator	Detector	Reference
12 toxic and priority PCBs	Clophen A50	LC-50 (10×0.15×0.10)× BPX-5 (0.4×0.1×0.10)	LMCS[a]	FID	[8]
Aroclor 1254	standard solution	SB-Smectic (24×0.20×0.15)× Ultra 2 (5.3×0.2×0.33)	two-stage TDM[b]	ECD	[10]
7 priority PCBs, toxaphene	Aroclor 1248, narwhal blubber	DB-1 (9.7×0.18×0.40)× CP-WAX-52 (0.8×0.10×0.20)	Sweeper	ECD	[11]
31 PCBs 17 PCDD/Fs	standard mixtures	DB-1 (9.0×0.20×0.33)× CP-WAX-52 (0.3×0.10×0.20)	Sweeper	μECD	[9]
12 toxic PCBs, 17 PCDD/Fs	cod liver	HP-1 (30×0.25×0.25)× HT-8 (1×0.1×0.10)	Sweeper	μECD	[12]
209 PCBs	seal blubber	DB-XLB (60×0.18×0.18)× LC-50 (2×0.15×0.10)	LMCS	μECD	[13]
12 toxic PCBs, 17 toxic PCDD/Fs	spiked milk	DB-XLB (30×0.25×0.25)× LC-50 (0.9×0.18×0.10)	loop modulator, CO_2	μECD	[14]
15 toxic and priority PCBs	sludge	HT-5 (15×0.25×0.10)× BPX-50 (1.0×0.10×0.10)	loop modulator, LN_2	μECD	[15]
12 toxic PCBs, 17 toxic PCDD/Fs	spiked milk, fish oil	DB-XLB (30×0.25×0.25)× LC-50 (1.4/0.9×0.15×0.10)	LMCS	μECD	[16]
12 toxic PCBs, 17 toxic PCDD/Fs	food and feedstuffs	DB-XLB (30×0.25×0.25)× LC-50 (0.9×0.15×0.10)	LMCS/loop modulator, CO_2	μCD	[17]
209 PCBs	standard solution	HT-8 (50×0.22×0.25)× BPX-50 (2.5×0.1×0.10)	quad-jet dual-stage modulator, LN_2	ToF MS	[18]

4 non-*ortho* PCBs, 17 toxic PCDD/Fs	fly ash, sediments, vegetation, fish	Rtx-Dioxin 2 (60×0.25×0.25)× Rtx-500 (2.0×0.18×0.10)	quad-jet dual-stage modulator, LN_2	ToF MS	[19]
17 toxic PCDD/Fs, 18 toxic and priority PCBs	foodstuffs	Rtx-500 (40×0.18×0.10)× BPX-50 (1.5×0.10×0.10)	quad-jet dual-stage modulator, LN_2	ToF MS	[20]
4 non-*ortho* PCBs, 17 toxic PCDD/Fs	standard solution	Rtx-Dioxin 2 (60×0.25×0.25)× Rtx-PCB (2.0×0.18×0.18)	quad-jet dual-stage modulator, LN_2	ToF MS	[21]
4 non-*ortho* PCBs, 17 toxic PCDD/Fs	spiked fish oil	Rtx-Dioxin 2 (60×0.25×0.25)× Rtx-PCB (3.0×0.18×0.18)	quad-jet dual-stage modulator, LN_2	ToF MS	[22]
38 PCBs, 11 OCPs, 12 PBDEs, 1 PBB	serum, milk	DB-1 (15×0.25×0.25)× HT-8 (1.2×0.10×0.10)	quad-jet dual-stage modulator, LN_2	ToF MS	[23]
17 PCDD/Fs	certified fly ash, flue gas	InertCap 5MS/Sil (60×0.25×0.10)× InertCap 17MS/Sil (1.5×0.075×0.10)	loop modulator, LN_2	HRToF MS	[24]
209 PCBs	standard solution	BP-1 (27×0.32×0.32)× VF-23MS (1×0.10×0.10)	home-made dual-jet modulator[c]	ECNI qMS	[25]
17 toxic PCDD/Fs	standard solution	DB-XLB (30×0.25 ×0.25)× LC-50 (0.9×0.10×0.10)	home-made dual-jet modulator[c]	ECNI qMS	[26]
125 PBDEs, 7 PBDE metabolites, 6 PBBs, HCBD, TBBP-A, Me-TBBP-A	standard solution, dust	DB-1 (30×0.25×0.25)× 007-65HT (1.0×0.10×0.10)	loop modulator, CO_2	μECD	[27]
17 PBDEs	fish (eel)	DB-1 (30×0.25×0.25)× 007-65HT (1.0×0.10×0.10)	home-made dual-jet modulator[c]	ECNI qMS	[26]
26 MBPs	dolphin blubber	Rtx-5 Crossbond(10×0.18×0.20)× BPX-50 (0.70×0.10×0.10)	quad-jet dual-stage modulator, LN_2	ToF MS	[28]

Table 1 (*Continued*)

Analytes	Sample	Column combination ($m \times mm\ ID \times \mu m\ d_f$)	Modulator	Detector	Reference
Toxaphene	technical mixture	DB-1 (10×0.25×0.25)× HT-8 (1.0×0.10×0.10)	LMCS	ToF MS	[29]
PCAs plus PCBs, PCDD/Fs, PBDEs, OCPs, PCDEs, PCNs, toxaphene, PBBs	dust	DB-1(30×0.25×0.25)× 007-6 5HT (1.0×0.10×0.10)	loop modulator, CO_2	μECD	[30]
PCAs	technical mixture (C_{10}, 65 wt% Cl)	DB-1 (30×0.25×0.25)× 007-65HT (1.0×0.10×0.10)	home-made dual-jet modulator[c]	ECNI qMS	[26]
PCAs	standard solution, technical mixtures, dust	DB-1 (30×0.25×0.25)× 007-65HT (1.0×0.1×0.10)	loop modulator, CO_2	ECNI ToF MS	[31]

[a]LMCS, longitudinally modulating cryogenic system.
[b]TDM, thermal desorption modulator.
[c]home-made dual-jet modulator [32].

could be used without significantly affecting the 2D analyte separation but promoting an interesting reduction in the analysis time, which was then completed in only 9 min. However, the limited loading capacity of the columns used in the study resulted in peak broadening for some more abundant congeners present in the technical mixture, that is, the indicator PCBs. De Geus et al. [10] avoided this problem by using a longer (24 m) and broader (0.20 mm) smectic ^{1}D column. But the price here was time: a 100-min chromatographic run was required to complete the GC×GC separation of the Aroclor 1248 components. In this early study, a conventional ECD was used as detector, something that partially explains the peak broadening observed in the contour plots. In a subsequent work, the authors reported a fourfold improvement (i.e., narrowing) in the peak widths by using a DB-1×CP-WAX-52 set of short narrow columns, and controlling the temperature of the ^{2}D means of an auxiliary GC oven [11]. Despite the remaining peak broadening associated with the large detector-cell volume, the general performance of the two-oven GC×GC system was demonstrated by satisfactory quantification of the seven priority PCBs (congeners No. 28, 52, 101, 118, 138, 153, and 180) in a cod liver reference matrix (certified concentrations in the range 42–1120 ng/g).

Further improvement was achieved when replacing the ECD (1.5-mL cell volume) by the μECD (150-μL cell volume) [9]. Although peak widths were still wider than those obtained with FID (10–20 μL cell volume), a significant peak shape improvement was obtained by the 10-fol reduction of the ECD cell volume. The narrowing of the peak widths also had the obvious consequence of peak sharpening, that is, enhanced sensitivity. In this pioneer study, LODs for individual PCB congeners in the 6–20 fg range (injected mass) were obtained. These rewarding results, combined with the selectivity of μECD for the analytes containing an electrophilic atom, justified its wide acceptation as GC×GC detector for the analysis of organohalogenated pollutants. Despite the short columns used in this study, the orthogonal nature of the selected column set (DB-1×CP-WAX-52) resulted in a wide spread of the targeted PCBs in the 2D plot and, even though the separation conditions were not really optimised, a first indication of an ordered pattern. The ability of GC×GC to generate structured chromatograms was demonstrated by Korytár et al. [12] in a wider study involving 90 PCB congeners. In this case, two nonpolar stationary phases, HP-5 (30 m×0.32 mm, 0.25 μm) and HP-1 (30 m×0.25 mm, 0.25 μm), were selected as ^{1}D columns and used in combination with phases of different polarity (BPX-50 and Supelcowax-10) and selectivity (HT-8). Two column sets, HP-1×HT-8 and HP-1×Supelcowax-10, allowed a complete separation of all 12 toxic non- and mono-*ortho*-CBs from the other congeners included in the mixture. HP-1× Supelcowax-10 provided the best separation of the PCBs investigated (84 congeners eluted free from interference with this combination), but little or no ordered structure was observed in the 2D contour plot. With HP-1×HT-8, 78 congeners eluted as resolved peaks in clearly structured chromatograms in which PCBs were grouped together according to the degree of chlorination, while within-group position was determined by the number of *ortho*-chlorine-substituents. This type of ordered structure allowed tentative identification of

PCB congeners for which the corresponding standard was not available. Consequently, this column set was preferred for further optimisation and application to the analysis of a purified cod liver extract spiked with the 17 toxic PCDD/Fs. Because of their planar structure, PCDD/Fs exhibited a strong retention in ^{2}D, eluting at retention times longer than those of the most retained PCBs, that is, the non- and mono-*ortho*-CBs, and separated from them. The only exception was one penta-CDD that co-eluted with PCB 169. Typical PCB structured chromatograms are shown in Figure 5 of Chapter 2 and Figure 1. In Figure 1, wraparound is also visible for the hepta- and octa-CDD/F congeners.

This first work was followed by several detailed studies involving more complete sets of either columns or analytes. Using DB-XLB as first dimension, Harju et al. [13] evaluated five column combinations and demonstrated that the DB-XLB×SP-2340 and DB-XLB×LC-50 sets provided the most satisfactory results allowing the separation (Rs$\geq$0.5) of, respectively, 176 and 181 of the 209 PCB congeners; and the detection of 126 of the 136 PCBs present in Aroclors 1232, 1248, and 1260 at concentrations greater than 0.05% (w/w). As a further illustration of the technique's potential for accurate determination of individual PCBs in complex mixtures, the DB-XLB×LC-50 set was used for analysis of a purified seal blubber extract. In this case, 64 PCBs were identified by applying a peak template, although use of an internal standard was mandatory to correct the observed small retention-time shifts. In another exhaustive study [14], the authors concluded that, under orthogonal conditions, using DB-1 as ^{1}D and a μECD as detector, the best simultaneous separation of the 17 2,3,7,8-substituted PCDD/Fs and the 12 dioxin-like PCBs was obtained using a relatively polar phase (VF-23;

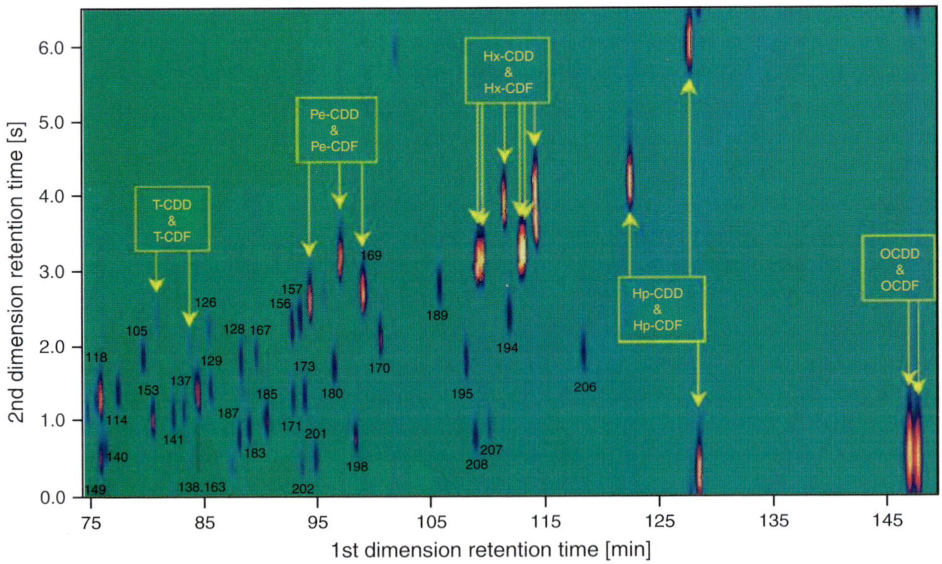

Figure 1 Structured GC×GC–μECD contour plot of a mixture of 90 PCBs and 17 PCDD/Fs with HP-1×HT-8 [12].

absolute cyano content 70–90%) or the shape-selective LC-50 column as ^{2}D. These columns sets had the extra advantage of providing structured chromatograms. However, they failed to separate the PCDD/Fs from the matrix constituents: the targeted congeners showed up in the matrix band when analysing a purified milk extract. This finding prevented use of these column sets for real-life applications at trace level. More importantly, it made evident that the matrix should also be considered during GC×GC method development. Complete separation of all 29 toxic analytes from each other and from the co-extracted matrix components was only achieved using DB-XLB×LC-50. This result confirmed previous observations from Harju et al. [13] and proved that, despite the identification potential derived from ordered structures, orthogonality should preferably be (partly) sacrificed to achieve the required separation from the sample matrix. Supporting this statement are the satisfactory quantitative data obtained with this column set (three orders of linearity, LODs below 70 fg for dioxin-like PCBs and in the 40–150 fg range for the 2,3,7,8-PCDD/Fs, and RSDs lower than 6.5% of all target compounds).

Further investigation on this topic [16,17] involved a variety of spiked and nonspiked food matrices and confirmed the main conclusions of this study. Probably the most interesting point here was that the use of VF-1 as first dimension allowed the elution of the analytes at a temperature ca. 20°C lower than that required for DB-XLB. In practice, that meant that the studied compounds had a higher retention in the ^{2}D column, LC-50. A 0.9-m-long LC-50 column sufficed then to achieve a separation for PCBs similar to that obtained with a 1.4-m-long column coupled to DB-XLB [16]. Unfortunately, it also resulted in co-elution of PCDD/Fs due to wraparound, even though a relatively large modulation period of 8 s was used. Because of the limitation in the maximum working temperature of the LC-50 phase, use of a faster temperature program and a secondary oven to hold the ^{2}D column did not represent practical solutions to the problem. As an alternative, the ^{2}D separation was speeded up by programming the flow of the carrier gas (1.3 mL/min (21 min), at 0.4 mL/min to 1.6 mL/min (5 min) and then at 0.4 mL/min to 1.3 mL/min). Experimental results showed that flow programming caused retention-time instability and different retention-time shifts for the analytes and the internal standard, which made it difficult to use templates for peak identification. In addition, the high flows adversely affected the separation in ^{1}D. The comparison of the total toxic equivalent of 2,3,7,8-TCDD (TEQs) of PCDD/Fs calculated using VF-1 (30 m×0.25 m×0.25 μm) as the first dimension and a 0.9 m×0.18 mm×0.15 μm LC-50 as the second column with those obtained by conventional GC–HRMS proved that, although this column set provided satisfactory results for toxic PCBs, the concentrations calculated for some PCDD/F congeners were overestimated. The separation obtained among the most toxic PCBs (i.e., congeners No. 77, 126, and 169) and the test 2,3,7,8-PCDD/Fs and (remaining) matrix components with DB-XLB×LC-50 is nicely illustrated in Figure 2, where the position of the selected internal standards (TCN and OCN) used for retention-time shift correction is also indicated. The improved separation yielded improved quantitation and demonstrated that, when properly tuned, GC×GC–μECD gives average concentrations comparable to those of the

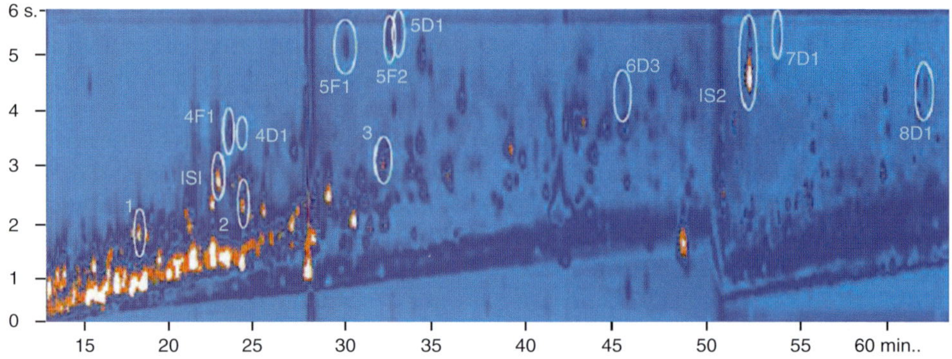

Figure 2 GC×GC–μECD contour plot of a fish oil (non-*ortho*-CB and PCDD/F fraction) with DB-XLB×LC-50 [16].

reference HRMS method. The (expected) concentration-dependent response of the μECD at concentrations close to the limit of detection (as those found when analysing nonspiked food), the baseline instability observed at those low concentrations, and the ca. tenfold higher sensitivity of GC–HRMS compared to GC×GC–μECD were suggested as possibly responsible for the somehow worse, though still acceptable (below 22%), repetitiveness of the latter technique.

All these results support GC×GC–μECD as a promising alternative for screening PCBs and PCDD/Fs in environmental samples. However, present software limitations regarding detection and quantification of compounds with low signal-to-noise ratio or close-eluting compounds (see also Chapters 2 and 4) made manual integration still mandatory in this type of analysis. The practical consequence is a significant increase in the final price (in terms of time) of the toxic PCB and PCDD/F analysis by GC×GC–μECD, which has been estimated as about double that involving the conventional HRMS procedure. Further technical development of μECD design is also desirable to avoid the deterioration of the chromatographic resolution caused by postcolumn band broadening in this detector.

Commercialisation of robust and fast-scanning mass spectrometers, and in particular of ToF MS, has decisively contributed to the increased use of this technique in combination with GC×GC to yield a powerful separation-plus-detection setup providing three-dimensional separation.

The first study reporting on the use of GC×GC–ToF MS for environmental application was published in 2004 [18]. This work was in line with those published at the time by other authors with μECD and reported on the relative merits of four column combinations — DB-1×HT-8, DB-XLB×HT-8, DB-XLB×BPX-50, and HT-8×BPX-50 — for the separation of the 209 PCB congeners. In all instances, relatively long columns (in the 50–75 m range for ^{1}D and of 2.5 m in the second dimension) and a slow ramp of temperature were used. This resulted in improved separation, but also in relatively long analysis times of up to 2.5 h. The use of thermally stable phases allowed the application of an offset temperature of

40°C to the secondary oven housing the ^{2}D column. Under these conditions, wraparound was minimized, and peak widths of 100–150 ms were obtained. In agreement with previous observations, the authors concluded that a satisfactory separation of the 12 toxic and 7 priority PCBs from any other congener present in the mixture was possible with several of the column sets assayed. However, they suggested HT-8×BPX-50 as the best alternative. This column combination resulted in highly ordered roof-tile structures, similar to those observed for DB-1×HT-8, but with a wider spreading of the compounds in the 2D contour plot. As a result, a clear separation of the homologue groups into subseries according to the *ortho*-substitution level was observed (Figure 5 of Chapter 2). A total of 188 PCBs were chromatographically separated with this column set. Use of the ToF MS allowed differentiation of four extra congeners (see Figure 5 of Chapter 2 and discussion on Section 4.2.1 of this chapter), yielding a total of 192 separated peaks in 146 min. This separation is somehow better than the previously mentioned 194 congeners in 240 min on 60 m DB-XLB×2.5 m BPX-70 [13] using GC×GC–μECD. However, the significant reduction of analysis time probably depended more on the substitution of DB-XLB by the high-temperature phase HT-8 as ^{1}D than on the use of ToF MS as detector.

The real potential of ToF MS, and again the need to consider the matrix effect during method development, were illustrated in a subsequent study dealing with the determination of the four non-*ortho*-PCBs and the 17 2,3,7,8-PCDD/Fs in a variety of environmental and biological samples, that is, fly ash, sediment, vegetation, and fish tissue [19]. Using a 40-m Rtx-Dioxin2 column as first dimension, a good separation was obtained among all tested PCDD/Fs, as shown in Figure 3A for the HxCDD/F congeners. However, the severe co-elution of these analytes, with matrix components not completely eliminated during sample preparation and having masses similar to those of the studied compounds, would have obscured their determination in 1D GC-ToF MS (see reconstructed 1D trace based on m/z 390 and 374 in Figure 3B). The satisfactory separation achieved among the target analytes and these isobaric interferences using a 2-m Rtx-500 column as second dimension solved the problem and allowed their accurate determination in as complex a matrix as a fish tissue, even though unit resolution was used. The selected column combination, Rtx-Dioxin2×Rtx-500, allowed satisfactory resolution of all test contaminants among them and from main matrix constituents in the purified extract, except for the critical pair 2,3,7,8-TCDD and PCB 126. This co-elution problem was solved by careful selection of the ion masses used for identification and quantitation of both compounds. (Later, Hoh et al. found an alternative chromatographic solution to this problem when replacing the Rtx-500 column used in these experiments by a 2- [21] or 3- [22] m-long Rtx-PCB.) Use of the isotopic dilution procedure based on ^{13}C-labeled standards helped increase the accuracy of both the identification and the quantitation processes with GC×GC–ToF MS. Thus, concentrations similar to those obtained using the conventional GC–HRMS method were reported for the 21 toxic analytes investigated. Despite the high detector voltage used in these experiments (1800 V), the lower PCCD/F levels detected in biological matrices, close to the instrumental LODs of the GC×GC–ToF MS (0.5 pg for 2,3,7,8-TCDD,

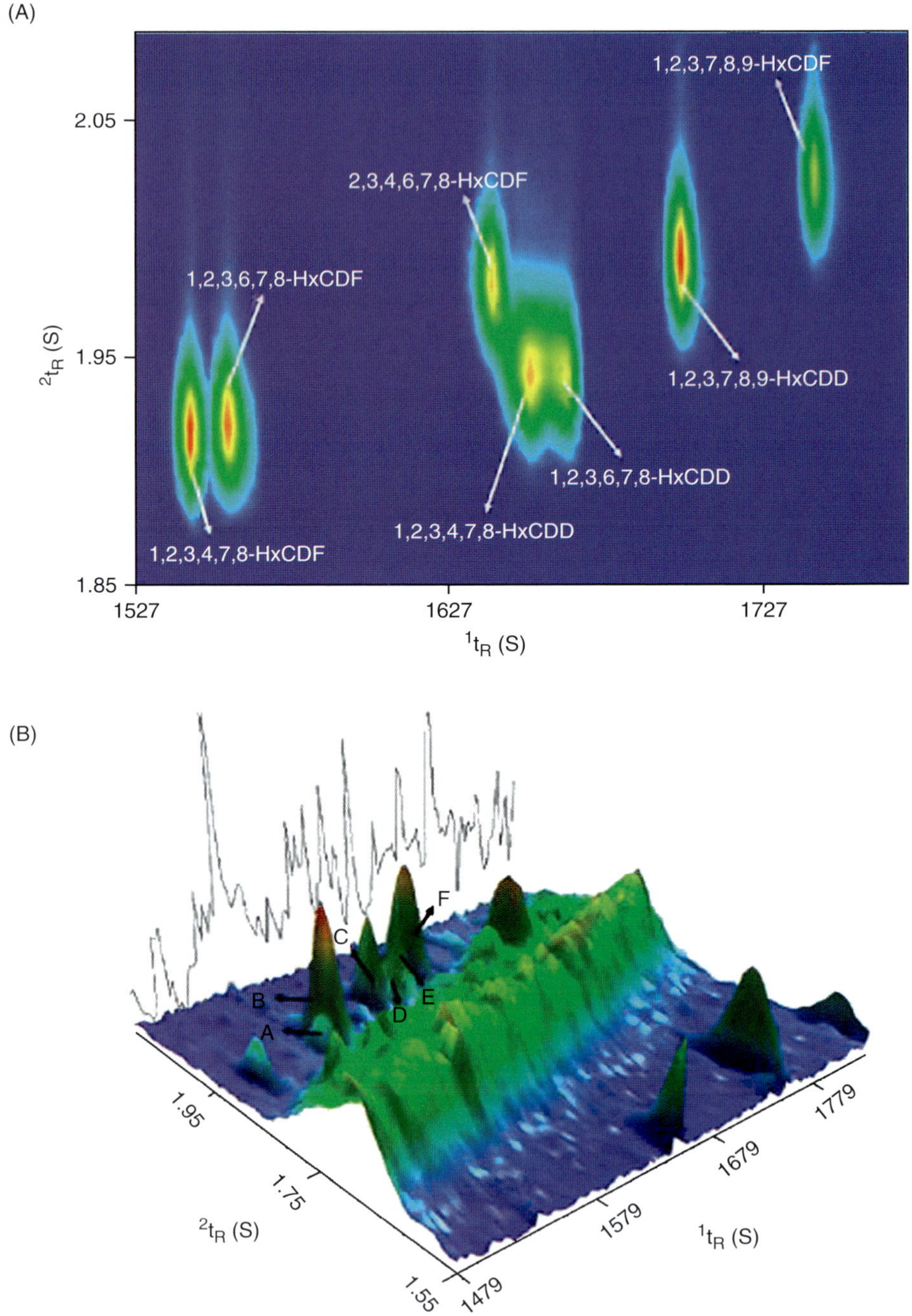

Figure 3 (A) Extended section of a GC×GC contour plot (*m/z* 374 + 390) of a standard solution of HxCDD/Fs. (B) Extended section of the HxCDD/F region of a GC×GC shade surface plot (*m/z* 374 + 390) of a purified fraction containing PCDD/Fs isolated from a fish sample [19].

while a ca. tenfold lower value of 0.04 pg is reported in GC–HRMS), compromised the application of the technique to the analysis of these particular families of pollutants, unless larger sample sizes were used. However, the higher concentration levels typically found for non-*ortho*-PCBs guaranteed their accurate determination even in these diluted and complex matrices. These two points were confirmed in a subsequent study involving fish, pork, and milk samples [20]. Though similar total TEQs were obtained for GC–HRMS, GC coupled to an ion trap working in the tandem MS mode, GC–ITD(MS/MS), GC×GC–ToF MS, and the dioxin-responsive chemical-activated luciferase gene expression (DR-CALUX) — the latter with the exception of milk — all three techniques were found to be less repetitive than conventional HRMS regarding the determination of individual congener concentrations.

Interestingly, in one of these works [19], the authors pointed out another interesting feature of GC×GC–ToF MS. Because of the nontargeted acquisition of the ToF MS, other closely related trace organohalogenated contaminants, such as certain organochlorinated pesticides (OCPs), polybromodiphenyl ethers (PBDEs), and phthalates, were identified in the same extract used for PCB and PCDD/F analysis. This observation suggested the potential suitability of GC×GC–ToF MS for simultaneous determination of all these relevant pollutants within the same run.

In a follow-up, the authors reported on the feasibility of GC×GC–ToF MS for simultaneous measurement of selected PCBs, OCPs, and relevant brominated flame retardants, such as polybrominated biphenyls (PBBs) and PBDEs, in a single chromatographic injection [23]. In this case, a shorter DB-5HT column (15 m) was used as ^{1}D to preserve the integrity of PBDEs, and a 1.2-m HT-8 was selected as ^{2}D for effective separation of the planar target analytes from other matrix components. The enhanced separation power provided by GC×GC, the possibility of ToF MS deconvolution, and the use of ^{13}C-labeled internal standards for quantitation, combined with the somehow higher concentrations at which these pollutants are detected in the human serum and milk samples compared to PCDD/Fs, resulted in accurate simultaneous determination of the 59 test pollutants and satisfactory comparison with results obtained with GC–HRMS. Good correlation coefficients spanned over three orders of magnitude (0.5–2000 pg/μL), instrumental LODs ranged between 0.5 and 10 pg/μL, and method LODs were in the 1–15 pg/μL range for all analytes. The reproducibility of the method was better than 11%, that is, almost as good as that of the standard HRMS method for analytes determined in nonspiked pooled human serum. These results demonstrated the suitability of GC×GC–ToF MS for environmental monitoring and, in particular, for human biomonitoring of these families of pollutants: While analysis of the several families of pollutants tested would require three separate runs with most 1D GC-based techniques today available, one single GC×GC–ToF MS run sufficed to obtain accurate simultaneous information regarding these trace micro-contaminant families.

To the best of our knowledge, at the time of writing, only one environmental application had been published involving GC×GC coupled to high-resolution ToF MS (HRToF MS) [24]. This study evaluates the feasibility of GC×GC–HRToF

MS for accurate determination of the 17 toxic PCDD/F congeners in a certified fly ash and several flue gas samples emitted from municipal waste incinerators using a rather conventional column combination — a 60-m 5% phenyl/phenyl-methyl silicone (InsertCap 5MS/Sil) coupled to a 1.5-m 50% phenyl/phenyl-methyl silicone (InsertCap 17MS/Sil), with a very narrow inner diameter (0.075 mm) for improved separation in the second dimension. A modulation period of 3 s was set to enable 75 data points per modulation since the minimum time for acquisition by the HRToF MS was 0.04 s. Using a mass resolution of 5000 at 500 *m/z*, the authors were apparently able to unambiguously determine all target compounds in a fly ash, even in the case of severe co-elution. As an example of the improved resolution power achieved with this instrument, Figure 4 shows the mass profile obtained at 40.28 min between *m/z* 335 and 346 for a fly ash crude (nonpurified) extract and demonstrated that the 337.8678 *m/z* [M^+] of PeCDF was separated from 337.3844 *m/z* and 338.2026 *m/z*, which are ions probably derived from other compounds. Despite these promising preliminary results and the low instrumental LODs reported (in the 0.4–5 pg range using ^{13}C-labeled compounds), comparison of the results found for real samples with those obtained using the reference method (i.e., GC–HRMS) indicated serious bias for specific congeners and suggested that further improvement might be necessary.

Two almost simultaneous papers reported on the feasibility of rapid-scanning quadrupole mass spectrometers (qMS) with an electron-capture negative ion (ECNI) option as detector for GC×GC analysis of PCBs [26] and PCDD/Fs [25]. In both studies, the instrument was operated in the single ion monitoring (SIM) mode. The limitation in the number of scanned ions resulted in the desired 33–50 Hz acquisition rate. However, the selected mass range should still ensure that

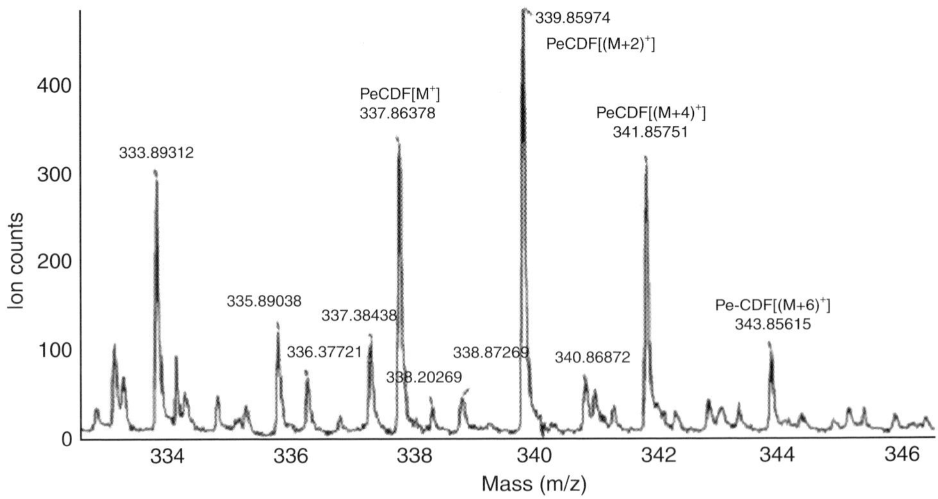

Figure 4 GC×GC–HRToF MS mass profile of a crude fly ash extract measured from *m/z* 335 to 346 at 40.28 min [24].

enough mass information was collected for the eluting peaks. In the case of PCBs, seven time-scheduled chromatographic windows were defined. Reported data proved that, despite the limited mass range selected, identification on the base of the mass spectrum was possible (matches of 98–99%). Satisfactory LODs below 2 pg injected mass were reported for a limited number of congeners, a value somehow higher than the 0.1–0.5 pg reported for ToF MS in [18], but still acceptable for a number of real-life applications.

PCDD/F determination with GC×GC–ECNI qMS required careful optimisation of the several experimental parameters affecting ionisation and data acquisition. In this application, only two selective ions, typically corresponding to the $[M]^-$ and $[M\text{–}Cl]^-$, were chosen to monitor the corresponding homologue group. Under these conditions, instrumental LODs below 100 fg, that is, in the range of those obtained using GC×GC–μECD [14] were obtained for the penta- to hepta-CDD/Fs. Unfortunately, the high LODs reported for OCDD (430 fg) and, especially, for 2,3,7,8-TCDD (710 fg) limited the applicability of the technique to analysis of real samples.

Next to PCBs and PCDD/Fs, other families of environmentally relevant aromatic organohalogenated pollutants rapidly attracted the attention of researchers working with GC×GC. Because of their widespread use and rapid increase of their environmental levels during the last decades, most attention was directed to flame retardants and, in particular, to PBDEs [23,27].

PBDEs are a numerous family of hydrophobic, persistent, ubiquitous, and rather nonvolatile pollutants with molecular masses in the range 482–950. According to their chemical structure, there are 209 possible PBDE congeners, which are identified following the same numbering used for PCBs. Technical mixtures contain a limited number of components, 20–25 congeners [34]. However, the instrumental analysis of these pollutants is complicated by thermal degradation of the higher brominated congeners and the frequent simultaneous presence in the purified extracts of other closely related analytes [23]. Korytár et al. [27] reported a detailed study on the GC×GC separation of 125 PBDE congeners and discussed the relative merits of six column sets regarding the notorious analytical problems associated with this type of analysis. Using DB-1 and DB-XLB as first dimension, 007-65HT was preferred as second dimension over VF-23 and LC-50 because (1) it provided a better resolution among the 125 studied PBDEs (co-elution persisted for 17 pairs, while 22 co-elutions were detected with 1D GC), and (2) it can stand with the high temperatures required to elute nona- and deca-BDEs. Regarding degradation, a severe decomposition of these four higher brominated congeners during the first-dimension run was observed, and DB-1×007-65HT then became the column set to choose. Figure 5 shows the apex plot of the 125 PBDEs on this column combination. Under finally proposed conditions, a satisfactory chromatographic separation was achieved for most relevant congeners in 80 min, both among them and from other relevant toxicants included in the study i.e., selected OH- and MeO-BDE metabolites, certain PBBs, and other relevant flame retardants commonly found in environmental samples, in particular hexabromocyclododecane (HBDE), tetrabromobisphenol-A (TBBP-A) and dimethyltetrabromobisphenol-A

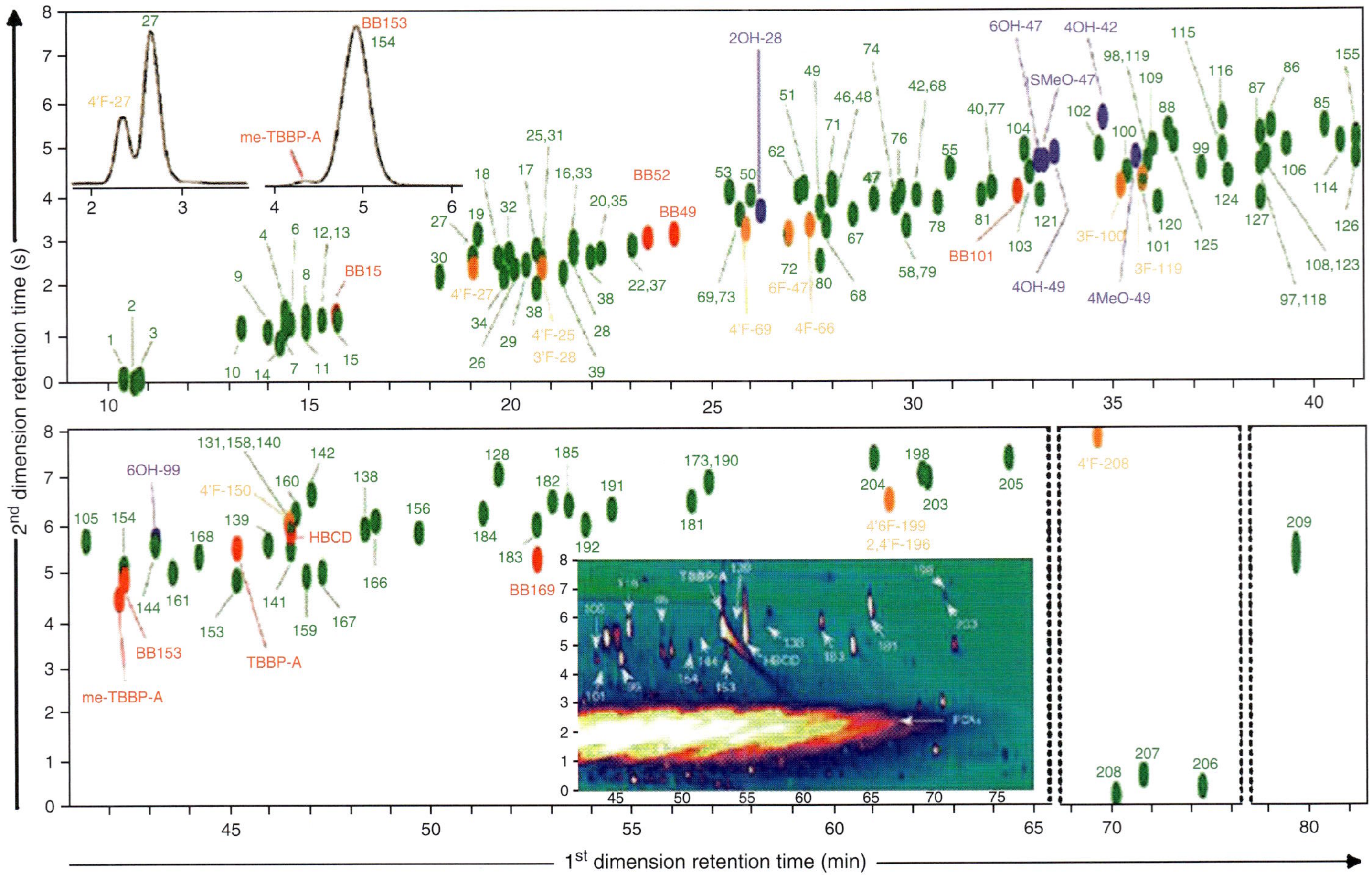

Figure 5 Overlaid GC×GC–μECD chromatogram on DB-1×007-65HT column combination of PBDES (green), fluorinated PBDEs (orange), other bronminated flame retardants (red), and PBDE metabolites (blue) [27].

(Me-TBBP-A). Wraparound was only observed for the nona- and deca-BDE congeners. The insert in the upper left-hand side of this figure demonstrates that a satisfactory separation among close eluting compounds was achieved through the whole chromatographic run. The insert in the lower right-hand side shows that successful separation was also obtained between PBDE 153 and TBBP-A and that accurate determination of PBDE 153 was possible even if a high concentration of TBBP-A was present. This study proved that μECD is an appropriate detector for this type of analysis. However, use of the most selective (and sensitive) ECNI qMS in the SIM mode has demonstrated that much more (unidentified) bromine-containing compounds can actually be present in real-life extracts [28]. The high separation demand posed by these substrates open a new working field for GC×GC.

This point has recently been illustrated by Pangallo et al. [28] in their work dealing with the identification of a set of 26 new halogenated 1′methyl-1,2′-bipyrroles (MBPs) in the dolphin blubber. MBPs are a family of halogenated natural products with physical and chemical properties similar to other persistent organic pollutants, which are of great interest to environmental chemists and toxicologists [28]. Bioaccumulative MBPs include more than 20 congeners of mixed halogenation (containing bromine and chlorine) in addition to the perchlorinated and perbrominated isomers. The source, biosynthethic pathway, environmental fate, and toxicity of these compounds remain unclear, but there is now evidence of geographic-dependent accumulation in marine mammals. One of the main present limitations regarding the analysis of this group of natural products is the lack of appropriate synthetic standards. In this study [35], four previously identified MBP isomers were isolated from dolphin blubber and, after purification, were used as reference and calibration standards. These four standards sufficed for tentative identification of 43 close eluting analytes as MBP isomers by GC–ECNI-qMS. Further evidence of the 28 partially halogenated MBP isomers was obtained by GC×GC–ToF MS using a 10-m Rtx-5 Crossbond column as ^{1}D and a 0.70-m BPX-50 housed in a secondary oven as ^{2}D. This orthogonal column set resulted in organised roof-tile structures in the 2D plane for the standard solution that extended to the newly found derivatives. MBP-Br_7 and its brominated congeners were found to align on none diagonal. Meanwhile MBP-Br_6Cl and congeners containing one chlorine aligned along a second, lower, parallel diagonal. Despite the (apparently) exceptional behaviour of MBP-HBr_5Cl and MBP-H_2Br_4Cl, the highly structured distribution of the target compounds provided support for tentative identification of the new analytes (Figure 6) and demonstrate the suitability of GC×GC–ToF MS for the identification of unknown compounds.

2.2 Nonaromatic organohalogenated pollutants

Technical toxaphene is a mixture of polychlorinated monoterpenes obtained by chlorination of camphene under UV irradiation. Introduced in 1945, toxaphene was mass-produced until the mid-1980 s and widely used as an insecticide, especially in cotton cultures. Currently, toxaphene is considered a worldwide

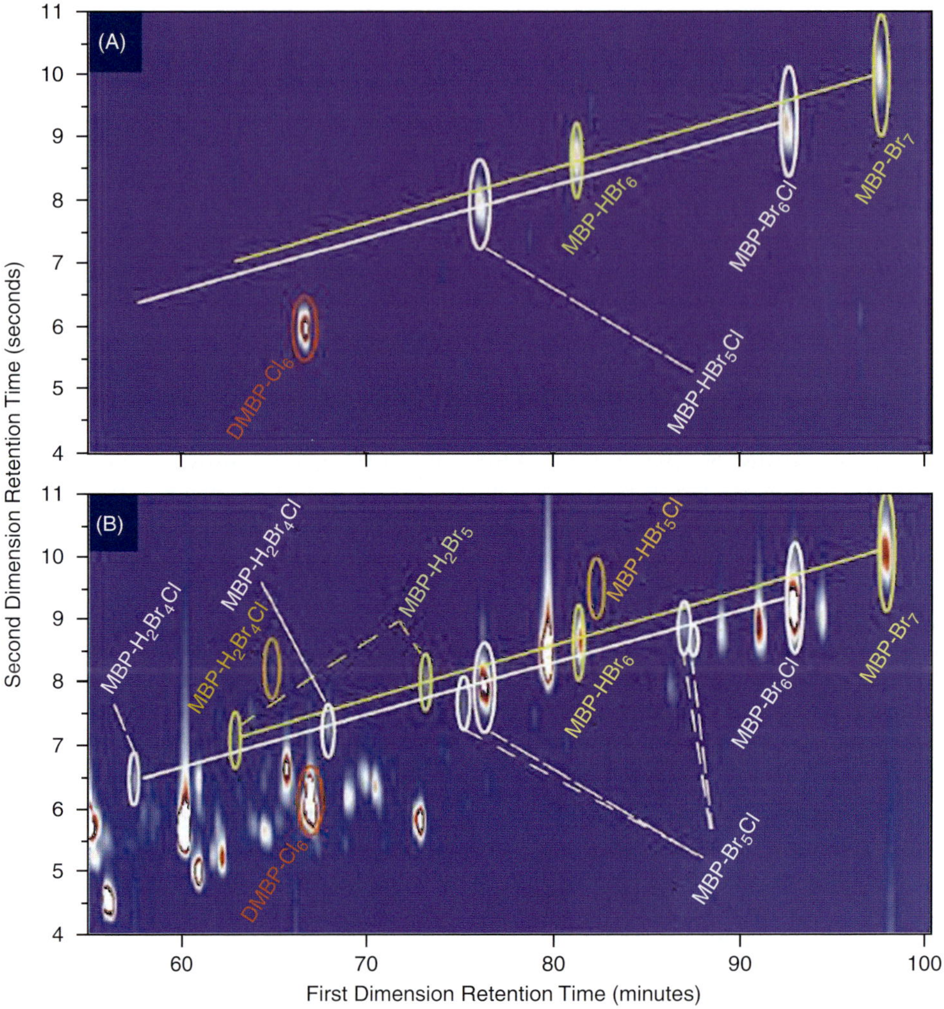

Figure 6 Section of the GC×GC–ToF MS contour plot of (A) the total in current of MBP standards and (B) the MBPs (*m/z*: 496, 540, 575, 620, 654, and 698) in a dolphin blubber extract. Detected MBP peaks are indicated by a coloured circle (yellow, white, and orange); the internal standard is indicated by a red circle; lines indicate the MBP containing bromine (top, yellow) and chlorine (bottom, white); peaks eluting above these lines are proposed to be MBPs (orange circle) [28].

distributed pollutant frequently detected at significant levels in fresh-water and marine biota as a complex mixture of compounds. The main constituents of toxaphene are chlorobornanes (for which 16,640 possible congeners have been calculated [36]) and chlorocamphenes (with 12,288 possible congeners). Chlorodihydrocamphenes (with 32,768 possible congeners) and chlorobornenes (or bornadienes) are considered minor components. The technical mixture also contains small amounts of other chlorinated and nonchlorinated hydrocarbons

[37]. Although most probably not all these theoretically possible congeners are present in technical mixtures, the figures help create an impression of their complexity and explain why the number of compounds estimated to be present on them has increased as the separation power of the instrumental approach selected for the analysis has increased [29]. Until the introduction of GC×GC, the best resolution (and highest figures, 675 congeners) was achieved by using multidimensional gas chromatography for the analysis of no less than 160 preseparated fractions of the mixture [38].

After a first attempt to use GC×GC, this time with ECD, for the separation of five bornanes [11], a much more detailed study on the composition of toxaphene was reported using μECD and ToF MS as detectors [29]. Using a 30-m HP-1 as first dimension and a 1-m HT-8 as second dimension, the authors obtained structured chromatograms reflecting the high complexity of the toxaphene mixture and the close structural relationship of the constituent compound classes. The total analysis time was 140 min, and the number of compounds present was estimated to be over 1000. The use of 23 individual congener standards (5 chlorocamphenes and 18 chloroboranes) and ToF MS as detector confirmed that these two classes of compounds do not separate from each other in the 2 D contour plot, and at the same time validated the fact that the several observed group peaks comprised congeners with the same number of chlorosubtituents with apparently no influence of the class of compound considered (Figure 7). Use of ToF MS also helped confirm that the mixture contained minor amounts of chlorodihydrocamphenes. In a subsequent study involving the same column combination but faster ramps of temperature, the authors also detected the presence of an

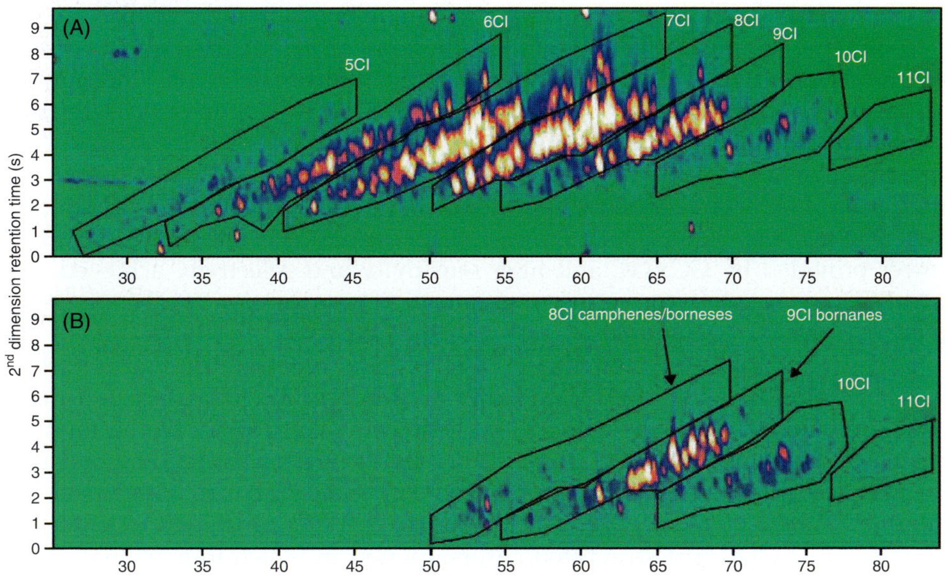

Figure 7 GC×GC–ToF MS chromatogram of technical toxaphene. (A) Total ion chromatogram (*m/z* 45-550), and (B) extracted ion chromatogram for *m/z* 413. [29].

extra group of compounds with roof-tile structure and eluting below toxaphene in the contour plot that was identified as hexa- to nona-chlorobornenes formed by thermal degradation of chlorobornanes [30].

Introduced in 1932, polychlorinated n-alkanes (PCAs) are complex mixtures obtained by chlorination of n-alkanes feedstock under uncontrolled conditions using molecular chlorine at temperatures of 50 to 150°C, elevated pressures, and/or under UV irradiation. PCA mixtures are classified according to their carbon chain length as short-chain PCAs (C_{10}–C_{13}), medium-chain PCAs (C_{14}–C_{17}) and long-chain PCAs ($>C_{17}$). These mixtures, with chlorination degrees in the range of 30–70%, have been used as additives in a large variety of industrial products and are found worldwide at variable concentration levels. Today, PCAs are classified as priority and toxic substances by the major international protection agencies.

The undetermined high number of components present in the PCA mixtures made impossible a congener-specific analysis with 1D GC regardless of the detector used, although examples involving ECD (because of its high sensitivity), qMS (mainly in the ECNI mode to avoid the excessive fragmentation produced by the EI mode), and ITD(MS/MS) (so far, probably the best alternative because of its higher selectivity) have been reported. At the time of writing, only a limited number of studies reporting on the GC×GC separation of PCAs can be found in the literature.

PCAs were included in a study [30] that evaluated the relative merits of five column combinations for group-type separation of 12 persistent organic pollutant (POP) families (see Section 2.4 below). The enhanced separation power and detectability provided by GC×GC–μECD, combined with the orthogonal character of most column sets assayed, resulted in structured chromatograms that allowed differentiation of several homologue bands in the technical mixture PCA-60. The complexity of this mixture became apparent through the several subgroups observed within these bands and, unfortunately, also through the partial overlapping among bands. DB-1×007-65HT was proposed as the best alternative for PCA analysis because it provided the clearest homologue band separation as well as the best separation between PCAs and other potentially interfering POP classes frequently detected in real environmental samples. Nevertheless, the results also showed that, despite the enhanced separation power provided by GC×GC and high sensitivity and selectivity achieved with the μECD used as detector in this work, a complete separation of all components present in the PCA mixtures was not possible [30].

Subsequent studies tested the feasibility of alternative detectors, such as ECNI-qMS [26] and ToF MS [31], to study the composition and characteristics of short-, medium-, and long-chain PCAs. Again, the complexity of the mixtures prevented a congener-specific analysis, but some features did become apparent. For instances, the analysis of a polychlorinated decanes mixture with an average chlorine content of 65 wt. % by GC×GC–ECNI-qMS using DB-1×DB-XLB as column set resulted in a structured chromatogram in which four parallel bands were separated. On the basis of the ECNI mass spectra, these bands were assigned, respectively, to hexa- and nona-chlorinated decanes, confirming that separation was based on the number of chlorine substituents.

Similar conclusions were obtained using ECNI-ToF MS as a detector, the same column combination, and a larger number of technical mixtures, as well as 35 individual PCA standards: compounds having the same chlorine substitution pattern but different carbon chain length were ordered as more or less parallel lines in the 2D contour plot [31]. The closely similar polarity was suggested as the most plausible explanation for this behaviour. It was also confirmed that, for compounds with the same carbon chain, congeners having chlorine substituents on only one end of the chain had shorter retention times in the ^{2}D column than those with chlorine substituents distributed over the entire carbon chain because of their lower polarity. In addition, use of individual standards proved that some congeners (e.g., 1,1,1,3,6,8,8,8-C_8Cl_8, 2,5,6,9-$C_{10}Cl_4$, 1,2,5,6,9-$C_{10}Cl_5$, and 1,2,5,6,9,10-$C_{10}Cl_6$) can exist as a number of diasteroisomers. A detailed inspection of the homologue group structures in technical PCA mixtures revealed the existence of subgroups within the homologue bands (see Figure 8 where there are no straight lines connecting the peak apices within each band). Overlay of the chromatograms obtained for technical short-chain (C_{10}-C_{13}) PCA mixtures (Figure 8E) indicated that compounds having the same number of carbon-plus-chlorine atoms showed up in the same diagonal line, but also that components with carbon chains differing in at least three carbons were efficiently separated with the DB-1×007-65HT column combination. In other words, this column set allowed partial differentiation of short-, medium- and long-chain PCAs, as it was demonstrated for the analysis of dust extracts.

2.3 Group-type analysis

The different toxicities of the individual microcontaminants made of analyte unambiguous determination a main requirement in the environmental field. As illustrated by previous sections, this need has made that the large majority of the studies dealing with GC×GC of pollutants focus on target analyte. In this type of determination, the main requirement is that the compound(-s) of interest be sufficiently separated from each other and from the sample matrix [39]. However, in many of these studies, the orthogonal nature of the selected column combinations result in organised chromatograms in which structurally related compounds typically elute as a band. This band-type organisation suggests the possibility of using the improved separation provided by GC×GC for fast screening of pollutant families using a group-type approach similar to that frequently used for characterising petrochemical (Chapter 7), fragrance (Chapter 9), and food (Chapter 10) samples. In this type of approach, the goal is to maximise the separation between the different component groups and among them and the sample matrix [39]. To achieve these requirements, if necessary, within-group separation could somehow be sacrificed because quantification is, for obvious reasons, not the main objective.

Despite the potential of this kind of group-type analysis for simultaneous fast screening of selected pollutant classes, for example, in monitoring analyses,until now the approach has somehow been overlooked. At the time of writing, only two papers have reported on the feasibility of this approach in the environmental

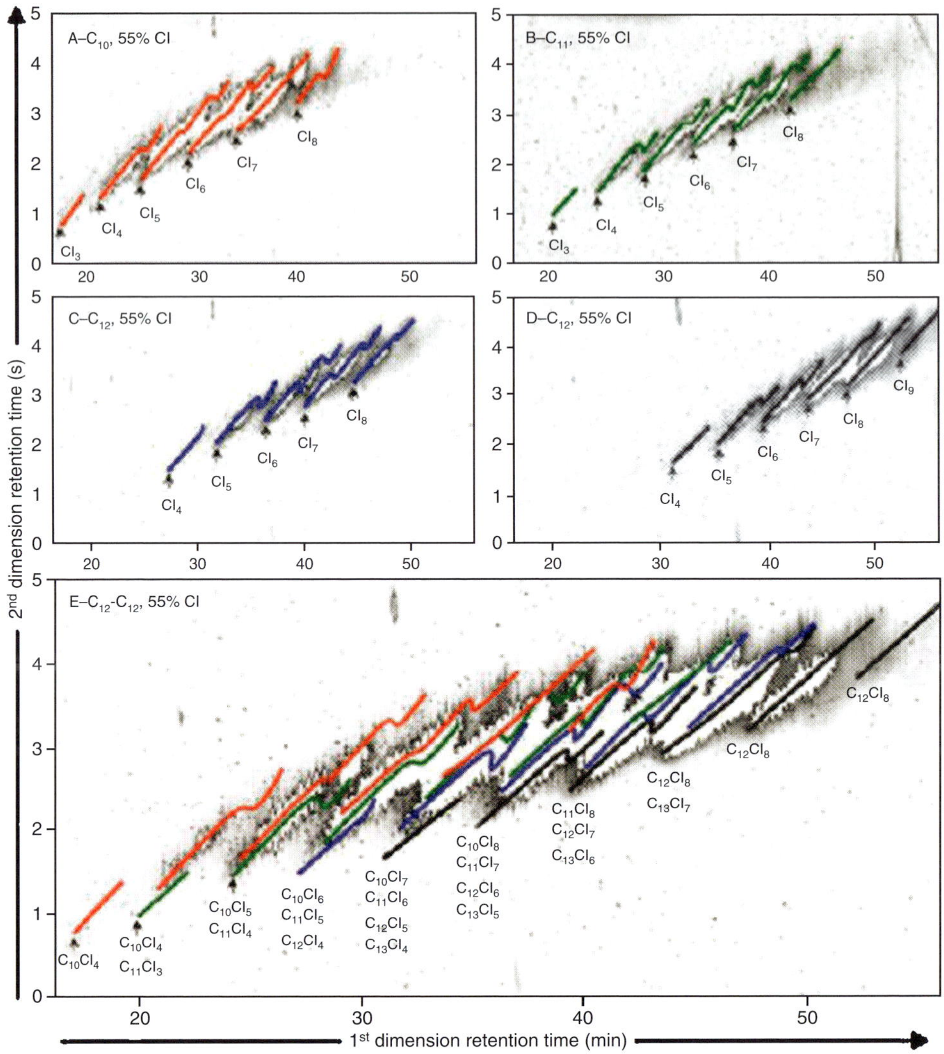

Figure 8 GC×GC–ECNI-ToF MS chromatograms of polychlorinated (A) decanes, (B) undecanes, (C) dodecanes, (D) tridecanes, and (E) C10-C13 technical mixture using obtained on DB-1×007-65HT column set. Lines indicate the positions of apices within the bands [31].

field [30,40]. In both cases, a μECD was used as detector, and closely related POP families were selected as test classes. Table 2 summarises the main conclusions of these studies.

None of the assayed column sets allowed the simultaneous and complete separation of all pollutant classes (12 POP families in [30] and 8 in [40]). However, some column combinations provided satisfactory separations among selected families and the rest of the pollutants investigated, indicating their suitability for fast isolation and detection of these particular group classes among all other POPs.

Table 2 Main features observed with different column combinations regarding group-type separation of aromatic and nonaromatic halogenated micropollutant families

Column combination	Analyte class separated	Reference
DB-1×007-210	–	[30][a]
DB-1×HT-8	–	
DB-1×LC-50	PCDDs, PCDFs, PCDTs, and PCNs from non-planar	
DB-1×007-65HT	PCAs and PBDEs Br- from Cl-substituted analogue classes	
DB-1×VF-23ms	–	
ZB-5×HT-8	–	[40][b]
ZB-5×BPX-50	PBDEs (except in marine samples)	
ZB-5×CW	–	
HT-8×BPX-50	PCDDs, PCDFs and PBDEs	
DB-17×HT-8	PCDDs, PCDFs and PCNs COPs (except in marine samples)	
DB-17×BPX-50	toxaphene (partialy from PCBs)	
DB-17×SW-10	–	
BP-10×HT-8	classes with one or two rings (PCBs, PCNs, toxaphene and OCPs) from those with three aromatic rings (PCDDs, PCDFs and PBDEs) PCBs from toxaphene (partial separation)	
BP-10×BPX-50	toxaphene	

Analyte classes included in the study:
[a]PCBs, PCDEs, PCNs, PCDTs, PCDDs, PCDFs, PCTs, PCAs, toxaphene, OCPs, PBBs and PBDEs.
[b]PCBs, PCNs, PCDDs, PCDFs, PCTs, toxaphene, OCPs and PBDEs.

As a typical example of the results obtained, Figure 9A shows the separation achieved among the several families investigated on DB-1×007-65HT [30]. This column set allowed a satisfactory separation of PCAs from all other studied POP classes, including some numerous groups such as toxaphene and PCTs, which were more retained in the second dimension (Figure 9B). Interestingly, this column combination was also found to provide a highly rewarding separation of PBDEs from other commonly overlapping POPs difficult to separate during sample cleanup, as well as a significant separation among Br-substituted compounds and the corresponding Cl-substituted class, as shown for PBBs and PCBs. Because of the orthogonal character of most of the column combinations assayed in these two studies and the high peak capacity provided by GC×GC, in many of the assayed column sets accurate within-group identification of target toxic compounds was still possible, giving the approach extra analytical potential.

The general practicability of this group-type analysis for fast environmental screening was demonstrated in both highly purified [30] and nonfractionated POP extracts [40] obtained from a variety of naturally contaminated samples, including sediments and dust [30] as well as biological tissues and foodstuffs [40].

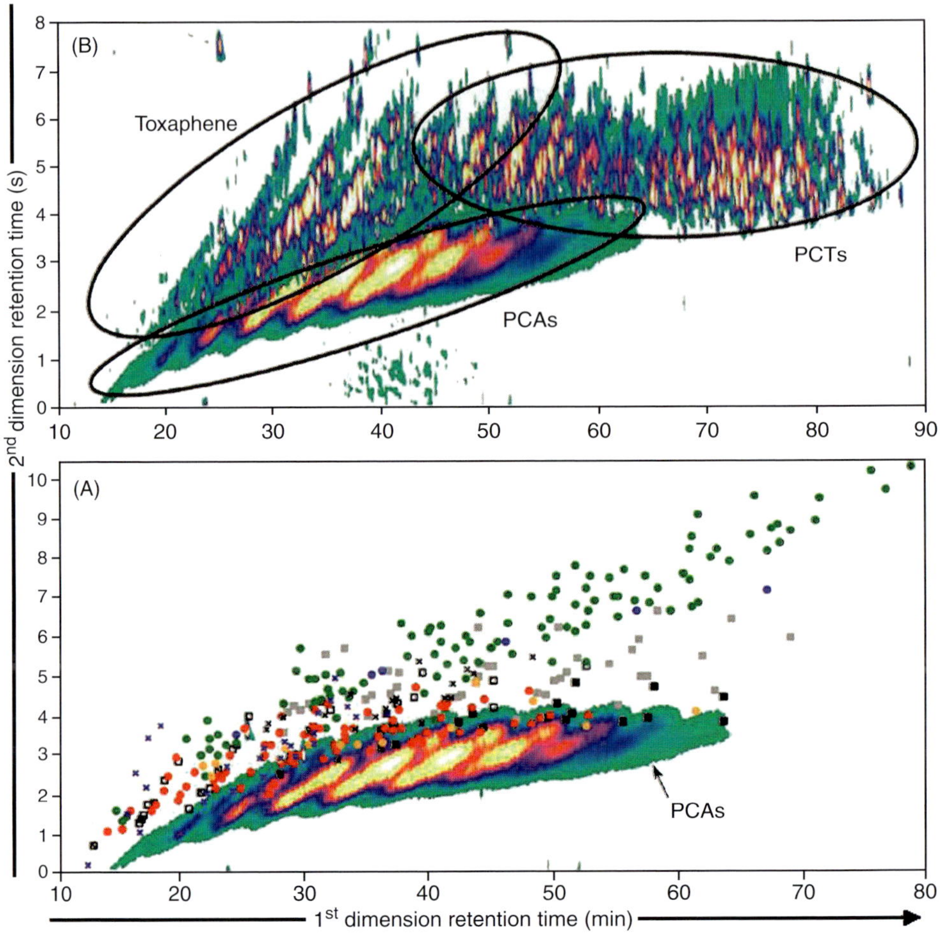

Figure 9 (A) Overlaid GC×GC–μECD chromatogram on DB-1×007-65HT column set of PCBs (red circle), PBBs (blue circle), PCDEs (orange circle), PBDEs (green circle), PCDTs (grey square), PCNs (white square), PCDD/Fs (black square), OCP (blue cross) individual toxaphene standrads (black cross), and PCAs (PCA-60). (B) PCA (PCA-60), PCTs (Aroclor 5442 + 5460) an toxaphene technical mixture [30].

3. NONHALOGENATED POLLUTANTS

3.1 Pesticides

Contrary to that observed for other classes of pollutants and for organohalogens in particular, the determination of (nonchlorinated) pesticides has attracted only limited attention (Table 3).

In some of the early studies reporting on the GC×GC separation of pesticides, these were chosen as model analytes rather than as real analytical targets [41,55]. However, impressive results were obtained, demonstrating the feasibility of this

Table 3 Selected GC×GC applications involving the analysis of nonhalogenated pollutants. For simplicity, only optimised experimental setups or those providing the most conclusive results have been mentioned (Acronyms as in Table 1)

Analytes	Sample	Column combinations (m×mm ID×μm d_f)	Modulator	Detector	Reference
17 Pesticides	human serum	DB-1 (2.0×0.25×0.25)× OV-1701 (0.8×0.1×0.05)	TDM	FID	[41]
9 Fungicides	Brussels sprouts	BPX-5 (30×0.25×0.25)× BPX-50 (1.0×0.15×0.15)	LMCS	NPD/μECD	[42]
33 Pesticides (OPPs[a], triazines, pyrethroids)	orange, pear, grape, apple	ZB-5 (30×0.25×0.25)× BPX-50 (0.8×0.10×0.10)	loop modulator, LN_2	μECD	[43]
92 Pesticides	spiked red grapefruit	SLB-5MS (30×0.25×0.25)× Omegawax (1.0×0.10×0.10)	loop modulator, LN_2	qMS	[44]
58 Pesticides	celery, carrot	CP-SIL5 CB (15×0.25×0.25)× BPX-50 (0.8×0.10×0.10)	LMCS	ToF MS	[45]
20 Pesticides	apple, peach	DB-XLB (30×0.25×0.25)× DB-17 (1.0×0.10×0.10)	quad-jet dual-stage modulator	ToF MS	[46]
51 Pesticides	grape	RTX-5MS (10×0.18×0.2)× TR-50MS (1.0×0.10×0.10)	quad-jet dual-stage modulator	ToF MS	[47]

Table 3 *(Continued)*

Analytes	Sample	Column combinations (m×mm ID×μm d_f)	Modulator	Detector	Reference
106 Pesticides	feed	RTX-CL (30×0.25×0.25)× BPX-50 (2.0×0.10×0.10)	quad-jet dual-stage modulator, LN_2	ToF MS	[48]
14 (OPPs and OCPs)	tobacco	Rtx-1 (30×0.25×0.25)× Rtx-200 (1.0×0.18×0.18)	quad-jet dual-stage modulator	ToF MS	[49]
36 Pesticides	tea	BPX-5 (40×0.18×0.18)× SupelcoWax (2.5×0.10×0.10)	quad-jet dual-stage modulato, LN_2	ToF MS	[50]
24 PAHs	soil	BPX-5 (30×0.25×0.25)× BPX-50 (1.2×0.10×0.20)	LMCS	FID	[51]
9 PAHs	sediment	HP-5MS (20×0.25×0.25)× BGB-1701 (0.5×0.1×0.10)	home-made semirotating cryogenic modulator[b]	FID	[52]
PAHs, nonylphenols	sediment	DB-5 (20×0.25×0.25)× BGB-1701 (1.0×0.10×0.10)	Dual-stage modulator, CO_2	ToF MS	[53]
12 Nonylphenol isomers	river water	DB-5 (30×0.25×1)× SP-Wax (1.0×0.10×0.10)	loop modulator, LN_2	qMS	[54]

[a]OPPs, organophosphorous pesticidas.
[b]home-made semirotating cryogenic modulator [33].

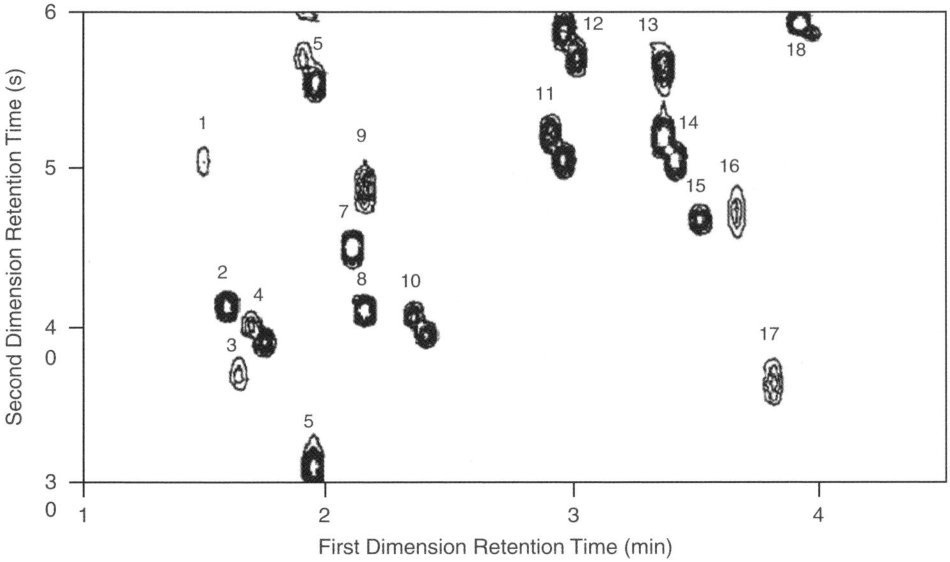

Figure 10 GC×GC separation of a pesticide mixture. Peak identification: 1, dicamba; 2, trifluralin; 3, dicloran; 4, phorate; 5, pentachlorophenol (internal standard); 6, atrazine; 7, fonofos; 8, diazinon; 9, chlorothalonil; 10, terbufos; 11, alachlor; 12, matalaxyl; 13, malathion; 14, metalachlor; 15, DCPA; 16, captan; 17, folpet; 18, heptadecanoic acid (internal standard) [41].

technique for this type of determination. As an example, Figure 10 shows the baseline separation of 17 pesticides and two internal standards, pentachlorophenol and heptadecanoic acid, achieved in 4.5 min using a 2-m DB-1 as first column and a 0.8-m OV-1701 as second dimension [41]. Using an FID as detector, on-column limits of detection in the range 2–4 pg were obtained. This result proved the practicality of the method for pesticide determination in relatively clean samples, such as human serum.

Khummueng et al. [42] compared the effectiveness of several column combinations for separation of N-containing fungicides in a Brussels sprout extract. With BPX-5×BPX-50 as the column set, LODs and limits of quantification (LOQs) below 74 and 250 ng/L, respectively, were obtained with the nitrogen-phosphorous detector (NPD) and even lower when using a μECD. These results, combined with the satisfactory repeatability and reproducibility of peak response proved the potential of the GC×GC method proposed for routine analysis of fungicides in vegetables. A similar column combination, ZB-5×BPX-50, was also found to provide the best separation among fruit matrix components and the 33 pesticides included in a recent study involving GC×GC–μECD [43]. As in the observations made in [42], analyte compression through the modulation process resulted in extremely low LODs that ensure accurate determination at the low MRLs set in current legislation.

Despite the satisfactory results obtained with these element selective detectors, in most of the studies concerning pesticide analysis, detectors providing

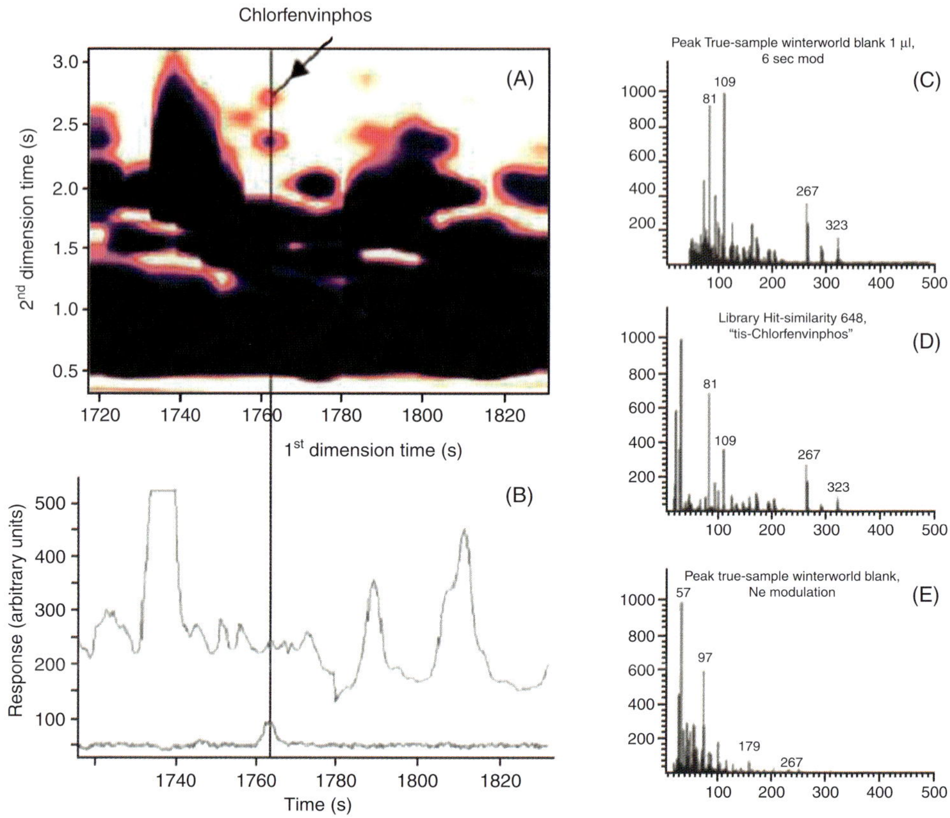

Figure 11 GC×GC–ToF MS vs 1D GC–ToF MS for the analysis of a carrot extract. (A) GC×GC–ToF MS contour plot. (B) 1D GC–ToF MS of the region selected in (A). (C) Mass spectrum obtained after GC×GC separation showing the characteristic *m/z* ions of chlorfenvinphos and comparison with (D) library spectrum and (E) spectrum obtained at the same retention time with 1D GC [45].

structural information (i.e., mass spectrometers) have been preferred (Table 3). Among them, ToF MS has been by far the most frequently used [45–50].

The potential of GC×GC–ToF MS for nontarget pesticide analysis was first illustrated by Dallügue et al. in 2002 [45]. The separation power of this three-dimensional technique was demonstrated through quite self-explicative chromatograms, as those reproduced in Figure 11. This figure compares the 1D GC–ToF MS mass spectra (Figure 11E) obtained for a carrot containing 0.02 mg/kg of chlorfenvinphos (corresponding to 10 ng/mL in the injected extract) with that obtained using GC×GC–ToF MS (Figure 11C). GC×GC can separate the quantification interferences from the matrix (Figure 11A and B), but the qualitative identification is provided through a combination of GC×GC and spectral deconvolution (peak true). In other words, while chlorfenvinphos identification was hampered by matrix components in the 1D GC approach, accurate determination at low concentration levels was possible with the comprehensive technique. The

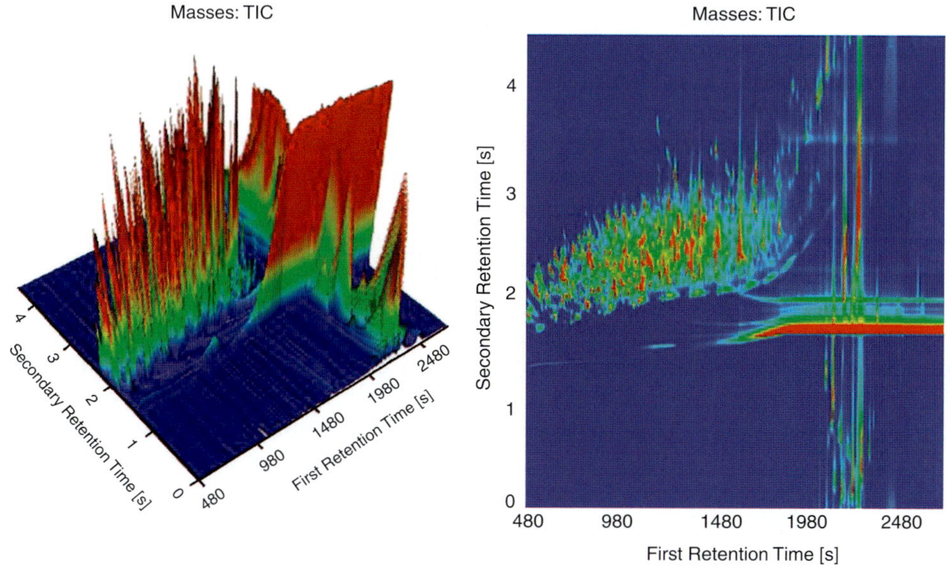

Figure 12 Typical 3D GC×GC–ToF MS image and contour plot obtained for a purified feed sample [48].

experimentally determined LODs of 10–30 pg for the N/P-containing pesticides included in the study lent support to this statement. In a somehow closely related study, Zrostlíková et al. [46] concluded that GC×GC–ToF MS can provide a 1.5- to 50-fold improvement in the LODs calculated for 20 modern pesticides as compared to the equivalent GC–ToF MS method. The combined effect of analyte compression on the modulator, improved separation from co-extracted sample material provided by the GC×GC, and deconvolution capabilities offered by the ToF MS explains these results [46,49].

A detailed study on the influence of the different experimental parameters affecting the modulation and acquisition data processes on the detectability and final identification of pesticides with GC×GC–ToF MS can be found in [47]. Results demonstrated that when pesticides should accurately be determined in complex extracts, a data-acquisition rate of 100 Hz or higher should preferably be used. Van der Lee et al. [48] used a data-acquisition rate of 200 Hz to deal with pesticide analysis in as complex a matrix as feed. Despite the laborious multistep procedure used for cleanup of the sample, analyte co-elution with matrix material was frequently detected (Figure 12). However, the accurate mass spectra definition achieved, thanks to the high data-acquisition rate, allowed automatic identification by the software, by comparison of the spectra of all individual compounds detected in the sample against a target library. Using a representative feed and setting a similarity threshold of 600, all 106 targeted pesticides were virtually detected through automatic screening at levels of 50 μg/kg. At a level of 10 μg/kg, 73% of the analytes were still fully automated, but at lower concentration levels the number of compounds detected decreased dramatically. The

GC×GC–ToF MS linearity was excellent in solvent and only slightly affected by the matrix, and the LODs were in general below 20 μg/kg. Furthermore, the acquisition of full-range mass spectra provided powerful confirmation of the pesticides in the sample, even for nontarget analytes [45,48,49]. This potential is in principle reinforced by the automated peak find and spectral deconvolution software capabilities, although in practice intensive manual data revision is often needed [45,48,50].

The new generation of high-speed qMS system represents a valuable alternative to the more expensive ToF MS also in this field. In these analyzers, the mass scanning range should still be limited to a relatively narrow range (at least compared to that typically used in ToF MS experiments), which limits the possibility of subsequent analysis of archived data for unknown identification. However, examples reporting satisfactory linearity (regression coefficients better than 0.9994 in the 1.0–15 ppm range with four-point calibration curves) and LODs low enough to ensure analyte detection at the low levels set in current legislation have already been described in the literature [44].

All previously revised studies focus on target analyte. Consequently, efficient separation of the investigated analytes from the sample matrix was the main concern during method development, especially when additional structural information was not available. This explains why group-type analysis has received limited attention in the pesticide field, despite the efficiency of this approach for fast screening proposes [43].

3.2 Other organic pollutants

Polycyclic aromatic hydrocarbons (PAHs) are a class of organic micropollutants containing two or more condensed rings. Apart from those cases associated with oil pollution, PAHs are produced mainly by anthropogenic combustion and are typically found in the environment as complex mixtures. However, not all components exhibited the same level of toxicity, and congener-specific determination has again become mandatory.

As shown in Chapter 7, PAHs are a relevant class of compounds in petrochemistry. However, up to now they have apparently received rather limited attention from environmental chemists working on GC×GC. This sharp difference among the number of studies concerning PAHs in these two research areas is probably a result of the different aims and analytical requirements for both types of analyses. In the area of petrochemistry, the main interest is group-type analysis and a FID suffices for accurate determination. Meanwhile, environmental studies are focussed on target analysis, and use of a nonselective detector, such as FID, can be problematic because of the low levels of the analytes and the typical high complexity of the extracts. On the other hand, although use of MS-based detectors can be desirable because they allow structural confirmation, their limited sensitivity can become a problem when analysing real-life samples with trace levels of these compounds.

The feasibility of GC×GC–FID as a screening method for the analysis of 24 environmental relevant PAHs in contaminated soils from a former gasworks site (concentrations in the 10–300 mg/Kg range) was first illustrated by Ong et al. [51]. Not surprisingly, structured chromatograms in which PAHs were grouped according to their number of aromatic rings were obtained using BPX-5×BPX-50 as the column set. All PAHs were separated among them and from other hydrocarbon classes present in the purified extract, with the exception of two pairs, benzo(*b*)fluorantene — benzo(*k*)fluorantene, and indeno(*c*,*d*)pyrene — dibenzo(*a*,*c*)anthracene. After careful optimisation of the sample preparation procedure, the clean obtained extracts allowed PAH identification by direct comparison of the ^{1}D and ^{2}D retention times in the soil extract with those found for the standards. Comparison of the GC×GC–FID method with the more conventional GC–MS from a quantitative point of view demonstrated that, although both sets of data compared reasonably well for low-molecular-mass PAHs, high-molecular-mass PAHs were underestimated with the comprehensive technique.

Similar conclusions were obtained in a subsequent study oriented toward the study of qualitative and quantitative aspects of GC×GC. In this case, PAHs were selected as model compounds and a FID as detector [52]. The quality and efficiency of the GC×GC separation were evaluated on the basis of several parameters, namely, the peak width at peak base, asymmetry, resolution, depth of the valley between PAH and preceding matrix peak, and total retention time, using both LC purified and nonpurified sediment extracts. For quantitation, two different approaches based on peak areas and volumes were tested. Results proved that increasing the matrix amount 16-fold compared to levels in the LC cleaned sediment did not affect the separation of PAHs in terms of peak width, asymmetry, and resolution. However, these large matrix amounts increased the depth of the valley between the considered PAH and the preceding matrix peak and reduced the repeatability of the retention times as well as those of the peak areas and volumes. Regarding quantitation, calibration based on peak area trended to underestimate the trace PAH levels in the sediment analysed, especially for high-molecular-mass components in nonpurified extracts. Volume-based calibration resulted in more accurate results, especially when using the external calibration procedure, for both cleaned and noncleaned extracts.

Nonylphenolpolyethoxilates are commonly used nonionic surfactants. In wastewater treatment plants, these compounds degrade to a number of products, including nonylphenols. Although nonylphenols are typically detected in environmental samples as a mixture of isomers due to branching of the C-9 group, only some of these isomers have been reported to exhibit estrogenic potential. The feasibility of GC×GC for unambiguous determination of some of these isomers was investigated by Ieda et al. [54] using a qMS as detector. To increase the data-acquisition rate, a limited scan range (m/z 105 to m/z 170), corresponding to 24.5 Hz, was set. As an example of the potential of the technique in this field, the authors reported on the separation of 102 peaks of nonylphenols in a technical mixture. The optimised method was then applied to quantification of selected nonylphenol isomers in water river with satisfactory results, that is, linear response in the 5–100 ng/L range with correlation coefficients better than 0.994; and LODs lower than 0.7 ng/L.

4. ANALYSIS OF CHIRAL POLLUTANTS

Previous sections have illustrated the complexity of most of the compound-specific analyses developed in the environmental field. It is easy then to figure out that the accurate determination of a possible enantiomeric enrichment of chiral pollutants is even more difficult owing to the many co-elution problems and low concentration levels of the analytes. This difficulty could explain the somehow limited research conducted on this topic. However, its interest is clear. Industrial contaminants, such as PCBs or toxafene, are released into the environment as racemates. Therefore, a nonracemic composition of these pollutants might be evidence of selective biotransformation and/or bioaccumulation. Some studies have also pointed to different biological and toxic behaviour for each of the enantiomers [56], something that can be especially relevant for pesticides exhibiting chiral properties.

Chiral analysis of trace pollutants in complex environmental samples has typically required laborious and time-consuming fractionation steps before instrumental determination of the target analytes [57]. Heart-cut multidimensional gas chromatography represents a valuable alternative to these approaches that efficiently contributes to reduce sample manipulation. However, it can also be rather tedious because only a limited number of target compounds can be transferred to the second column in a single run (Chapter 1).

The first attempt to use GC×GC for chiral separation of organic micropollutants was reported by Harju and Haglund in 2001 [58] (Table 4). Nine out of the 19 atropoisomeric PCBs (PCB Nos. 45, 84, 88, 91, 95, 131, 132, 135, 136, 139, 144, 149, 171, 174, 175, 176, 183, 196, and 197) [63] were used as test compounds. In particular, those that can be separated into enantiomers on the permethylated β-cyclodextrin column (Chirasil-Dex) selected as first-dimension column. Using a shape-selective column as second dimension, LC-50, six out of nine studied atropisomeric PCBs were resolved (PCB Nos. 91, 132, 135, 136, 149, and 176) and two more (PCBs 84 and 174) were partially separated from co-eluting congeners in a mixture of 144 congeners using a μECD as detector. In a follow-up [59], the authors concluded that the use of VF-23 ms as second dimension instead of LC-50 yielded more satisfactory results when analysing real-life samples. With this column set and a single injection, the authors reported on the enrichment factor (EF) of five atropisomeric PCBs (PCB Nos. 91, 95, 132, 149, and 174) and simultaneously determine the concentrations of the seven priority and twelve toxic PCB congeners in grey-seal samples. Results demonstrated that the EFs of some PCBs deviated strongly from racemic (results confirmed by GC×GC–ToF MS) and that ratio deviations were higher in liver than in blubber, suggesting a possible enantioselective metabolism.

Using the same column combination but with a longer first dimension (25 instead of 10 m Chiralsil-Dex column), Bordajandi et al. [60] were able to elute free from interferences seven out of the nine atropisomeric PCBs that can be resolved into enantiomers with this phase (i.e., PCBs 84, 91, 95, 136, 149, 174, and 176). However, PCBs 135 and 132, the latter being one of the most abundant in food samples, co-eluted with PCBs 82 and 141, respectively. With

Table 4 Selected GC×GC applications involving the analysis of chiral pollutants. For simplicity, only optimised experimental set-ups or those providing the most conclusive results have been mentioned (Acronyms as in Table 1)

Analytes	Sample	Column combination (m×mm ID×μm d_f)	Modulator	Detector	Reference
9 Chiral PCBs	standard solution	Chirasil-Dex (10×0.10×0.10)× LC-50 (1.0×0.10×0.10)	LMCS	μECD	[58]
9 Chiral PCBs (+7 indicator and 12 toxic PCBs)	144 PCBs, seal liver and blubber	Chirasil-Dex (10×0.10×0.10)× VF-23MS (1.5×0.10×0.10)	LMCS	μECD	[59]
9 Chiral PCBs	milk, cheese	Chirasil-Dex (25×0.25×0.25)× VF-23ms (1.0×0.1×0.10) Chirasil-Dex (25×0.25×0.25)× Supelcowax-10 (0.9×0.1×0.10)	loop modulator	μECD	[60]
19 Chiral PCBs (+7 indicator and 12 toxic PCBs)	milk, cheese, salmon	BGB-172 (30×0.25×0.18)× Supelcowax-10 (1.0×0.1×0.10)	loop modulator	μECD	[61]
5 Chiral toxaphenes	fish oil	BGB-172 (30×0.25×0.18)× BPX-50 (2.0×0.10×0.10)	loop modulator	μECD	[62]

Chiralsil-Dex×Supelcowax, PCBs 91, 95, 132, 135, 149, 174, and 176 eluted without interference from a test mixture of 95 PCBs, but congeners 84 and 136 co-eluted with PCBs 56 and 85, respectively. The combined use of these two column sets allowed accurate EFs determination for the nine atropoisomeric PCBs in milk and cheese samples. In a subsequent study [61], three β-cyclodextrin-based columns (Chirasil-Dex, BGB-172 and BGB-176SE) were combined with HT-8, BPX-50 and Supelcowax-10 as second dimension and evaluated for simultaneous determination of the 19 chiral PCB enantiomers, the seven priority and twelve toxic PCBs. In general, the best results were obtained with those column sets involving Supelcowax as second dimension. In addition, these column combinations allowed priority and toxic congeners to be detected in real food samples free from interferences. Regarding chiral PCBs, BGD-172×Supelcowax-10 provided the best overall separation allowing all enantiomers to be determined free from interference (Figure 13), with the only exception of PCB 91. Accurate determination of the EF for this particular congener was only possible on BGB-176SE×Supelcowax-10. Further confirmation of the results obtained with the comprehensive approach by heart-cut multidimensional gas chromatography analysis of the extracts proved the feasibility of the proposed method even if structural information was not available.

BGB-172 and BGB-176SE in combination with HT-8, BPX-50 and Supelcowax-10 were also evaluated for the enantiomeric separation of five chiral toxaphenes typically found in real-life marine samples, Parlar 26, 32, 40, 44, and 50, in nonfractionated extracts containing other POPs and using a μECD as detector [62]. Under these conditions, BGB-172×BPX-50 provided the best results allowing the unambiguous determination of the EFs of the five studied toxaphenes with satisfactory repeatability and reproducibility values (RSDs lower than 11%) and with adequate LODs of 2–6 pg/μL. Again, the EF values calculated for real

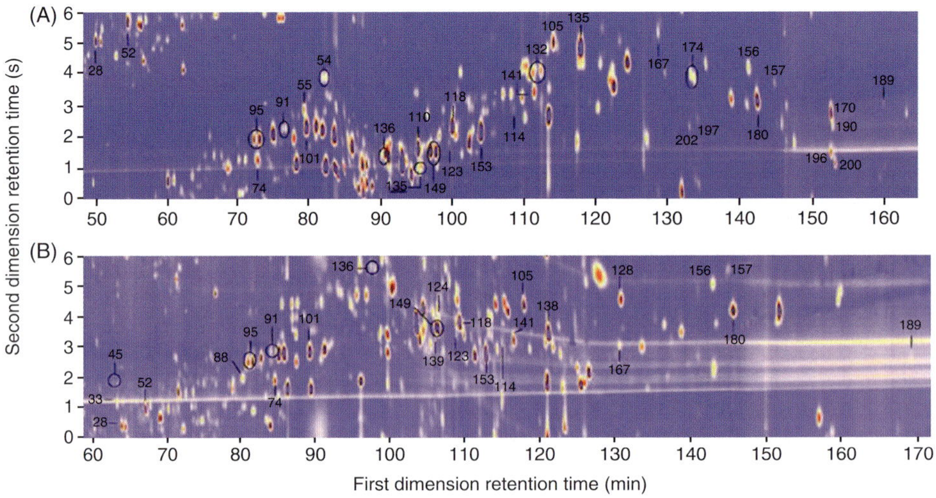

Figure 13 GC×GC–μECD contour plots of a salmon extract on (A) Chirasil-Dex× Supelcowax-10 and (B) BGD-172×Supelcowax-10 [61].

samples, in this case fish oil, agreed with those determined by using heart-cut multidimensional gas chromatography.

5. CONCLUSIONS

The complexity of most of the micropollutant mixtures and the constant demand for enhanced separation in this research area made GC×GC immediately attract the attention of the environmental analytical chemists. Early attempts to use GC×GC to unravel the composition of numerous classes of contaminants, such as PCBs, demonstrated the potential of the technique for accurate target analysis, as far as an adequate detector was available. Such a detector was μECD, as demonstrated by the many studies reporting on this subject since its introduction in 1990. μECD allowed a convenient data-acquisition rate of 50 Hz and, more importantly, provided enough selectivity and sensitivity to allow detection of organohalogenated pollutants at the low levels typically found in the environment. Many column combinations and experimental conditions were then assayed to determine the best experimental conditions for chromatographic isolation of the target toxic congeners from other compounds belonging to the same class and to closely related chemical families that cannot completely be separated by sample preparation procedures in use. Because of their toxicity, most attention was devoted to PCBs and PCDD/Fs, and, due to the nonpolar nature of these toxicants, nonpolar×(semi-)polar columns sets providing highly structured chromatograms were initially preferred. However, subsequent application of optimised methods to analysis of real-life samples showed the relevance of sample matrix in this type of determination. The complexity of many of the extracts analysed in this field frequently resulted in the co-elution of matrix components with the target compounds and the consequent overestimation of the concentrations of the latter. Ongoing investigations demonstrate that the use of shape-selective columns, and in some cases reverse configuration, can help solve this problem. At present, GC×GC–μECD is considered a valuable analytical alternative for accurately determining individual congeners in classes such as PCBs, PBDEs, PCNs, and OCPs. Depending on the concentration levels, that is, in the case of highly contaminated samples, it can also provide satisfactory results for PCDD/Fs however, in all cases, only after an intensive, that is, mainly manual, data processing step that seriously limits the applicability of the technique for routine and monitoring analyses.

For other more numerous classes, such as PCAs, toxaphene, and PCTs, the enhanced sensitivity and separation power provided by GC×GC–μECD have contributed to gathering useful information about the several subgroups of structurally related components simultaneously present in these families. However, isomer-specific analysis is not yet possible. The introduction of ToF MS as a GC×GC detector added an extra separation dimension to that of GC×GC by incorporating structural information and deconvolution capabilities. GC×GC–ToF MS indeed added extra insight into the composition of these complex mixtures, but further research is needed to unravel their composition.

Interestingly, GC×GC–ToF MS also contributed (albeit slowly) to extending the application field of GC×GC to other classes of pollutants that previously had received only limited attention, with PAHs and pesticides as prominent representative classes. Once more, the main goal in these studies has been target analysis, and consequently, separation of the tested analytes forms the sample matrix has been the most urgent demand. The feasibility of this three-dimensional approach for accurate automated detection of a complete set of pesticides at acceptable concentration levels of 50 μg/kg in complex matrices has been demonstrated. However, at lower levels (i.e., below 10 μg/kg), the efficiency of the method decreased dramatically, and manual integration was again required. Unfortunately, the sophisticated nature of ToF MS, together with its high price, prevents the introduction of this powerful detector in many laboratories. In these cases, much less expensive and more user-friendly rapid-scanning qMS instruments provide satisfactory results using limited mass ranges of 200–250 Da — which is a sufficiently wide range for most of these target-type applications.

Regarding future trends, a specific research field in which ToF MS offers unsurpassed capacities is in the preliminary identification of new (i.e., unknown) pollutants. Here, the continuous (and complete!, m/z range, 5–999) structural information provided by ToF MS through the entire GC×GC chromatogram, combined with the powerful deconvolution algorithms incorporated in its advanced software, represents a distinguished feature that no doubt will be explored in coming years. The scripting capabilities recently introduced in commercial MS-based software packages could help in the use of GC×GC for fast screening of selected families of pollutants in monitoring studies. Finally, the application of chemometric approaches for efficient data interpretation and pattern recognition could also contribute to the practical implementation of this technique in environmental laboratories.

ACKNOWLEDGMENTS

Authors acknowledge MICINN for financial support via grant CTQ-2006-14993/BQU. JJR and MPA thank MEC for FPI and FPU grants, respectively.

REFERENCES

1 M. Adahchour, J. Beens, R.J.J. Vreuls and U.A.Th. Brinkman, *Trends Anal. Chem.*, 25 (2006) 726.

2 M. van den Berg, L. Birnbaum, A.T.C. Bosveld, B. Brunström, P. Cook, M. Feeley, J.P. Giesy, A. Hanberg, R. Hasegawa, S.W. Kennedy, T. Kubiak, J.C. Larsen, F.X.R. van Leeuwen, A.K.D. Liem, C. Nolt, R.E. Peterson, L. Poellinger, S. Safe, D. Schrenk, D. Tillitt, M. Tysklind, M. Younes, F. Waern and T. Zacharewski, *Environ. Health Perspect.*, 106 (1998) 775.

3 U.S.EPA Office of water engineering and analysis division (4303), Method 1613: Tetra- through octa-chlorinated dioxins and furans by isotope dilution HRGC-HRMS, USA, (1994).

4 D.G. Patterson, J.R. Barr, E.S. di Pietro, J. Granger, V.E. Green, C.R. Lapeza, V.L. Maggio, P.C. McClure, S. Sirimanne and W.E. Turner, *Organohalog. Compd.*, 27 (1996) 309.

5 J.D.M. Dimandja, J. Grainger, D.G. Patterson, W.E. Turner and L.L. Needham, *J. Expos. Anal. Environ. Epidem.*, 10 (2000) 761.

ANNEX

Troubleshooting

Lourdes Ramos and **Jesus Sanz**

Contents

1. INTRODUCTION

Previous chapters in this volume have shown the impressive separation power of comprehensive two-dimensional gas chromatography (GC×GC). Compared with conventional monodimensional gas chromatography (1D GC), GC×GC provides a better overall separation among analytes and between these analytes and matrix components, which results in more reliable compound determination. Contour plots illustrating extra benefices associated with this technique, such as the possibility of obtaining structured chromatograms and the enhanced sensitivity achieved as compared to 1D GC have been shown throughout this volume. However, as explained in Chapter 4 and illustrated in many examples of this book, visualization can efficiently show other relevant characteristics of GC×GC data, including chromatographic features related to both the first (^{1}D) and second (^{2}D) dimension separation, as well to the modulation process.

As in 1D GC, simple inspection of the peak shape, in this case in contour plots or in nonconverted 1D (i.e., raw) chromatograms, suffices to experienced GC×GC users to identify chromatographic problems such as analyte tailing and column overloading, in the ^{1}D and/or ^{2}D. In GC×GC, however, other problems

Comprehensive Analytical Chemistry, Volume 55
ISSN: 0166-526X, DOI 10.1016/S0166-526X(09)05512-3

inherent to this comprehensive technique also become apparent. The reasons for these problems have been explained in detail at different points of the text, and the most plausible practical solutions have been described. Therefore, this annex does not attempt to be an exhaustive revision of these already discussed points but simply tries to provide some typical examples of chromatograms and contour plots where some of the most common problems can visually be identified. Our hope is that these examples will serve as a complement to previous theoretical descriptions for newcomers in the field and as a help during the optimisation of their GC×GC separations.

2. TROUBLESHOOTING IN GC×GC

As previously discussed, the particular features of GC×GC make this technique very attractive for anyone dealing with the analysis of complex extracts. However, as for any other chromatographic technique, the final degree of separation achieved will depend on the careful optimisation of a number of experimental parameters. Parameters related to the chromatographic process will produce chromatographic responses in a GC×GC system similar to those observed in any other GC-based separation. Therefore, knowledge of the cause of GC problems is directly applicable here for trouble identification and correction. Parameters related to the transfer of the narrow ^{1}D fractions to the ^{2}D column are partially shared with other multidimensional techniques. However, because of the high speed of this process and the special requirements of the 2D separation, the GC×GC transfer process becomes particularly delicate or, in other words, prone to problems during method development. Peak shape and eventually GC×GC resolution are affected by both categories of parameters, as illustrated below through some selected examples.

2.1 Chromatographic peak distortion and analyte degradation

2.1.1 Peak Tailing

Low programming temperature rates of the ^{1}D column can result in a widening of the chromatographic peak, which in addition can show a typical asymmetry (tailing peak) derived from extra column effects and from active points in the column. On the other hand, peak front tailing can originate from overloading or from too fast carrier gas velocities.

In GC×GC, peak tailing can similarly be recognised by simple inspection of both the 2D contour plot and the chromatographic raw data. The peak highlighted in Figure 1, which corresponds to a trimethylsilyl oxime ether (OTMS) carbohydrate, shows simultaneously a frontal asymmetry in the first dimension and a tail in the second dimension, which can, respectively, be identified by the asymmetric 1D peak and the tailing modulated peaks observed in the raw data. In the contour plot, these result in deformation of the typical ellipse shape of GC×GC peaks in the corresponding directions, that is, to the left (in the first case) and to longer retention times in ^{2}D (in the second one).

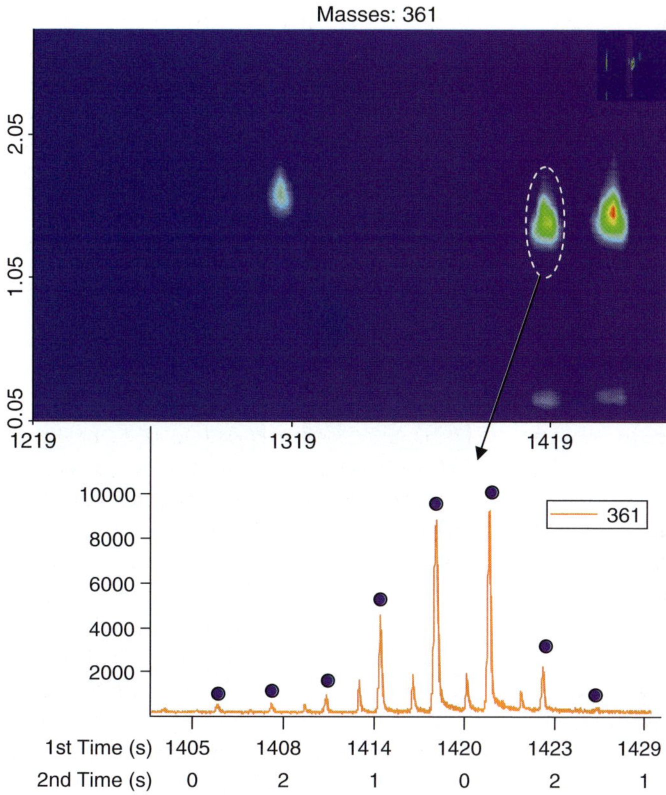

Figure 1 Typical contour plot and chromatographic raw data obtained for a peak tailing in both chromatographic dimensions. In the nonconverted chromatogram (lower figure), blue points indicate the several modulations obtained for the peak highlighted in the contour plot (upper figure).

2.1.2 Analyte degradation in the injector and in the ^{1}D column

Degradation of labile compounds is a well-known phenomenon for chromatographers that can also be visually identified on 2D contour plots.

Chromatograms on the left side of Figure 2 show 11 chromatographic peaks obtained for 9 fungicides analysed in BPX-5×BPX-50 with GC–NPD and GC×GC–NPD (nonconverted –raw– data) [1]. In the latter, the baseline rise between peaks 9A and 9B is an indication of a degradation process (see expansions), in which iprodione (peak 9 B) yields in the injector a degradation product, causing peak 9A. The same degradation process continues through the 1D elution, producing the baseline rise.

This degradation is clearly visualized in the contour plot (right side) by the band linking peaks 9A and 9B with GC×GC–NPD. A severe decomposition of co-extracted matrix components eluting before peak (1) in the ^{1}D column (white band eluting from 15 to 25 min) is graphically shown. Similar results appear when analysing the extract with GC×GC–μECD, as shown in the expanded region.

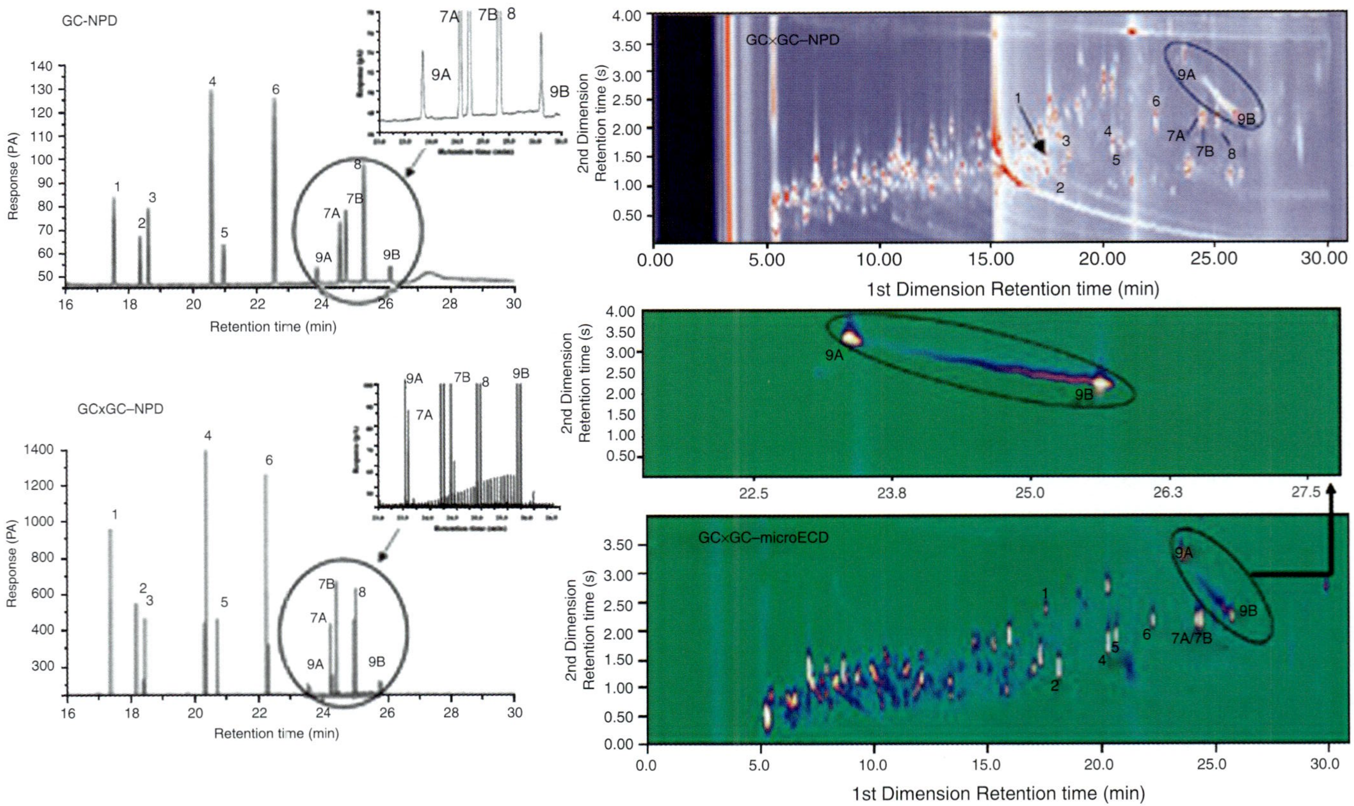

Figure 2 Visual identification of analyte degradation in the injector and/or in the [1]D column in the contour plot (right side) and in nonconverted chromatograms (left side) [1]. Peak identification: (1) chlorothalonil, (2) vinclozolin, (3) metalaxyl, (4) penconazole, (5) procymidone, (6) myclobutanil, (7A, 7B) propiconazole diastereomers, (8) tebuconazole, (9B) iprodione, and (9A) iprodione degradation product.

Another example of degradation in the ¹D column during the chromatographic run is presented in Figure 3, which shows sections of the GC×GC–μECD contour plots obtained for polybrominated diphenylethers (PBDEs) on two column sets [2]. On DB-1×007-65HT (upper contour plot), the BDE congener 209 elutes at 88.9 min from the ¹D column and is visible as a vertical yellow band rather than as the typical ecliptic spot because of the high amount injected. The decomposition products (BDEs 206, 207, 208, and 198+203) formed by degradation of BDE 209 in the injector and in the 1D column are clearly identifiable as well as the characteristic bands connecting BDE 209 with these particular congeners (e.g., band A for BDE 208 and band B for BDE 207).

A more severe degradation pattern was observed on DB-XLB×007-65HT (lower contour plot), where a complete decomposition in the injector and/or first dimension of deca-BDE 209 into compounds of lower molecular weight is observed. This result prevents use of DB-XLB as ¹D column in this type of study.

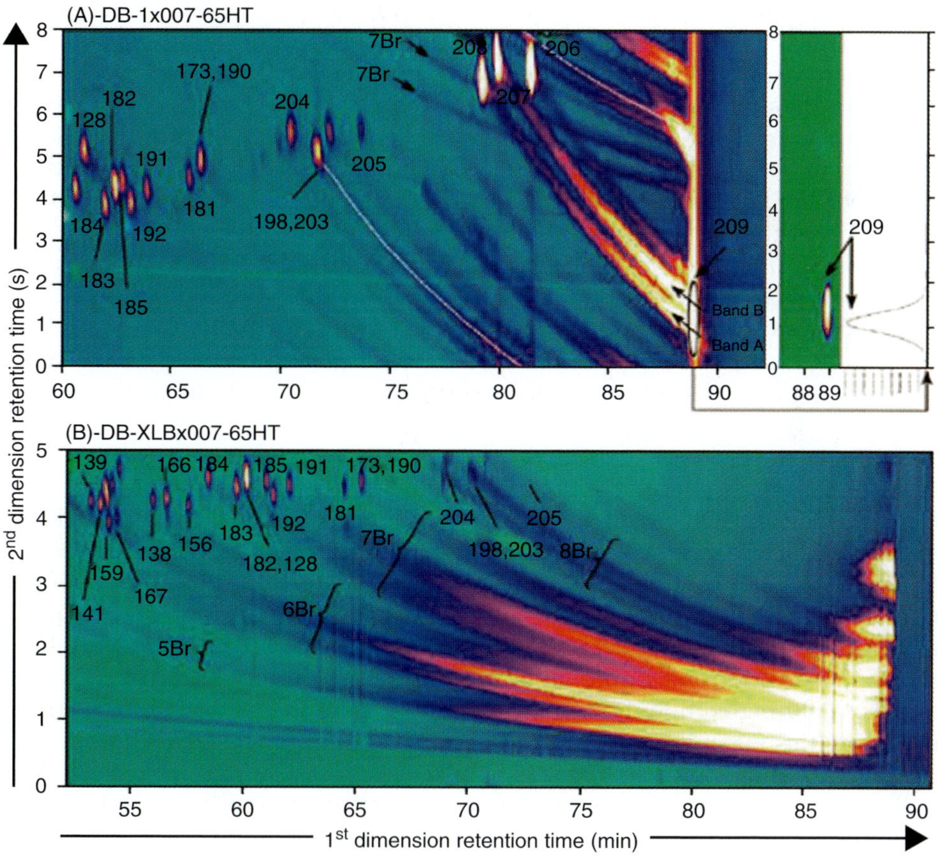

Figure 3 Severe degradation of BDE-209 in the injector and/or ¹D column as observed in the GC×GC–μECD contour plots obtained using DB-1×007-65HT and DB-XLB×007-65HT [2]. In both cases, a high concentration of BDE-209 was injected to ensure proper visualisation of the process.

2.1.3 Analyte degradation in the modulator and in the ^{2}D column

Some models of thermal modulators include, apart from the cold jet(s) for cryotrapping of the analytes eluting form the ^{1}D, hot jet(s) for subsequent fast release of the trapped compounds to the ^{2}D column (Chapter 2). In these setups, hot jet pulses at temperatures above those of the main oven are applied to ensure the transfer of the analytes concentrated in the modulator as a narrow band to the ^{2}D column. In the case of thermolabile compounds, wrong selection of this modulator temperature can result in partial degradation, which is visualised in the 2D contour plot by the presence of two peaks with the same ^{1}D retention time but different ^{2}D retention. In spite of the short elution times, labile compounds can also degrade in the ^{2}D column. In this case, a baseline rise joints the eluting points of the original compound and its degradation product.

Contour plots in the left side of Figure 4 show the chiral separation of five selected toxaphene congeners on BGB-172×BPX-50 (Figure 4A) and on BGB-172×Supelcowax-10 (Figure 4B) using GC×GC–μECD and the same temperature program [3]. A comparison between both plots indicates that the use of polyethyleneglycol-type stationary phases for analysis of this family of pollutants is not recommended. In the upper plot (Figure 4A), chromatographic peaks show a correct shape in both ^{1}D (chiral) and ^{2}D (nonchiral) columns. In Figure 4B, however, the front bands (such as those observed for Parlar 26) and the long tails stretching in ^{2}D between peaks from original and altered analytes (e.g., for Parlar 40 and 50) can be caused by degradation in the ^{2}D column. Column bleeding is also visible at the upper part of the contour plots as a white band.

Figure 4C shows the presence of a similar effect in the contour plot obtained with GC×GC–μECD for some pesticides on ZB-5×Carbowax 10 [4]. Degradation in the modulator is shown by the double peaks observed in the ^{2}D column for γ-HCH (peak b), *p,p*′-DDD (peak i) and *p,p*′-DDT (peak j). But degradation also occurs through the ^{2}D elution, as shown by the vertical bands connecting the two peaks.

2.1.4 Overload

In GC×GC, column overload can occur in both dimensions. Similarly to 1D GC, overloading in the ^{1}D column results in non-Gaussian peaks exhibiting a characteristic peak front tail. This effect can be identified graphically in both the contour plot and the nonconverted (raw) 1D chromatographic profile, in the same way as explained for peak tailing in Figure 1. Severe overloading will result in broadened ^{1}D chromatographic peaks that can eventually lead to overlapping with close eluting compounds. Apart from the obvious loss of chromatographic resolution, this effect can ruin the separation, especially when trace components elute close to the overloaded peak.

Because of the requirement of fast separation in the second dimension in GC×GC (Chapter 2), ^{2}D columns are essentially short and present a moderate to low film thickness. From a practical point of view, that means that both separation power and loading capacity of the ^{2}D column are rather limited. These considerations support the need for careful optimisation of the mass transfer conditions from the ^{1}D to the ^{2}D in GC×GC and explain why overloading in the

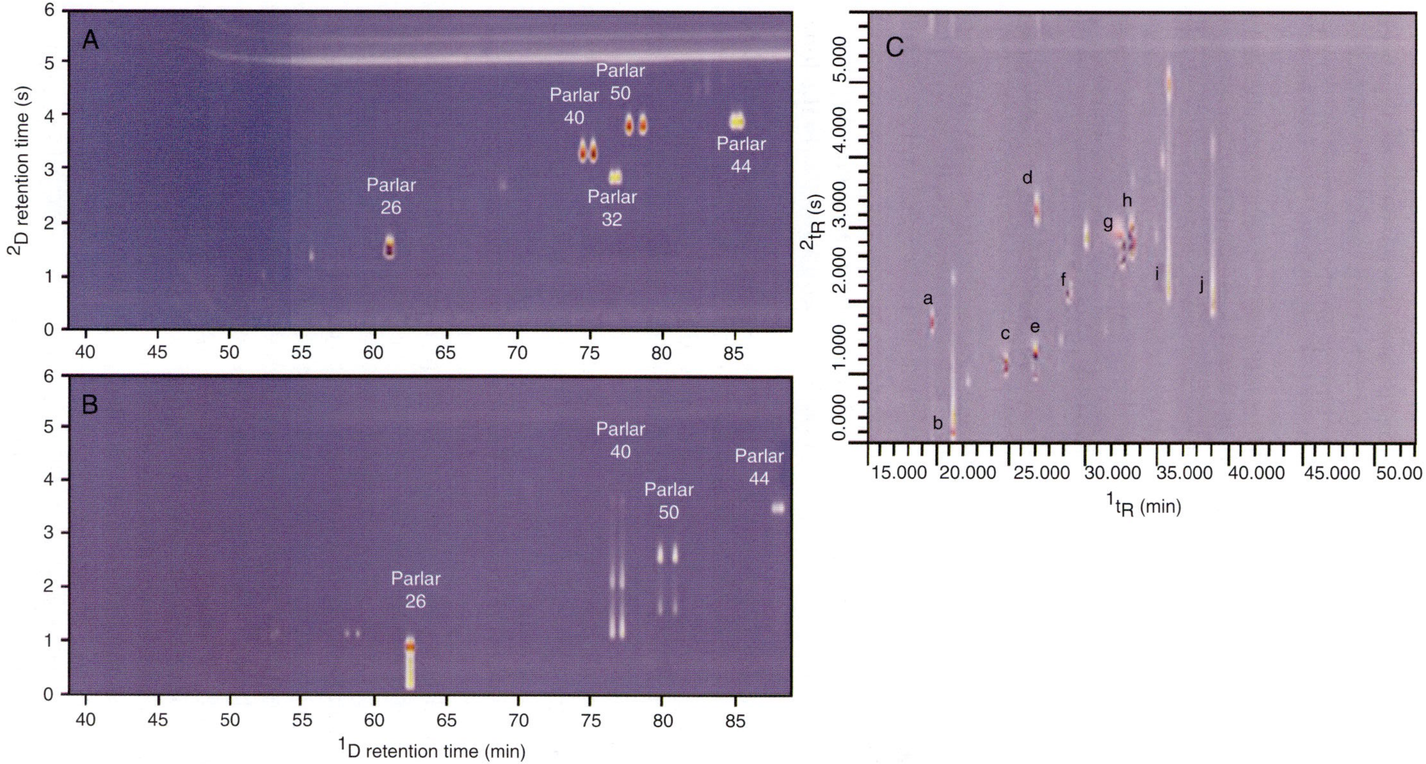

Figure 4 Chiral separation of selected toxaphene congeners on (A) BGB-172×BPX-50 and on (B) BGB-172×Supelcowax-10, where analyte degradation in the ^{2}D column is visible for Parlar 26, 40, and 50 [3]. Degradation in the modulator and ^{2}D column is also evident in (C) for γ-HCH (peak b), *p,p′*-DDD (peak i) and *p,p′*-DDT (peak j) on ZB-5×Carbowax 10 [4]. Other peak identification: (a) α-HCH, (c) heptachlor, (d) DBF, (f) heptachlor-epoxide, (g) *p,p′*-DDE, and (h) dieldrin.

^{1}D will typically result in overloading in the ^{2}D, leading to front tails and broadened chromatographic peaks in this dimension as well, as illustrated in Figure 5. A typical example of overloading can be found in Figure 3B of Chapter 8 (see broadened chromatographic peaks eluting in the last part of the contour plot).

Severe overloading, which can be identified in the contour plot by characteristic continuous vertical bands in the ^{2}D, is observed in Figure 3 (see previous discussion) and on the left part of the GC×GC–NPD contour plot in Figure 2.

2.2 Programmed temperature separations

In GC×GC, slow temperatures ramps should be preferred when working with oven programmed temperature because the application of fast oven temperature ramps can adversely affect the possible 2D structure and peak shape visualization in the contour plot as well as the separation itself.

Figure 6 shows the GC×GC–μECD contour plot obtained on BPX-172×BPX-50 for the chromatographic components of a fish oil sample, in which a waving band was produced when a relatively fast temperature ramp was applied to the ^{1}D column [3]. In this study, the temperature program for the first dimension was: 90°C (2 min), at 25°C/min to 190°C (2 min), at 2°C/min to 210°C (27 min), and at 10°C/min to 240°C. Meanwhile, the temperature program for BPX-50 was as follows: 210°C (8 min), at 2°C/min to 230°C (27 min), and at 10°C/min to 260°C. The trend toward a faster elution observed for analyte peaks between min 45 and 48 (delimited with dotted white lines in the figure) corresponded to the 10°C/min ramp.

Figure 4 in Chapter 8 shows another example of waving bands: the negative consequences of this effect on the structure of the 2D chromatograms are evident.

When the temperature of the ^{2}D column changes significantly between consecutive modulations during the elution of an analyte, the apex of a ^{1}D peak in consecutive ^{2}D runs may shift from one modulation to another and result in a mistaken assignation as two or more different compounds (see Section 4.3 in Chapter 4 for further details).

The analysis of mixtures of compounds of different volatility by GC×GC can easily result in relatively broad peaks for low-volatility compounds in the first dimension, which are sampled a high number of times during modulation. This effect, combined with the application of a relatively fast temperature ramp to ensure their fast elution, can make consecutive modulated fractions of a peak to elute at increasingly earlier elution times from the ^{2}D column. In an extreme case, as that shown in Figure 7A for OTMS-laminaribiose (highlighted peak), multiple peaks can apparently be differentiated in the contour plot. However, non-converted 1D chromatographic data inspection proves that only one peak was present in the mixture, as demonstrated for the analysis of the corresponding individual standard (Figure 7B). As illustrated in Figure 7C, this problem can easily be solved by using a slower temperature program. Alternative solutions are the use of a shorter modulation period (P_M) or a slower sampling rate.

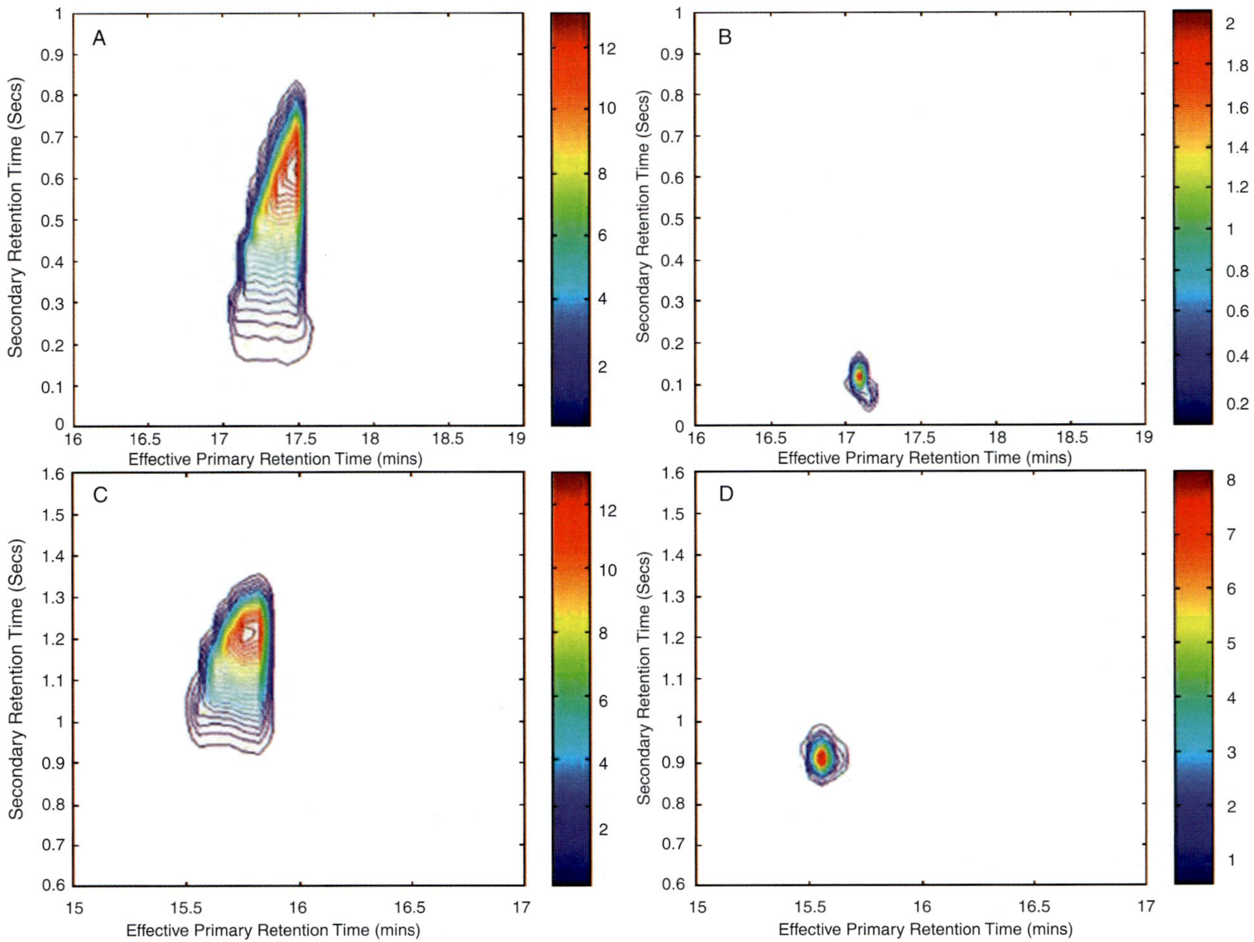

Figure 5 Typical peak shape obtained for an analyte injected (A and C) at very high concentration and (B and D) after a 1000-fold dilution using an (A and B) 100 μm and a (C and D) 250 μm ^{2}D column. Note that the analyte actually overloads the two column sets in both dimensions [5].

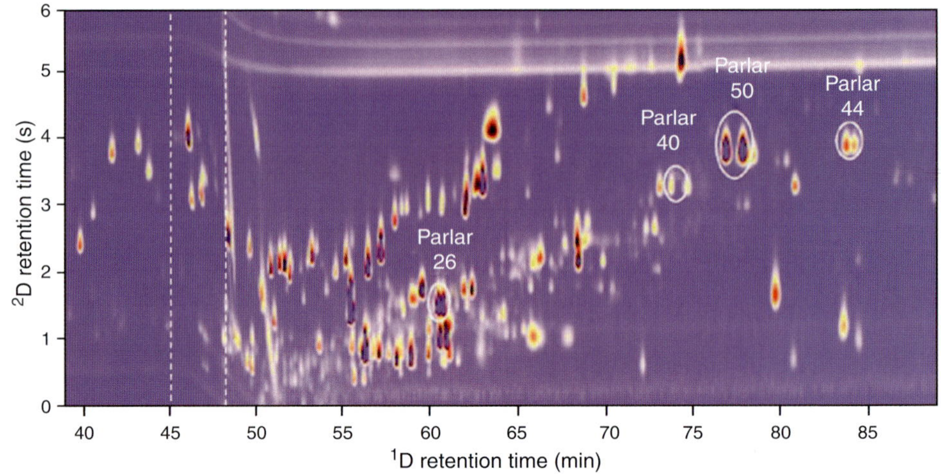

Figure 6 Adverse effect observed in the 2D structure of a chromatogram obtained for chiral separation of toxaphene in fish oil on BPX-172×BPX-50 with GC×GC–μECD when applying a fast ^{1}D oven temperature ramp [3].

The application of fast temperature ramps can also adversely affect the separation process itself. To take advantage of the GC×GC separation power, it is advisable (and, in some cases, even mandatory) to work under optimum separation conditions also in the first dimension. This is especially true when analysing very complex mixtures of structurally related compounds, especially if limited separation is expected in the second dimension. As a typical example, Figure 8 compares the 1D GC separation (A–C) and the reconstructed ^{1}D GC chromatograms (D–F) obtained for polychlorinated biphenyls (PCBs) nos. 169, 196 and 203 using temperature ramps of (A,D) 0.5°C/min, (B,E) 1.5°C/min and (C,F) 3°C/min [6]. A P_M of 4 s was used in the GC×GC experiments. PCBs 169 and 196 were not resolved in any of the runs. However, as expected, the resolution between this pair and PCB 203 increases as the temperature gradient decreases. Similar results were only obtained at 0.5°C/min. Figure 8E compares the 1D reconstructed chromatograms obtained using a 2 m ^{2}D column and a P_M of 4 s (continuous line), and a 0.9 m ^{2}D column and a P_M of 2.5 s (dotted line). This modification decreased the peak capacity of the ^{2}D while increasing the number of modulations per peak from three to five, and resulted in a resolution that was only slightly poorer than that of the original 1D GC separation (Figure 8A).

2.3 Wraparound

Wraparound occurs when the retention time of a compound in the ^{2}D column exceeds the modulation period. It can be identified in the 2D profile by the elution of the affected compound during the subsequent modulation cycle(s). Figure 9 shows the GC×GC–μECD contour plot obtained on ZB-5 x HT-8 for a

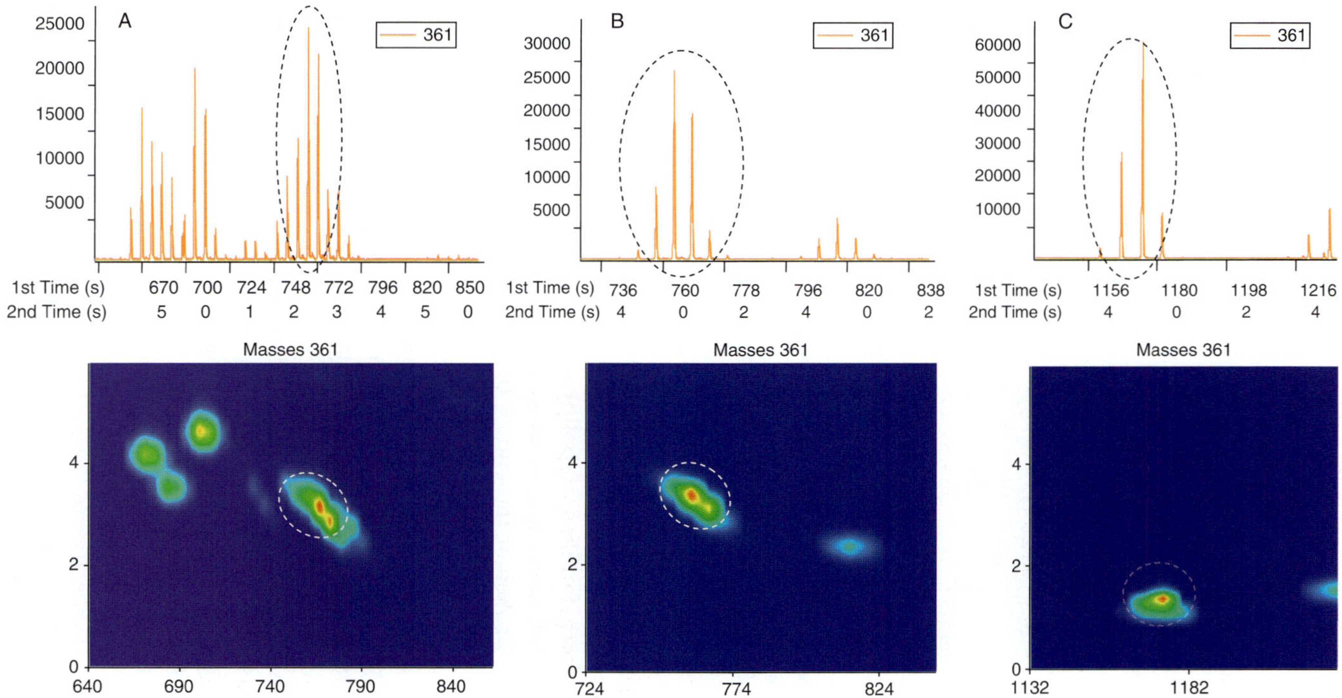

Figure 7 (A) Adverse effect observed in peak shape of OTMS-laminaribiose (highlighted peak) when the ^{2}D column temperature changes too rapidly relative to the modulation period and (B) confirmation by analysis of the corresponding standard under similar conditions. ^{2}D oven temperature ramp: 230°C (12 min) at 7°C/min to 270°C (1 min); P_M: 6 s; hot jet time: 1 s. The problem was solved by (C) reducing the speed of the temperature ramp applied from 7°C/min to 2°C/min until complete elution of the target compound.

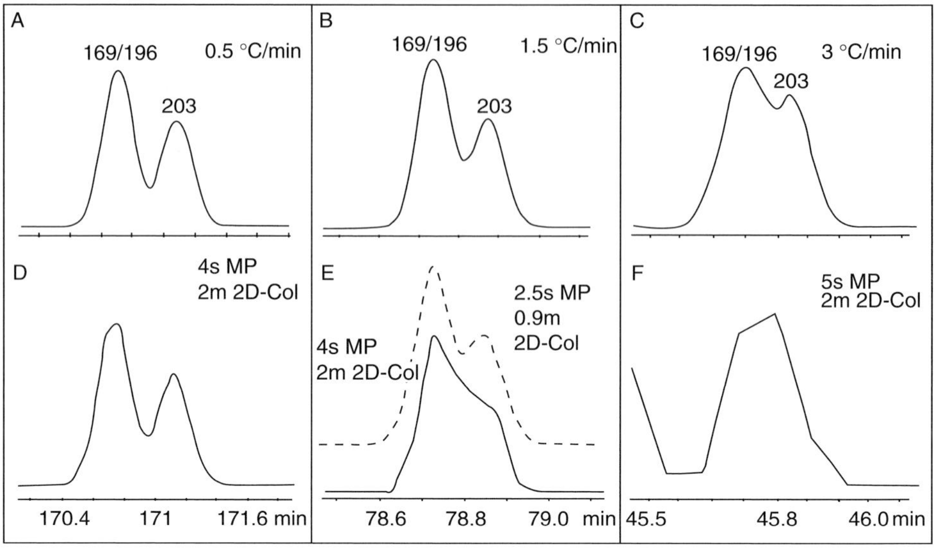

Figure 8 Comparison among the (A–C) 1D GC and the (D–F) reconstructed ^{1}D chromatograms obtained for PCBs 169, 196, and 203 at (A,C) 0.5°C/min, (B,E) 1.5°C/min, and (C,F) 3°C/min using DB-XLB (1D GC experiments) and DB-XLB×LC-50 with a P_M of 4 s in the GC×GC experiments. (E) shows the influence of the ^{2}D column length and P_M in the final separation [6].

technical mixture of polychlorinated terphenyls, PCTs (Aroclor 5460), also containing PCBs, whose position is indicated with a white circle [4]. Orderly structures were obtained for both families of pollutants. But while no wraparound was observed for PCBs, this effect was evident for less volatile PCTs, for which the structure appears as broken at the end of the ^{2}D elution time.

Another example of wraparound is observed in Figure 2, in the GC×GC–NPD contour plot, for the peak eluting at 5 min and with an apparent retention time in the second dimension (^2t_R) of 2.50s. Broadening the peak in the ^{2}D because of wraparound is clearly visible, too, when comparing the ^{2}D width of this peak with those of any other close eluting compound.

2.4 Modulation process

The enhanced separation power provided by GC×GC can partially be lost if the resolution achieved in the first dimension is not preserved during modulation and separation in the second dimension (Chapters 2 and 3).

Apart from the previously discussed need for careful selection of the maximum working temperature in thermal modulators to prevent degradation of thermolabile analytes (Section 2.1), some other experimental parameters related to the modulation step should be considered (and optimised) to ensure appropriate transfer of the ^{1}D fractions to the ^{2}D column and a satisfactory separation in the latter. Among them, selection of a suitable P_M and, in the case of

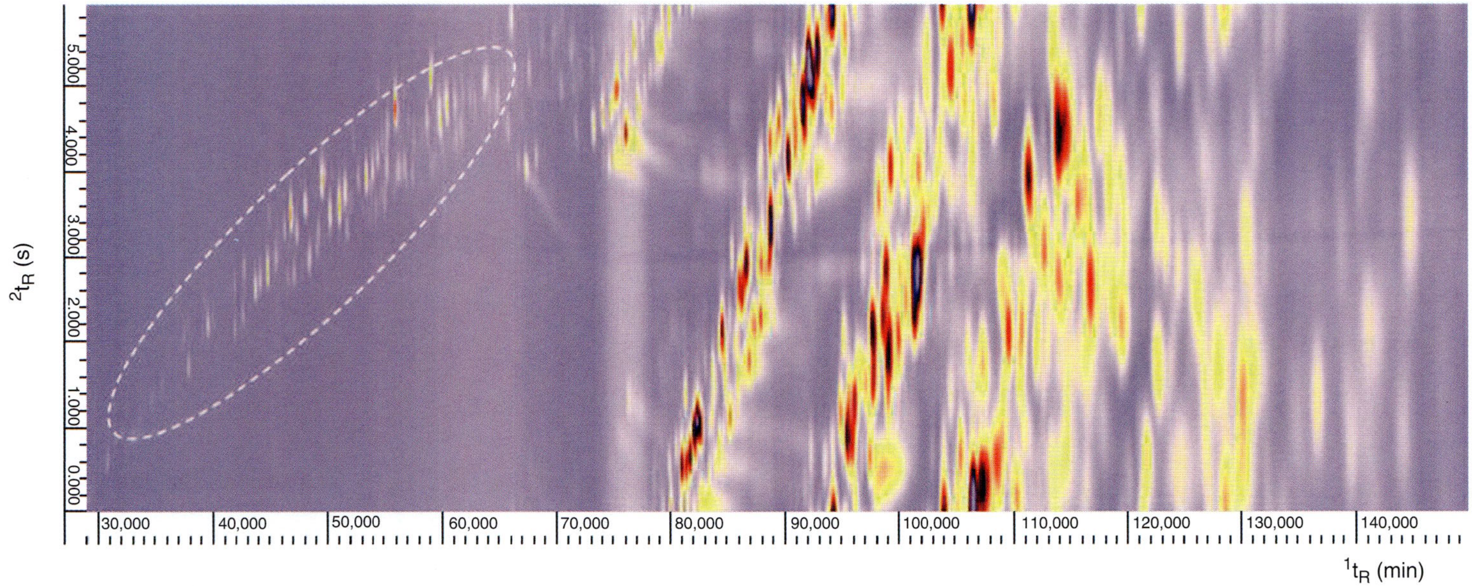

Figure 9 GC×GC–μECD contour plot obtained on ZB-5×HT-8 for a technical mixture of PCTs (Aroclor 5460) showing a structured chromatogram with no wraparound for PCBs (position is indicated with a white circle) but where less volatile PCTs showed a clear wraparound [4].

thermal modulators, a proper fitting among this parameter and those affecting the cold trapping and subsequent quantitative release of the concentrated analytes are probably the most relevant.

2.4.1 Inadequate modulation period

It is generally accepted that peaks eluting from the ^{1}D column should ideally be modulated at least four times (Chapter 2). Failure to select an adequate P_M can ruin not only separation in the ^{2}D column, but also separation previously achieved in the first dimension. Figure 10 compares the separation achieved on DB-5×BPX-50 for two selected polycyclic aromatic hydrocarbons (PAHs), indene[1,2,3-*cd*]pyrene and dibenzo[*a,h*]anthracene, analysed under similar chromatographic conditions but using different modulation periods. A quad-jet dual-stage modulator was used in the study, and the hot jet pulse was set as 600 ms in all instances. As shown in Figure 10A, DB-5 provided some degree of separation between indene[1,2,3-*cd*]pyrene and dibenzo[*a,h*]anthracene, while both analytes exhibited similar retention times on BPX-50. A (too) long P_M (7 s) resulted in recombination of the already separated peaks in the ^{1}D modulator and their joint injection in the ^{2}D column (Figure 10C). Decreasing the P_M to 6 s (Figure 10B) contributed only partially to solving the problem: co-elution remains, although peaks are now modulated three times. Further reduction of the modulation time to 5 s (Figure 10A) resulted in four modulations per peak, preserved the separation achieved in the first dimension, and allowed an adequate separation of the two analytes, which appeared in the contour plot as two separate spots.

Although shortening of the modulation period is an efficient way to solve this problem, depending on the application, wraparound of less volatile analytes can then become the price to pay (see analyte 3 in Figure 10A). As suggested by Adahchour in [7], use of narrower ^{2}D columns (if overloading is not a problem) or of lower ^{1}D temperature programming rates (resulting in an increase of the ^{1}D peak widths) is a valuable analytical alternative to consider.

2.4.2 Incomplete trapping or release of analytes in thermal modulators

At the time of writing, thermal modulators without moving parts are the most widely marketed and used. In these systems, analytes eluting from the ^{1}D are

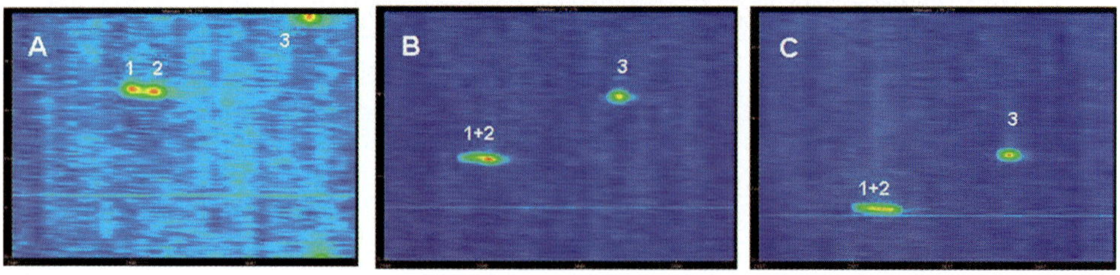

Figure 10 Reconstructed GC×GC–ToF MS contour plots (*m/z* 276 + 278) obtained for (1) indene[1,2,3-*cd*]pyrene, (2) dibenzo[*a,h*]anthracene, and (3) benzo[*ghi*]perylene on DB-5×BPX-50 using a P_M of (A) 5 s, (B) 6 s, and (C) 7 s. All other experimental parameters were kept constant throughout the study (see text).

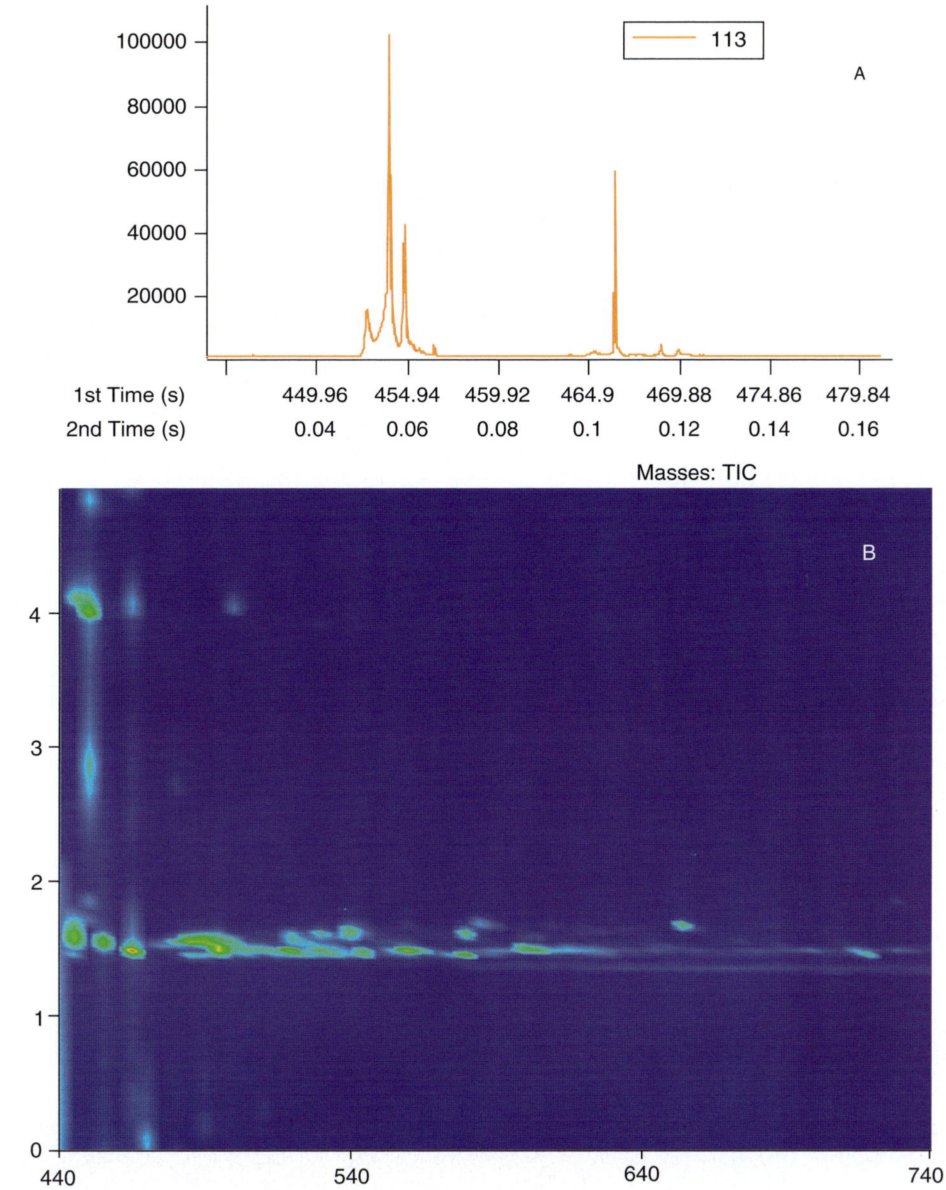

Figure 11 Improper trapping of early eluting analytes in GC×GC using a quad-jet dual-stage modulator as visualised in (A) the nonconverted 1D data (fragmentogram expansion corresponding to *m/z* 113), and (B) the contour plot (TIC).

trapped by cryo-focusing at low temperatures and subsequently released, ideally as a narrow band, to the ^{2}D column by rapid increase of the temperature. The efficiencies of the trapping and reinjection steps have a profound influence on the ^{2}D separation process, as explained in [7] for different types of modulators.

Incomplete trapping is typically associated with insufficiently low temperatures in the modulator and results in breakthrough of the otherwise cryo-accumulated analytes. Breakthrough in the modulator can visually be recognised in the contour plot by characteristic vertical bands that are rather similar to those derived from severe overloading of the ^{2}D column (see Section 2.1).

Figure 11 shows a not so extreme, and consequently much more frequent, example of incomplete retention of the analytes in the modulator, in this case a quad-jet dual-stage modulator. Improper trapping of early eluting analytes can be identified in the contour plot (Figure 11B) by double peaks with essentially similar retention times in the ^{1}D but separated in the ^{2}D by 2.5 s (note that this value is equal to half the P_M, 5 s). In this particular example, the application of a relatively long hot pulse (0.6 s) at high temperature can be responsible for this effect, which became evident when inspecting the peak shapes in the non-converted chromatogram (Figure 11A, *m/z* 113, for simplicity).

REFERENCES

1 W. Khummueng, C. Trenerry, G. Rose and P.J. Marriott, *J. Chromatogr. A*, 1131 (2006) 203.
2 P. Korytár, A. Covaci, P.E.G. Leonards, J. de Boer and U.A.Th. Brinkman, *J. Chromatogr. A*, 1100 (2005) 200.
3 L.R. Bordajandi, L. Ramos and M.J. González, *J. Chromatogr. A*, 1125 (2006) 220.
4 L.R. Bordajandi, J.J. Ramos, J. Sanz, M.J. González and L. Ramos, *J. Chromatogr. A*, 1186 (2008) 312.
5 J. Harynuk, T. Górecki and J. de Zeeuw, *J. Chromatogr. A*, 1071 (2005) 21.
6 M. Harju, C. Danielsson and P. Haglund, *J. Chromatogr. A*, 1019 (2003) 111.
7 M. Adahchour, Troubleshooting in GC×GC in www.chromedia.org, (2008).

SUBJECT INDEX